Technology Of Machining Systems

[Including New Technology]

Technology Of Machining Systems

[Including New Technology]

Dr Bimal Kumar Mullick

MME, PhD (Eng.), FIE (Ind.), CE(Ind.), CEng, MIProdE (Lond.)

Professor, Production Engineering Dept, Jadavpur University

Dr Amitabha Bhattacharyya

ME (Gold Medalist), MS (Illinois), PRS, PhD (Eng.), FIE (I), FNAE, FNA, FNASc, CEng (UK), CEng (I), Mowatt Medalist, MIProdE (Lond.)

Professor, Production Engineering Dept, Jadavpur University
Former Director, IIT, Kanpur

NEW CENTRAL BOOK AGENCY (P) LTD

DELHI • KOLKATA • PUNE • ERNAKULAM

TECHNOLOGY OF MACHINING SYSTEMS

Publisher

NEW CENTRAL BOOK AGENCY (P) LTD

- 8/1 Chintamoni Das Lane, Kolkata 700 009
- 4262/3 1st Floor, Flat Nos. 105 & 106
  Ansari Road, Daryaganj, New Delhi 110 002
- Shop Nos. 3 & 4, Vinayak Towers
  681/B Budhwar Peth, Appa Balwant Chowk
  Pune 411 002
- GSS Shopping Complex (1st Floor)
  Opposite College Post Office
  Convent Road, Ernakulam 682 035

Typesetter
Graphitech India Ltd
216/3 A J C Bose Road, Kolkata 700 017

Printer
New Central Book Agency (P) Ltd
Web-Offset Division, Dhulagarh, Sankrail, Howrah

Cover Printer
Biswanath Studio
3 Garanhatta Street, Kolkata 700 006

Diagrams
Biren Das

Technical Editor
Dr B N Lahiri

Project Team
Prabhat Jas, Swapan Sarkar

ISBN: 81-7381-330-2

First Published: May 1990
Reprinted: January 2009

Price: Rs 100.00 [Rupees One Hundred only]

PREFACE

"Technology of Machining Systems" is a synthesis of two very interactive fields of the mechano-production scenario — the "Metal cutting" on one side and the "Machine Tools" on the other.

The need for such a synthesised text based on the system concept — the intellectual order of the era — was long felt and well established. That 'need' was the 'drive' which, however, took a long gestation period to present this system based synthesis in its current form. The idea has now been translated into a reality.

It is a matter of great sorrow that the co-author of this book, Prof B. K. Mullick, who tirelessly worked for many long years to synthesise this concept by judicious integration of the two separate subjects of "Metal cutting" and "Machine tools" into one, could not see this book in print form. The hard reality of bringing this book could not be achieved except for the initiatives put by him in teaching and documenting the subject in the Department of Production Engineering.

The contents of this book have been arranged in such a manner so as to cater to the requirements of Undergraduate and Post-graduate degree courses of all Indian universities, all IITs, Post-graduate diploma courses, as well as Professional diploma courses like AMIE (India), AMI Mech.E (Lond.), AMI Prod.E (Lond.) being carried out in the country.

The task of printing a text book requires different kinds of expertise. To take care of so large a number of illustrations and to set the proof right is a tedious job. Shri Brajen Roy volunteered to help me in this respect, for which I am thankful to him.

Discussions with my students and now colleagues, particularly Dr. Jyotirmoy Saha, Dr. Sanat Kumar Mukherjee, Shri Kamal Kumar Dutta and Shri Subir Kumar Sanyal, Readers of the Department of Production Engineering, have helped me in enriching the contents of this synthesis. The editorial help has been rendered by Dr. B N Lahiri, Reader, of this Department. It is difficult for me to express in language my indebtedness to them.

Lastly, it is proper to thank my publisher, New Central Book Agency, and printer, Books & Allied (P) Ltd., for taking up this task.

15th April, 1990

Production Engr. Dept. **A. Bhattacharyya**

Jadavpur University **B. K. Mullick**

CONTENTS

CONTENTS

Chapter 7 Machines of Machining

Chapter 8 Cutting Tools

Chapter 9 Machining of Surfaces

CONTENTS

CHAPTER 1

Introduction

1.1. TECHNOLOGY

In various forms the word 'Technology' appears in academic and everyday speech. In past decades the term 'technology' had a very specific and limited meaning. Some employed the term is 'practical art' or 'the study of practical arts', etc. In the literature of the 18th and 19th centuries such meanings were unique and clear with no other connotation for further analysis or deliberation. Many a people described technology directly in terms of machines, tools, factories, industries, crafts and engineering and did not bother about technology as a distinctive phenomenon.

In the 20th century, linguistic conventions gradually changed. The word technology expanded rapidly in both its denotative and connotative meanings. Now this word of speech has expanded to an unbelievable diverse collection of phenomenon—tools, instruments, machines, organizations, methods, techniques, systems and the totality of all these and similar things. The shift of the meaning from relatively precise, limited and unimportant has changed to something vague, expansive and highly significant. The definition in Webster's Second International (1909) of this word meant 'Industrial science, the science or systematic knowledge of the industrial arts, especially of the more important manufactures.' In the 1961 edition, Webster blossomed the definition into: 'The totality of means employed by a people to provide itself with the objects of material culture.' To many, meaning of technology in this form is uncomfortable. J Ellul defined *la technique* as 'the totality of methods rationally arrived at and having absolute efficiency (for a given stage of development) in every field of human activity.' Today, technology points to a vast, diverse, ubiquitous totality that stands at the centre of modern culture. New tendency to describe technology at modern times can finally be described as 'Technology is everything and everything is technology.' Technology has thus become a staggering collection of phenomena and frequently refers to a variety of organizational sets—factories, workshops, bureaucracies, armies, research and development teams, etc., in forms of various rational productive arrangements.

1.2. MAPPING TECHNOLOGY

The concept of 'mapping' is a projected manifestation from one form into another form subjected to several criteria like

(i) equality,
(ii) ortho-morphism or conformality, and
(iii) equivalence, etc.

These or similar criteria are basic and need not be mutually exclusive.

The process of mapping of technology can be such attributes as

(i) momentum and motive of technology,
(ii) technological evolution,
(iii) technological determinism,
(iv) technological drift,
(v) technological imperative, etc.

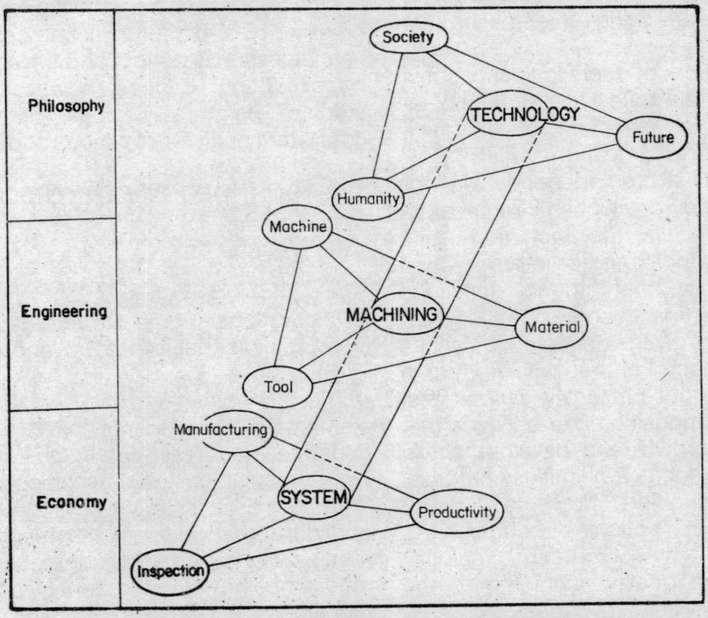

Fig. 1.1: Thematic likage amongst technology, machining and systems

Some people think that technological development goes by a process of inertia having a character of self-propelling, self-sustaining ineluctable flow.

On the other hand, the thermo-organization theorists attribute a certain dynamism and ineluctability to this process of technology associated change. Industrialization, modernization and growth, etc., contain concepts as speed, momentum, force and acceleration. The semantics of such ideas inextricably bring forward such expressions as 'dynamics of modernization', 'thrust of technical change', 'accelerated pace of industrial development', etc.

Behind the massive processes of momentum transfer, one always finds a realm of human motives and conscious decision in which the actors at various levels determine which kind of apparatus, technique and organization are going to be developed and applied. Behind modernization are always the modernizers; behind industrialization, the industrialists. Science and technology do not grow of their own momentum but advance through the mapping of creative individuals following paths of discoveries, inventions and productive innovations.

A mapping through Darwinian evolution has been noticed by Karl Marx. In *Capital,* he commented, 'Darwin has aroused our interest in the history of natural technology, that is, in the formation of the organs of plants and animals, as instruments of production for sustaining life'. However, Marx did not believe that the evolution of technology was a self-generating process. Biological evolution has given way to a far more rapid process—technological evolution. Examining the progress of scientific discoveries and technological innovations, history encounters something resembling an evolution of forms.

Karl Marx's conception of technical and social changes is another attribute for mapping of technology. The principal aspect of this theory is rooted to deterministic historical materialism. He states, 'What individuals are.....depends on the material conditions of their production.' To speak of 'determinism', Marx recalls the fact that each generation is strongly conditioned or informed by a technological inheritance that in no sense is 'chosen'. The socio-technical context into which people are born simply is given. Marx states, 'Means of production do not depend on free will.' Further in Marx's view, needs are not present in any simple, finished form in any composition. He believes that the tendency towards growth and development in human motives through needs is a limitless process. Marx's mapping of technology presents a direct challenge to voluntarist notion of choice. As technology advances, human needs actually 'expand' and change in quality according to the deterministic theory of Marx.

The premise for mapping for technological imperatives is : 'Technologies are structures whose conditions of operation *demand* restructuring of their environment.' The technological imperative contains a logic that accounts for much of the change in society. This logic is not that of the syllogistic inference, rather in the pragmatic rationale of necessary action. The force of technological imperatives is reinforced by their connection to the necessities of life. Certain technologies stand at the very roots of human survival and without certain technologies mankind suffers discomfort. Thus technological change is actually a logical corollary of technical rationality.

1.3. DEFINITION OF TECHNOLOGY

Technology has been defined in many ways. The following definition is suggested: "Technology is created capability: it is manifested in artefacts the purpose of which is to augment human skill."

Key concepts include:

– *Creation.* Technology is not a free gift of nature: it does not come about by itself. It is the product of deliberate action. If technology is to be employed as a resource it has to be cultivated, nurtured and supported.

– *Capability.* This concept refers to a particular type of handiness, namely that of manipulating aspects of the physical world.

– *Artefact.* This is the generic term for all devices, tools, instruments or machines. Artefacts are the repositories of capability.

– *Augment.* This concept is used to convey two meanings; on the one hand, enhancing human ability, such as adding instrumentation to human activity, and on the other hand, replacing human ability, by substituting it with competent artefacts.

–No definition is perfect and the above formulation is no exception. It does, however, provide a good starting point and does permit an internally consistent analysis of various aspects of technology.

1.4. TECHNOLOGY COTYLEDON

(a) Technology

A technology is a set $\mathcal{T}$ not empty of elements, configurations or structures, closed under isomorphism, that is, if $Z \in \mathcal{T}$ and Z' is isomorphic to Z, then $Z' \in \mathcal{T}$.

The technology is again represented by a set of elemental components and this technology generates another cotyledon, the cotyledon of structures which are buildable in the technology. That is, the technology cotyledon consists of configurations that are the resultants of coupling recipes, each elemental component of the coupling recipe being a component available in the technology set. Thus, it can be shown that each

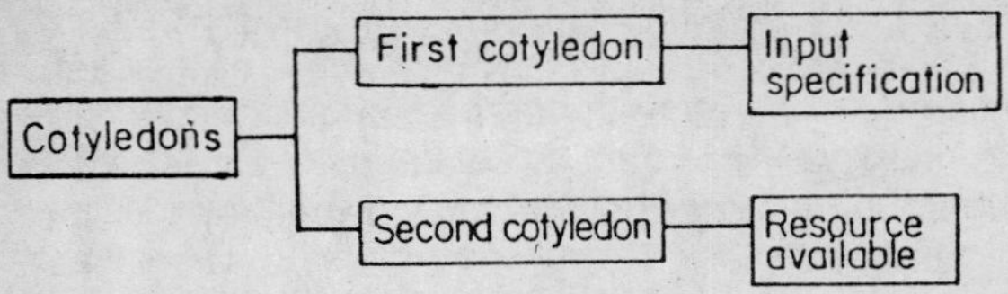

Fig. 1.2: Establishing different cotyledons

point of the technology cotyledon is determined by a coupling recipe and the resultant of the coupling recipes. Such a point is labelled by the pair ($\mathscr{H}$, Z).

The *first cotyledon* is determined by the input/output specification, which, in turn, is a formal description of what the ultimate system might do basically, and the *second cotyledon* is determined by the resources available to solve the problem, Fig. 1.2. In many cases, of course, the resources available to solve the problem will simply be defined as including any structure that can be built in the technology of the society in which the design process is taking place.

(b) Buildability

A structure Z is buildable in the technology $\mathscr{T}$ if and only if there exists a coupling recipe $\mathscr{H}$ such that $\mathfrak{X} \subset \mathscr{T}$ and $Z = RES(\mathscr{H})$.

(c) Technology Cotyledon

If $\mathscr{T}$ is a technology, then the technology cotyledon determined by $\mathscr{T}$ is denoted as $\mathscr{C}(\mathscr{T})$ and is defined as follows: $\mathscr{C}(\mathscr{T})=\{(\mathscr{H},Z) : \mathscr{H}$ is a coupling recipe, $\mathfrak{X} \subset \mathscr{T}$, Z is a structure, $Z=RES(\mathscr{H})\}$.

(d) Feasibility Cotyledon

A structure Z is implementable in a technology $\mathscr{T}$ if and only if there exists a system Z^r buildable in the technology $\mathscr{T}$ such that Z^r simulates Z.

Let $\mathscr{X}$ be an input/output specification and let $\mathscr{T}$ be a technology. Then the feasibility cotyledon determined by $\mathscr{X}$ and $\mathscr{T}$ is denoted $\mathscr{C}(\mathscr{X}, \mathscr{T})$ and is defined as follows:

$$\mathscr{C}(\mathscr{X}, \mathscr{T})=\{(Z, x, \zeta, U, \mathscr{H}, Z^r, Z^s, \rho, \mu, \zeta, U) \in \mathscr{C}(\mathscr{X}), (\mathscr{H}, Z^r) \in \mathscr{C}(\mathscr{T}), Z \text{ is simulated } (Z^s, \rho,\mu,\theta) \text{ by } Z^r\}.$$

An input/output specification $\mathscr{X}$, a technology $\mathscr{T}$, and the input/output cotyledon and technology cotyledon will be generated therefrom. When these artefacts are brought together, it will be discovered that there is really a third cotyledon namely, the feasibility cotyledon is generated by *both* the input/output specification $\mathscr{X}$ and the technology $\mathscr{T}$. The feasibility cotyledon consists of all structures that not only satisfy the input/output specification $\mathscr{X}$ but also are buildable or implementable in the technology $\mathscr{T}$. It is important that each point in the feasibility cotyledon is determined by a structure Z, and initial state x of the structure Z, an output function ζ for the structure Z, and a subset U of the time scale of the structure Z. It should be noted that such a list of artefacts determines an assertion that the structure Z satisfies an input/output specification. The next two items on the list defining a point in the feasibility cotyledon are a coupling recipe $\mathscr{H}$ and its resultant structure Z^r. Again, it should be noted that such a pair of artefacts determine a point in the technology cotyledon. The other artefacts on the list, namely Z^s, ρ, μ, and θ are necessary to assert that the structure Z is implementable in the technology $\mathscr{T}$ through the coupling recipe $\mathscr{H}$ and its resultant Z^r. This is extremely technical language which will be clarified later; perhaps it is better now to think simply of the feasibility cotyledon as consisting of all structures or configurations each of which satisfies the input/output specification $\mathscr{X}$ and is imnlementable in the technology $\mathscr{T}$.

CHAPTER 2

Concept of System

2.1. DEFINING SYSTEM

A system S is represented by a set, Fig. 2.1,
$S=\{N, G, U, A, X, O, Y, T\}$
where N represents the set of environment objects, $G{:}N \rightarrow U$ is the mapping of environment states into system inputs, U represents the set of admissible system input functions. Let, X represents allowable state space for the system when $A : U \rightarrow X$ is the state transition function for the system. Let the operational process of the system be,

$$O : U \otimes X \rightarrow Y$$

where Y is the set of possible output functions from the system. T represents the time scale.

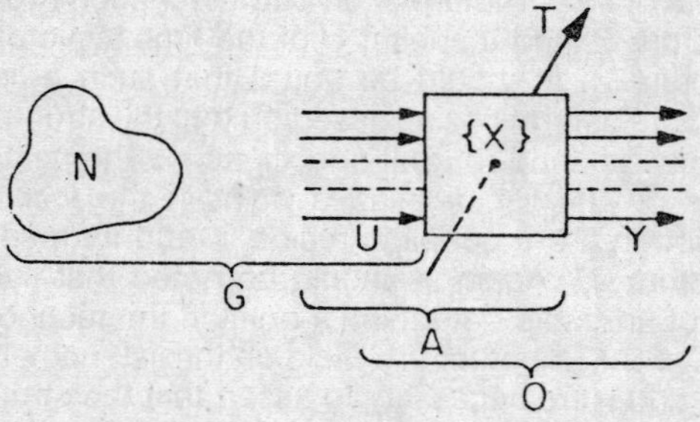

Fig. 2.1: Set representation of a system

2.2. MODELLING OF SYSTEMS

A system is a mathematical abstraction that is devised to serve as a model for a dynamic phenomenon. It represents the dynamic phenomenon in terms of mathematical relations among three sets of variables known as the *input*, the *output*, and the *state*.

A system therefore can be either an element, a configuration, or a structure that is responsive (outputs) to stimulus (inputs).

The *input* represents, in the form of a set of time functions or sequences, the external forces that are acting upon the dynamic phenomenon. In similar form, the *output* represents the measures of the directly observable behaviour of the phenomenon. Input and output bear a cause-effect relation to each other; however, depending on the nature of the phenomenon.

In the operation of a system thus three types of variables can be distinguished namely, input, output and intermediate (or state). If $u(t)$, $y(t)$ and $x(t)$ denote the vectors corresponding to the above variables in the order stated and the system is describable by a differential equation, the following two relationships hold

$$\frac{d}{dt}(u(t)) = \phi(u(t));\ x(t)$$
$$\frac{d}{dt}(y(t)) = \psi(u(t);\ x(t)) \qquad (1)$$

where ϕ and ψ are single valued functions of their arguments.

The assignment of any variable to the input or output vector depends on the interactions occurring on the interface between the system and its environment, on the directions of the existing causal links and on the purpose for which the description of the system is required.

The general relations among the input, output, and state of a system are illustrated in Fig. 2.2

Since the input, state, and output in general consist of sets of variables, they may be represented by vector quantities; for example, an m-variable input is written as

$$\{u\} = \begin{bmatrix} u_1 \\ u_2 \\ \cdot \\ \cdot \\ \cdot \\ u_m \end{bmatrix} = [u_i], \qquad i=1, 2,\ldots, m$$

The state and output are denoted by the vectors $\{x\}$ and $\{y\}$ respectively where $\{x\}$ is of dimension n and $\{y\}$ of dimension p.

2.3. PERSPECTIVES OF PRODUCTION SYSTEMS

Production is a process of transformation of a set of input elements into a set of desired output elements for increasing the utility of goods and services. The word, 'process' means the engineering methods or techniques for performing technological transformation.

Consider a set of input elements $\{u_i\}$ in Fig. 2.3 which undergo a technological transformation to yield a set of output elements $\{v_i\}$ in order to optimize an objective function B which not only includes the

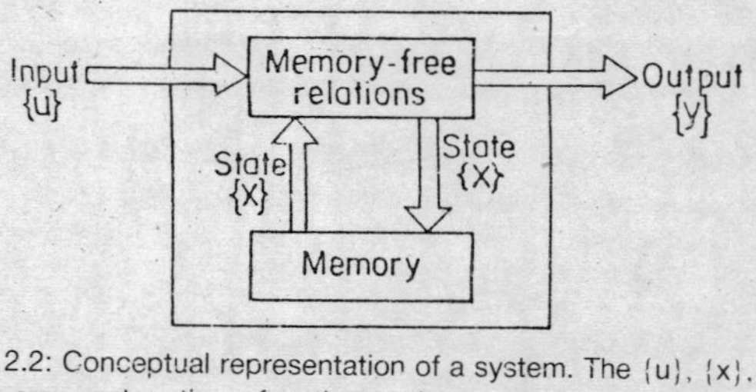

Fig. 2.2: Conceptual representation of a system. The {u}, {x}, and {y} are vector time functions of dimensions m, n, and p, respectively.

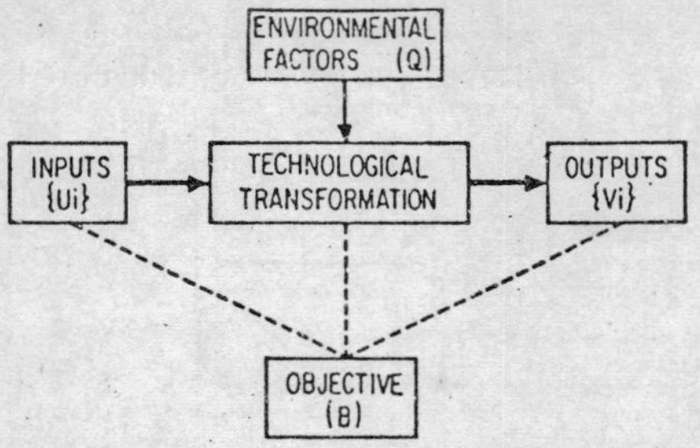

Fig. 2.3: Production process as an input-output system.

input and output variables, but also a factor Q dependent on the characteristics of the environmental conditions.

The optimality will, however, be governed by a set of constraints involving parametric combinations of input and output variables.

Thus

$$T\{u_i\} = \{v_i\}$$

in order that

$$B = B_{\text{optimum}} = f\,[u_i\ v_i\ Q]$$

subject to a set of constraints

$$C_j[\{u_i\}, \{v_i\}] < 0.$$

The criteria of optimality can be

(i) increased *utility* of goods and services,

(ii) enhanced *productivity* which is defined as the ratio between the value of outputs and the value of inputs.

The index of utility can be any of the economic criteria used in common practice like gross revenue, profit or rate of return, etc.

In a production or technological transformation system, the enhancement of productivity characterized by a production function essentially depends not only on the simple parameters of $\{u_i\}$ or $\{Q\}$ or $\{v_i\}$ but should also be viewed as a vector function of time (t). This factor enters not only in a given train of events from the time the input elements come into play until transformed to outputs of required objective function, but also plays an important role through historicities of scientific breakthroughs in the form of 'technological innovation and change'. A state has now reached where implementation of a production process is accomplished due to close-loop interaction of a large number of elements and events involving innovations, controls and changes of technological means of the production system at various levels of operating hierarchy. Fig. 2.4 depicts such a scheme of production system.

The products have got to conform with the required specifications, such as size, shape, surface finish, surface integrity and dimensional accuracy. Control functions are generated from any deviation from the specification of the products and are required to monitor and control the production system at its various stages.

In order to identify the problems in the future development of manufacturing technology and discuss the possibilities of application of advanced scientific tools of synthesis, analysis and optimization, a conceptual model of a production system with interconnected subsystems is proposed by Peklenik.

Fig. 2.5 indicates a possible structure of a production system with individual blocks representing subsystems such as managerial decisionmaking and control, product development, design for production, central information processing and integrated manufacturing systems. The input is specified by the motivation and ideas to manufacture a product for the market representing the output as a result of combined efforts of the system. The *major* feedback of the system is the market analysis affecting managerial decision-making and control. Internally, the *minor* feedbacks

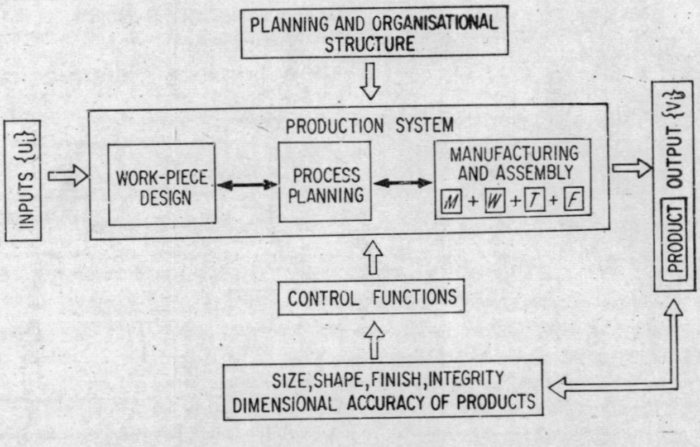

Fig. 2.4: Close-loop production system.

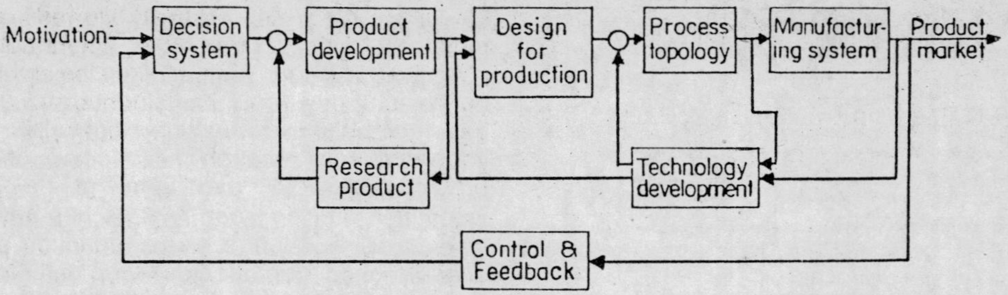

2.5: Conceptual model of production system with subsystems

are established for the subsystems: 'Product development', 'Design for production', 'Data processing' and 'Manufacturing systems'.

The data processing of geometrical and technological information for control and optimization of the integrated manufacturing systems are a fast expanding area, linking the design and manufacturing subsystem, and deserves a great deal of attention. New heuristic approaches in programming and information gathering, pattern analysis, graphics, system simulation, etc., are opening new vistas in approach and procedure.

2.4. MANUFACTURING SYSTEM

Manufacturing is a broader subsystem wherein unprocessed material brought as input is technologically transformed into a useful and desired product. During this process, the raw material may be modified physically or geometrically as shown after McAdams in Fig. 2.6. At a given time, *t*, the condition of the work piece can be defined by a vector of dimensions $\{d\}$ and a vector of material properties $\{p\}$ upon which, the manufacturing process can be viewed as a vector function of time (t).

The generalized term 'manufacturing' therefore describes the entire sequence of events from the time the raw material is brought into the production system until it is changed into the desired characteristics of a product. Fig. 2.7 represents the manufacturing process as the path generated when the work piece from its initial condition A changes to its final condition B.

A special class of manufacturing process is the 'machining system' where a *workpiece (W)* with the aid of a *'machine tool' (M)* and *'cutting tool' (T)* is transformed into a *product (P)*. Both the workpiece and cutting tool need *fixtures (F)* for proper holding, guidance in order that required size, shape and accuracy are obtainable. The machining system comprising of such elements is commonly described as *M-W-T-F* system and is schematically shown in Fig. 2.7.

The elements of the manufacturing system are identical with the machine tool's components. These components are connected with each other due to necessary functions. Within a machine tool there are seven complexes of functions to be distinguished. Each of them is necessary for performing the machining process. The components employed in performing these functions form seven subsystems, respectively. The subsystems as shown in Fig. 2.8 are:

1. workpiece system,
2. tool system,
3. dimension system,
4. kinematic system,
5. energy system,
6. information system,
7. component system.

The workpiece system consists of the workpiece

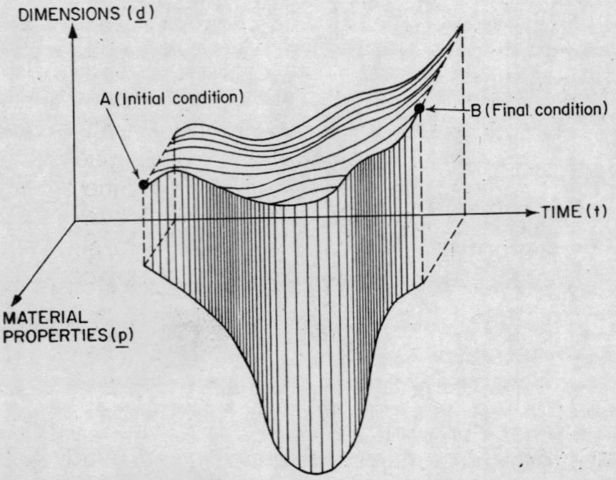

Fig. 2.6: Generalized concept of manufacturing process

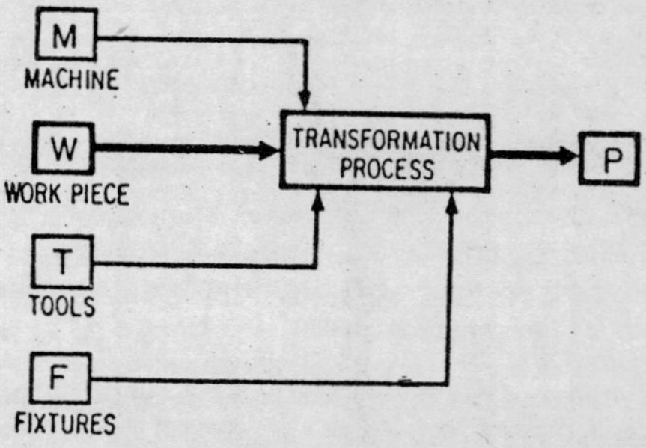

Fig. 2.7: *M-W-T-F* machining system

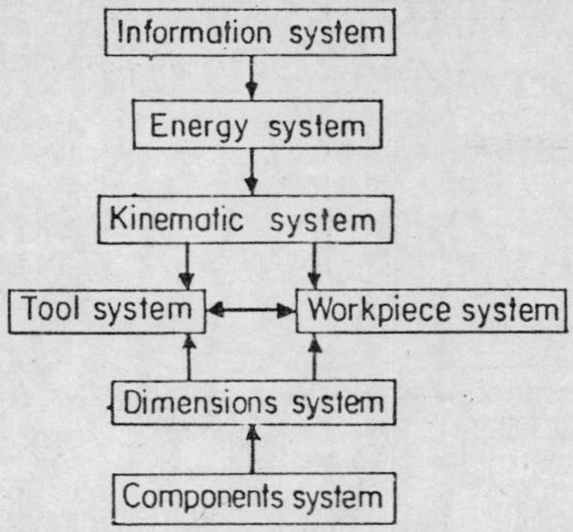

Fig. 2.8: Block diagram of manufacturing system

and all components of the machine tool necessary for clamping and changing the workpiece.

The tool system includes the tool and all components necessary for clamping and changing the tool.

The dimension system describes the geometrical relations within the machine tool.

The kinematic system yields all the components necessary for the relative motions of workpiece and tool.

In the energy system of a machine tool, all components transforming energy are considered.

The information system deals with the organisation responsible for programming the process.

The component system considers the configuration of the entire machine tool, except the workpiece and the tool. It includes the way of organization by which the design has been materialised.

2.5. CONCEPTUAL FRAME OF MANUFACTURING SYSTEMS

2.5.1. SPUR Model

By one model the manufacturing system can be defined consisting of a machine tool, the tools and the machining process.

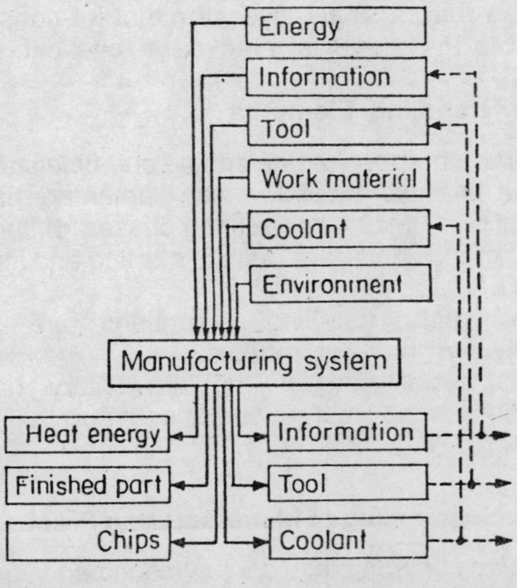

Fig. 2.9: Schematic diagram of analyitical manufacturing system

The further considerations will be concerned with the analysis of the manufacturing system. In order to optimize manufacturing systems, an evaluating algorithm is required to describe the technical value of a machine tool by means of various mathematical methods. The algorithm should contain all the parameters, Fig. 2.9, which determine the behaviour of the manufacturing system. Such a mathematical model has not yet been developed because of the machine tools' complexity.

2.5.2. Peklinik's Model

The basic idea in developing a new concept of so-called "manufacturing system" was to integrate the manufacturing process (*M*-process) with the machine tool (*M*) in a closed loop system for surface generation. It is impossible to analyze and optimize the manufacture of components without considering both major units of the system. In addition by introducing this concept, the powerful tools and methods of system engineering may be employed for synthesis, analysis and optimization. The integration of the manufacturing system with the computer, provides means for sophisticated automation representing a base for future development of the automatic factory.

Figure 2.10 reveals the basic principle of a manufacturing system of the first order. the manufacturing process, such as turning, boring, milling, grinding, etc., is placed in the forward loop. The machine tool with its subsystems for positioning (*P*-system), kinematics (*K*-system) and energy (*E*-system), is in the feedback loop and is aimed to provide the relative positions between the tool and workpiece, various speeds, such as cutting and feed speeds, and the required energy for removing a volume of material in a time unit. The combination of geometrical and kinematic elements yields the condition for generation of a workpiece with the required shapes and dimensions.

A manufacturing system of the first order may be manually controlled. However, according to a proposed classification the level of automation and control may be increased. The *P*-system and/or *K*-system are numerically controlled. In addition, the on-line identification of the manufacturing processes and the following optimization yields an adaptive controlled manufacturing system with substantially increased and sophisticated levels of control.

2.6. TOPOLOGY OF MANUFACTURING PROCESSES

2.6.1. Principal Concepts

Manufacturing engineering is a study of the processes required to make parts and to assemble these.

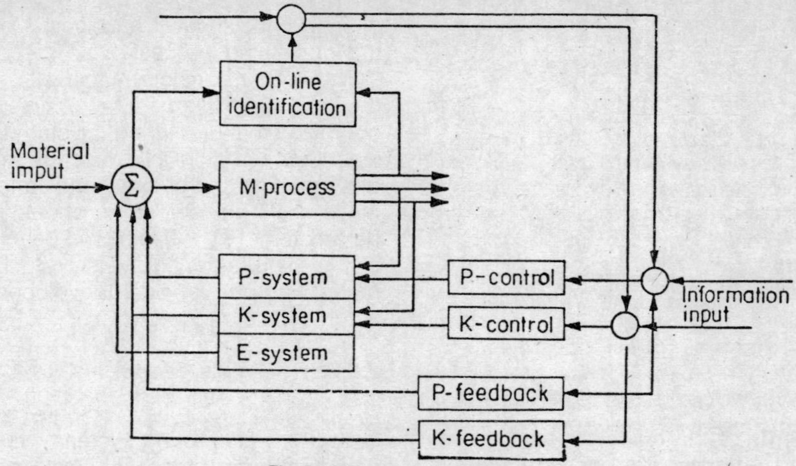

Fig. 2.10: Model of the process

The process of producing machines and mechanisms consists of the complex of operations for making the blanks, machining the blanks or stock to obtain finished parts, assembling the parts into units, and finally assembling the units and certain separate parts into machines.

The sum total of separate processes involved in the conversion of raw materials and semi-fabricated products into blanks, finished parts, units and mechanisms is the production process. The production process is thus the procedure followed in a plant for converting semi-finished products or raw materials into the finished product.

2.6.2. Machining Operations

Machining operation is the part of the production process directly concerned with changing the form and dimensions of the part being produced.

More elaborately, an operation is a part of the manufacturing process for machining a single part (or group of parts) performed at one workplace continuously until starting to repeat the same machining procedure for the next part in the same lot or batch.

Thus, the whole sequence of actions constituting a definite finished work cycle, and performed by the operator and machine tool continuously in reference to one part (or group of parts) is called an operation.

Therefore, the basic subdivision of the machining process or machining cycle is the operation.

Transporting parts, their handling and storage, as well as other elements of the production process, *not* directly concerned with changes in form and state of the part, are not considered to be components of the manufacturing process.

2.6.3. Operation Elements

In connection with machining processes, the terms station', 'setting' and 'operation element' are often employed.

A *station* is each of the different positions of the workpiece relative to the machine during one setting.

A *setting* is that part of an operation, that is, performed with a single clamping of the workpiece.

An *operation element*, as referring to machining, is a part of an operation in which the work surface, cutting tool, and speed and feed remain unchanged.

The machining of different sections of the workpiece surface simultaneously by several cutting tools is indicated in operation and instruction sheets as a single operation element. A number of similar operation elements, directly following each other and performed at one station, are also listed as a single operation element.

2.6.4. Cuts/Passes

Operation elements are subdivided into cuts or passes.

A *cut* or *pass* is the part of the operation element performed by one working movement of the cutting tool in the direction of the feed.

As a rule, a single layer of metal (of constant or variable thickness) is removed in one cut.

2.6.5. Handling Elements

For the purpose of rate setting and time study, the labour process is divided into elementary handling operations which are distinct phases of the work performance and have definite starting and stopping points.

Elementary handling operations are further subdivided into unit motions.

This subdivision of the elementary handling operation is the smallest component part that can be measured in a time study.

2.6.6. Sequencing of Manufacturing Processes

A blank or workpiece can be machined in different ways. The manufacturing process may consist of a small number of operations performed in a correspondingly small number of machine tools. The same blank can, however, be machined in a great many machine tools, in many operations, each consisting, as a rule, of one simple kind of machining.

The sequence of manufacturing operations for producing a part may be planned in different ways in accordance with the prevailing conditions. The classification of the processes is shown in Fig. 2.11.

For example, a stepped shaft may be rough turned in one operation consisting of a large number of operation elements performed consecutively. The same shaft may be turned in several operations, each consisting of a small number of operation elements. A multiple-tool lathe may be employed for performing several operation elements simultaneously.

(A) *Differentiation*

If the manufacturing process is broken down into simple operations, each containing a small number of operation elements, this may be called a *differentiated* manufacturing process, Fig. 2.12.

The manufacturing process is differentiated into the simplest single-element and two-element operations in certain stages for machining a part in large-lot and mass production. This is usually associated with the machining of various types of surfaces in operations requiring small labour input (for example, centreing shafts, chamfering in finishing up parts after machining on an automatic lathe, etc.).

The degree of differentiation is determined by a number of factors, such as the size and weight of the part, required output, type of machining, conditions and required labour input for setting up the workpiece on the machine tool.

(B) *Consecutive Concentration*

In cases when a large number of consecutive operation elements are performed on a machine tool, such an arrangement of the manufacturing process is called consecutive concentration, Fig. 2.13.

The number of operation elements in a given operation characterize its complexity and comprehensiveness or, as it is called, 'degree of concentration'.

For lot production, process planning is usually along the lines of consecutive concentration of the machining operations. Only one workpiece is machined and the operator consecutively performs the work required for all operation elements that can be done conveniently on the given machine tool. For example, the machining operation by a single tool may be done in a machine by the tracing of a template.

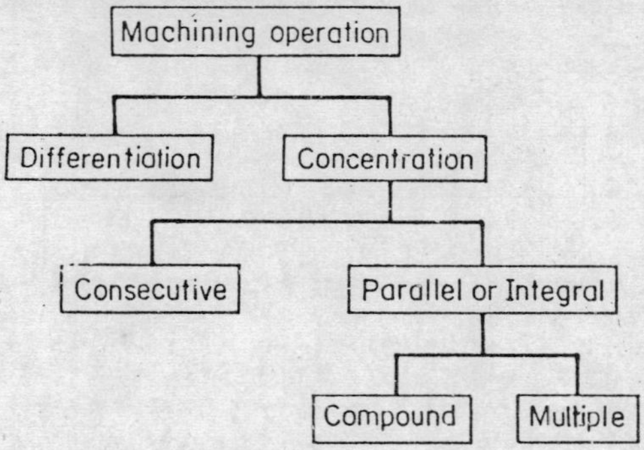

Fig. 2.11: Classification of the manufacturing processes

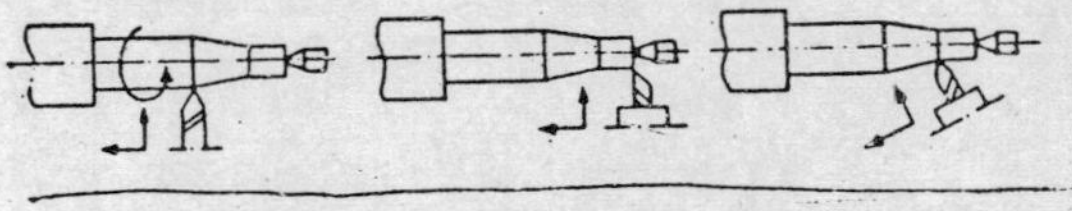

Fig. 2.12: Differentiating a manufacturing process into simple operations

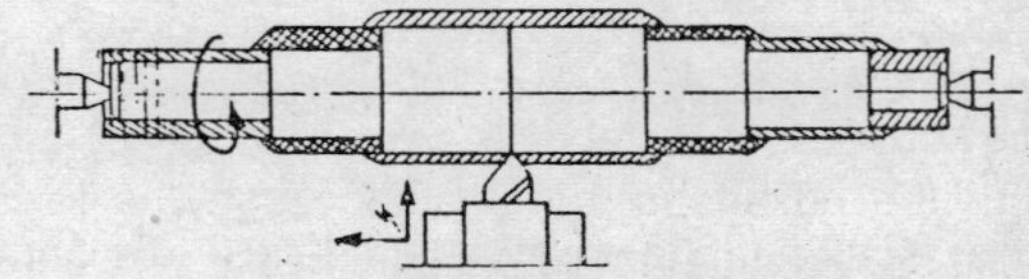

2.13: Consecutive concentration of the maching operations

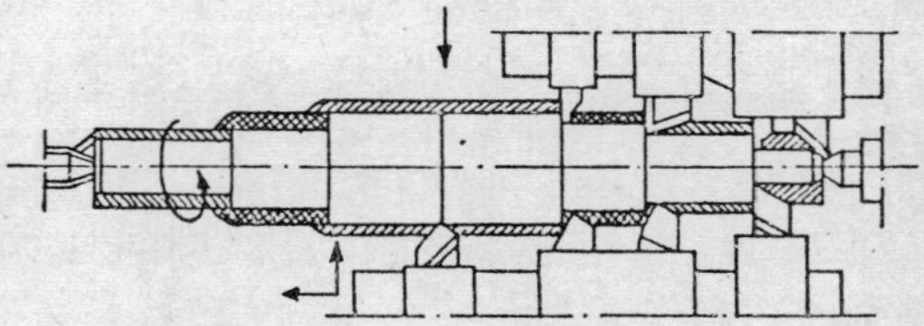

Fig. 2.14: Integral or parallel concentration of operations with multiple cutting-tools in simultaneous action

(C) *Integral or Parallel Concentration*

If a considerable number of operation elements are performed in parallel, simultaneously in a single operation, such an arrangement is called *integral* or *parallel concentration* of the manufacturing process, Fig. 2.14.

As a rule, parallel concentration of the manufacturing process is associated with the application of multiple cutting-tool machines (multiple-tool, multiple-spindle, etc.) which provide for a high-production capacity. The application of such machine tools, however, may be economically feasible only when a large output of the particular parts are required.

(i) Compound Cuts

This is a method by which cuts can be taken both

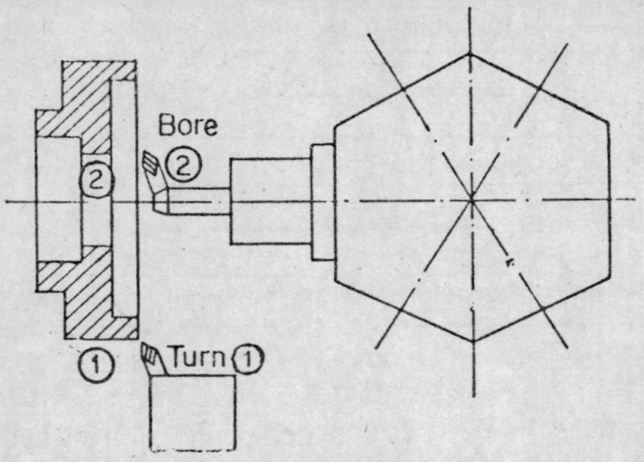

Fig. 2.15(a): An example of compound cuts

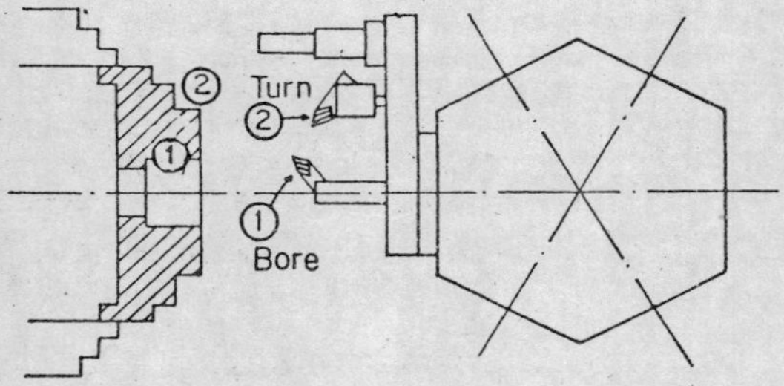

Fig. 2.15(b): An example of multiple cuts

from the hexagonal turret station and the square turret station at the same time. This overlapping helps to reduce actual cycle time of machining. Fig. 2.15 (a) shows an example of compound cut.

(ii) Multiple Cuts

The taking of more than one cut from the same station is called multiple cutting. An example is turning of the outside diameter and boring of a hole at the same time with tools held in one turret station. A typical example of multiple cut is shown in Fig. 2.15 (b).

2.6.7. Assembly Process

The *assembly process* for machines and mechanisms is the part of the production process concerned with the consecutive joining of the finished parts into assembly units and complete machines, of a quality that meets the manufacturing specifications.

Assembly tasks thus result from the requirement to fit together certain individual parts, dimensionless substances and sub-assemblies into assemblies of final products of higher complexity in a given quantity or within a given unit of time. Assembly therefore represents a cross-section of the problems within the whole of production engineering, but very different activities and assembly processes are performed in the individual branches of industry. Assembling can thus be thought of series of actions as follows:

ASSEMBLING means

- bunkering
- feeding into magazin
- forwarding
- branching-off
- organizing
- rotating
- turning
- allocation
- feeding-in
- taking output
- positioning
- gripping
- releasing

CHAPTER 3

Fundamental Machining Principles

3.1. WORK-TOOL CONTACT

The motions of a cutting tool with respect to workpiece produce the machined surface which occupies an entire series of consecutive positions. The envelope of a family of surfaces $f(X, Y, Z, t)=0$ is expressed by a system of equations

$$f(X, Y, Z, t) = 0$$

$$\frac{\partial f (X, Y, Z, t)}{\partial t} = 0$$

The surface, $\bar{t}_s$, tangent to a position M at the machined surface of Fig. 3.1 must contain the initial tool surface. The cutting edge of the tool must lie on the tangent plane. If the cutting edge project beyond the boundaries of surface $\bar{t}_s$, this will cut into the machined surface during machining operation and an excess material will be removed from the workpiece. Thus, the workpiece and cutting tool form two conjugate links which contact each other at the plane $\bar{t}_s$ at point M in course of machining. A condition of contact can be developed when $\bar{V}$ is the velocity in the tangential plane in order that point M of the cutting tool profiles its conjugate point on the work piece.

The conditions of contact can be written as the dot product of velocity vector $\bar{V}$ in the tangent plane and vector $\bar{N}$ in the normal plane when

$$\bar{N} \cdot \bar{V} = 0$$

This condition determines the points of contact of conjugate surfaces at successive intervals of time. The set of contacting points in a system related to the workpiece is the surface of the workpiece and the set of contact points in a system related to the cutting tool gives the initial tool surface. For cases of $\bar{N} \cdot \bar{V} \neq 0$, two cases may emerge as shown in Fig. 3.2. The first case is a situation of mutual penetration or interference. For the cases

$$0 < \frac{\bar{N} \cdot \bar{V}}{|N||V|} \leq 1$$

penetration of profiles is observed and this must be avoided. A condition of machining is to provide contact of the initial tool surface with the workpiece without interference. There are two types of interferences. The first kind consists of mutual penetration of the conjugate surfaces within the contact zone. The second kind is the mutual penetration of conjugate surfaces beyond the contact zone.

However, when a situation arises such that

$$-1 \leq \frac{\bar{N} \cdot \bar{V}}{|N||V|} < 0$$

where the condition emerges for 'no contact' or relief. In these cases, there is a departure of the tool surface from the work surface.

In Fig. 3.3, normal to the tangent plane at M_0 is perpendicular to both $\bar{t}'_B$ and $\bar{t}''_B$ and hence parallel to $\bar{t}'_B \times \bar{t}''_B$, the cross-product. The normal to the surface at P is the same as the normal to the tangent plane at P.

If a trajectory is a boundary line or a set of points located at the conjugate surfaces of adjoining parts, it is necessary that the velocity $\bar{V}$ of the points on the boundary line at the moment of profiling be directed tangentially to the line or be equal to zero. Considering the machining of the work piece configuration shown in Fig. 3.4, the boundary of W_1 and W_2 will be the line $\bar{AB}$. Two normals $\bar{N}_1$ and $\bar{N}_2$ can be drawn and for compliance of contact conditions

$$\bar{N}_1 \cdot \bar{V} = 0 \text{ and } \bar{N}_2 \cdot \bar{V} = 0$$

This implies that the velocity vector is perpendicular to both the normals indicating that the velocity is directed along a tangent to the boundary AB between W_1 and W_2.

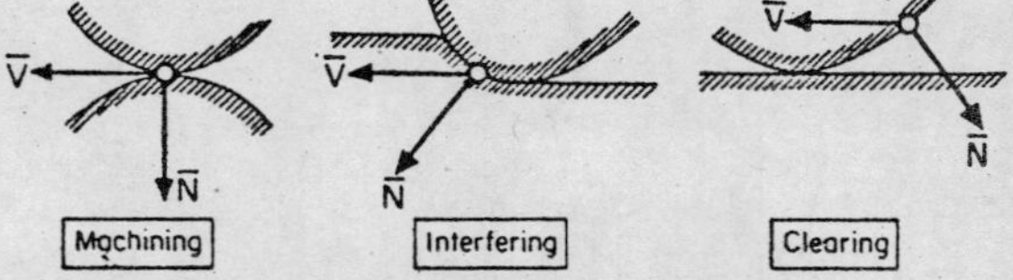

Fig. 3.2: Condition of contact N̄·V̄ = 0, N̄·V̄ > 0 and N̄·V̄ < 0 respectively

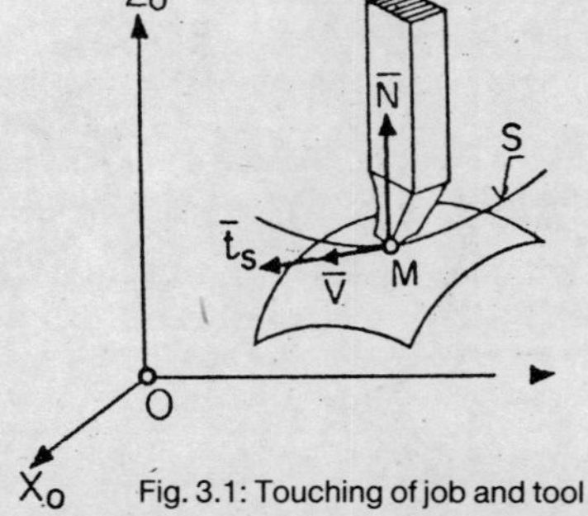

Fig. 3.1: Touching of job and tool surfaces

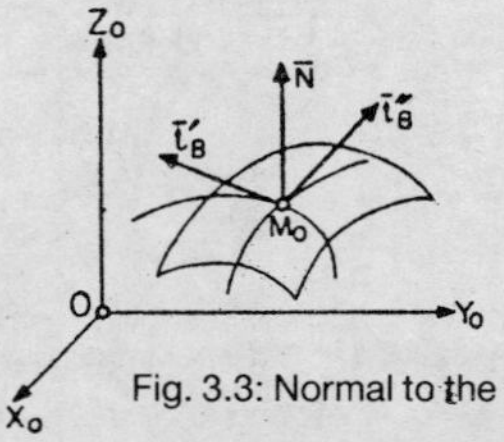

Fig. 3.3: Normal to the tangent plane

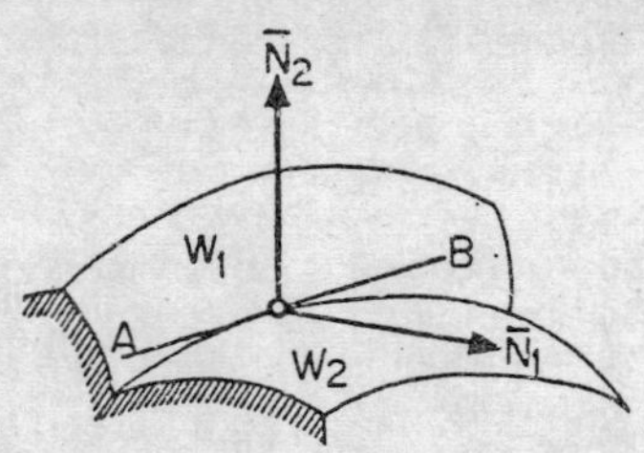

Fig. 3.4: Trajectory as a boundary line between two surfaces

If the velocity V is vector sum of two components such that

$$\bar{V} = \bar{V}_1 + \bar{V}_2$$

then for machining condition

$$\bar{N}_1 \cdot \bar{V} = 0 \text{ and } \bar{N}_2 \cdot \bar{V} = 0$$

The normal vector to any plane to be machined can be systematically determined. Consider a profile B of power screw as shown in Fig. 3.5. The parameter p is defined by

$$p = \frac{s}{2\pi}.$$

A longitudinal cross-section of the screw reveals the profile where a point M is located. The points M and M_0 lie on the same helical profile and the co-ordinates for M will be

$$r = x_0, \quad \phi = \phi_0$$
$$Z = Z_0 + p\hat{\phi}_0$$

In cartesian form

$$x = r\cos\phi_0, \quad y = r\sin\phi_0, \quad Z = Z_0 + p\hat{\phi}_0.$$

Further

$$\tan\beta = \frac{2\pi r}{s} = \frac{r}{p}$$

$$\tan\xi = \frac{s}{2\pi r} = \frac{p}{r}$$

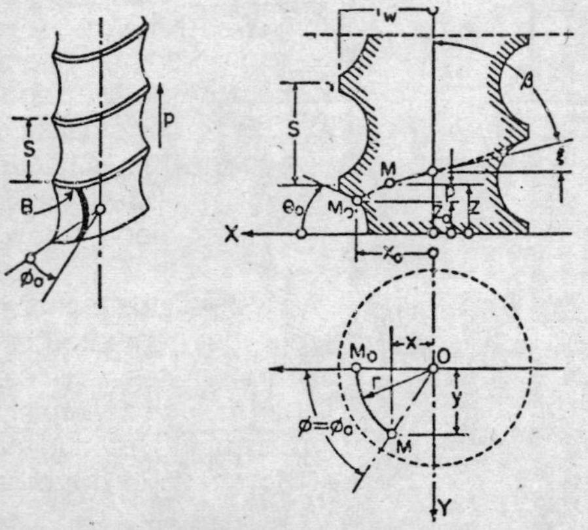

Fig. 3.5: Profile of a power screw

In Fig. 3.6 a transverse section of the screw profile B is shown on which a point M is located such that

$$\phi = \phi_c + \phi_0$$

$$Z = p\hat{\phi}_0$$

ϕ_c measures the location of M with respect to line of symmetry while ϕ_0 is the shift of line of symmetries of x-axis.

The tangent vectors $\bar{t}\,'_B$ in x-y plane and $\bar{t}\,''_B$ drawn at the point M are given by

$$\bar{t}'_B = i\cos(\theta_c + \phi_0) + j\sin(\theta_c + \phi_0)$$

$$\bar{t}''_B = -i\sin\beta\sin(\phi_c + \phi_0) + j\sin\beta\cos(\phi_c + \phi_0) + k\cos\beta.$$

Hence

$$\bar{N} = [\bar{t}'_B \times \bar{t}''_B]$$

$$= \begin{vmatrix} i & j & k \\ -\sin\beta\sin(\phi_c+\phi_0) & \sin\beta\cos(\phi_c+\phi_0) & \cos\beta \\ \cos(\theta_c+\phi_0) & \sin(\theta_c+\phi_0) & 0 \end{vmatrix}$$

$$= -i\cos\beta\sin(\theta_c+\phi_0) + j\cos\beta\cos(\theta_c+\phi_0) - k\sin\beta\cos(\theta_c - \phi_c)$$

If the velocity vectors in two-orthogonal direction are known, then

$$\bar{N}\cdot\bar{V}_1 + \bar{N}\cdot\bar{V}_2 = \bar{t}'_B\cdot\bar{t}''_B\,\bar{V}_1 + \bar{t}'_B\cdot\bar{t}''_B\,\bar{V}_2 = 0$$

This is the condition of work-tool contact at point M.

3.2. MACHINABLE SURFACES

A surface is usually defined as a continuum of consecutive geometrical positions obtained in the motion of one geometrical generating line called the *generatrix (gx)* along a path called *directrix (dx)*. Fig. 3.7 shows the generalized from of machinable surface where the *gx* line G is made to traverse a defined *dx* path B. Over this directrix a point P_B is determined by chosen co-ordinate system. Normal section to the surface will reveal the path of B, as

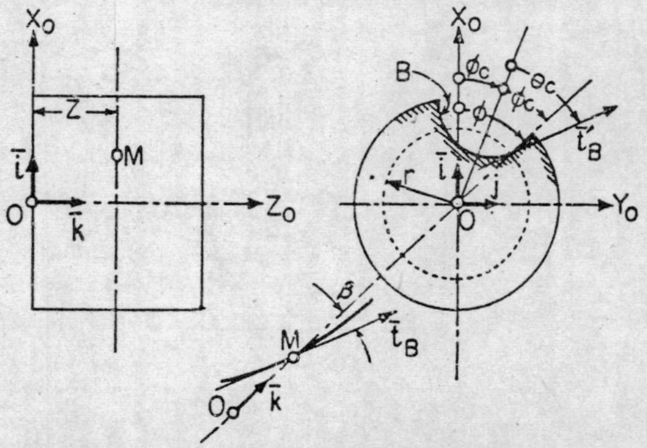

Fig. 3.6: Transverse section of the screw profile

shown in the Fig. 3.7. Depending on the manner in which *G* is traversed along *B* rotational or non-rotational surfaces will be produced (Fig 3.8). The normal plane will reveal that the workpiece has edges bourne out of several elementary surfaces.

Sometimes, a surface is conditionally broken into a series of elementary surfaces each of which is separately machinable [Fig. 3.9 (a) and (b)]. Same surfaces may be enveloping or internal (Fig. 3.10).

Most of the geometrical surfaces can be shaped by two geometrical generating lines with appropriately chosen relative motions. The general scheme is shown in Fig. 3.11.

The surfaces produced by geometric generatrixes are classified according to the variability of generating lines. A number of combinations showing how surfaces of revolutions or flat surfaces are produced have been shown in Fig. 3.12 and Fig. 3.13.

The machining of a surface involves the production of the geometrical lines as a result of whose formative motions the required surface is produced.

3.3. KINEMATICS OF WORK-TOOL INTERACTION

All motions producing or resulting in a path or profile may do so by either of the following techniques:

(i) tracing,

(ii) tangent tracing or enveloping,

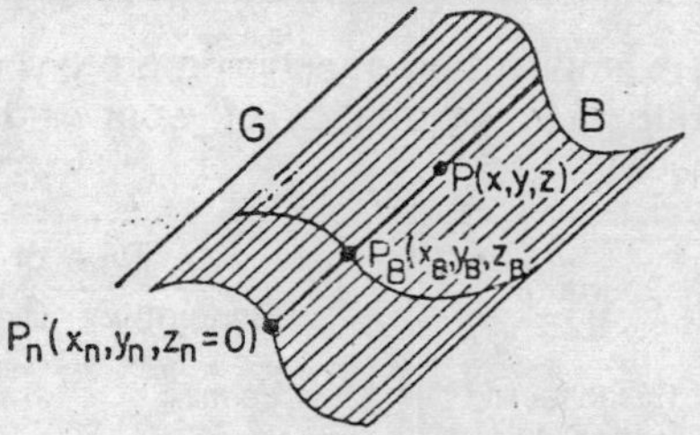

Fig. 3.7: Generalized form of machinable surface

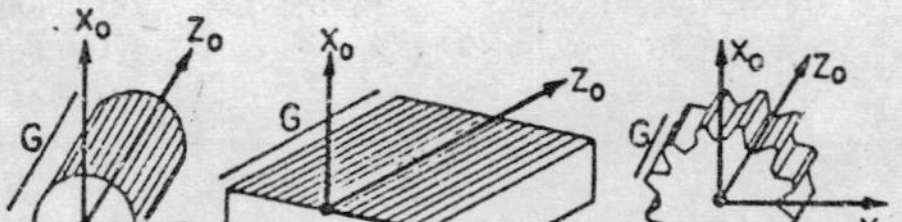

Fig. 3.8: Rotational & non-rotational surfaces

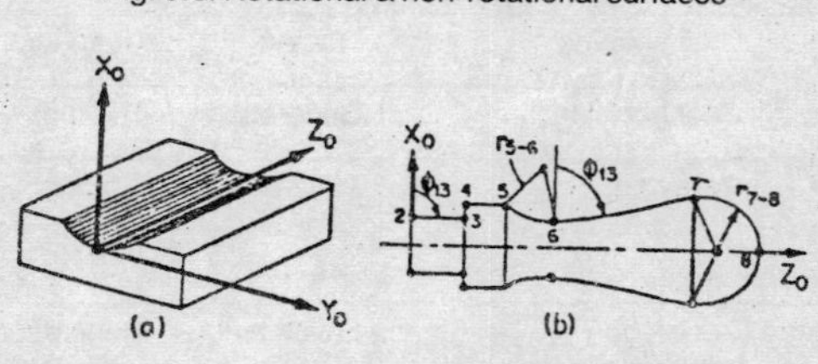

Fig. 3.9: Mixed surface

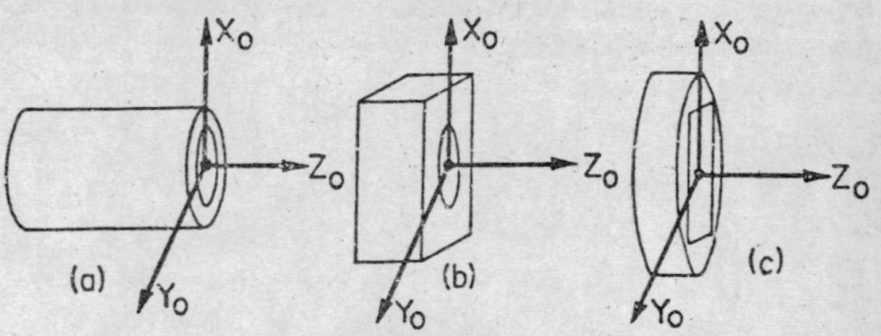

Fig. 3.10: Internal surfaces

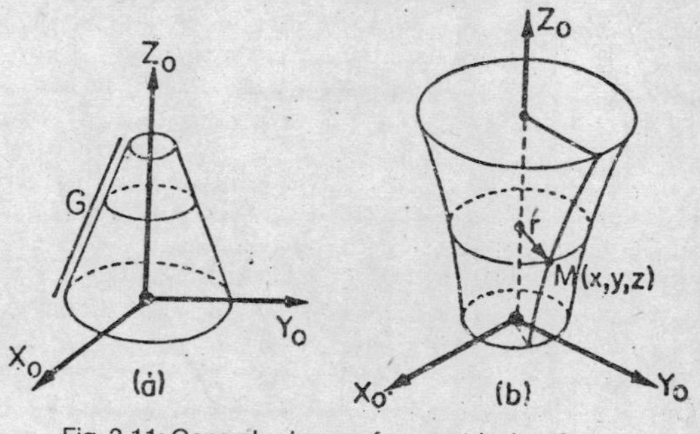

Fig. 3.11: General scheme of geometrical surfaces

(iii) forming,

(iv) generating.

In the *tracing method,* a short length of cutting edge of the tool considered as a point produces a line to be formed as the trace left behind (Fig. 3.14).

When the trace of a point continuously changes direction with the addition of extra formative motion, a series of traced paths as shown in Fig. 3.15 (a) and (b) results whose tangent trace or envelope will be the produced line or surface. An example is shown in Fig. 3.16.

Sometimes, the produced profile or surface is a true replica in form and extent of the trace of cutting points (i.e. the work surface becomes a conjugate surface of the tool edge). This method is known as 'forming' technique (Fig. 3.17). Condition of conjugate surface at any point of the profile requires that the contact condition $\bar{r}_B'\ \bar{r}_B''.\ \bar{V} = 0$. This, in turn, means that the relative velocity vector must be tangent to this contact surface. It also implies that there is no formative motion in the direction of common normal.

Examples of producing profiles by this forming principle are indicated in Fig. 3.18 (a), (b) and (c).

When a line or a surface, curved or straight, *rolls* along another line or surface which is obtained by the envelope of the continuously mapped conjugate points of the rolling element, the method is known as 'generation principle'.

In other words, the envelope of the continuous positions of a rolling element is a generated line or surface. The word rolling connotates the relative motion of two lines or surfaces in such a manner that the consecutive points or elements of one come successively into contact with those of the other in their order.

In the first example, shown in Fig. 3.19 (a), the

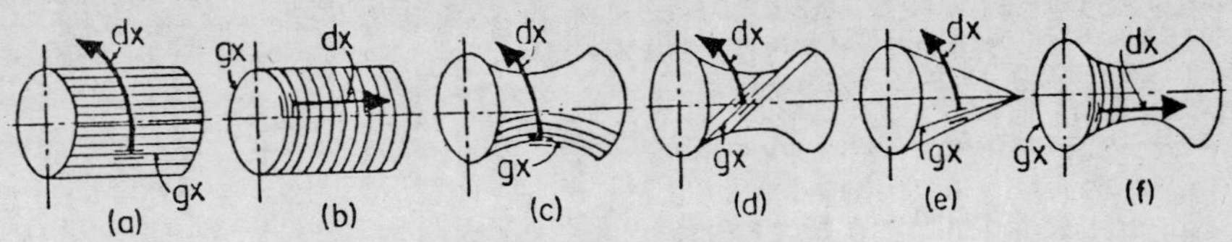

Fig. 3.12: Surfaces of revolution

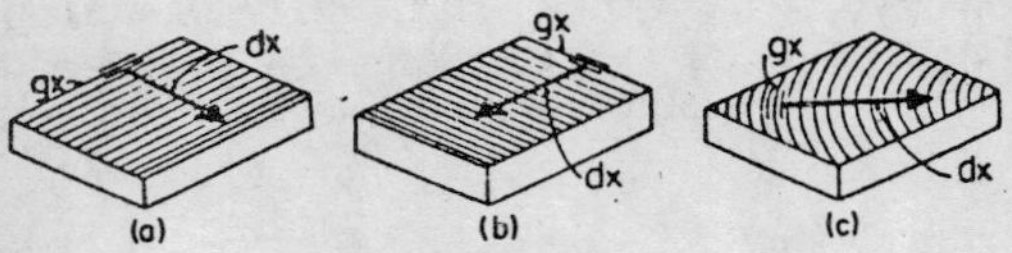

Fig. 3.13: Flat surfaces

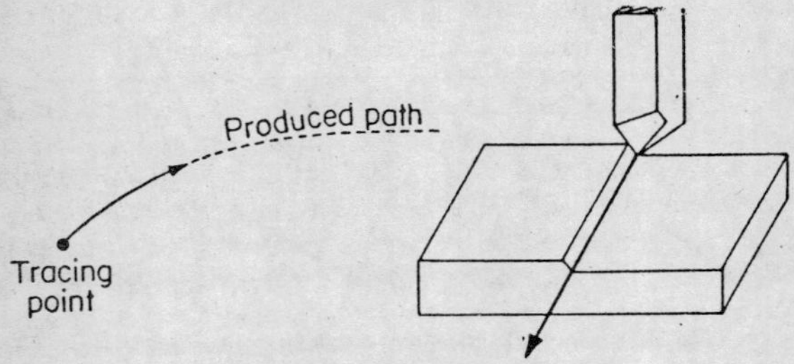

Fig. 3.14: Tracing method

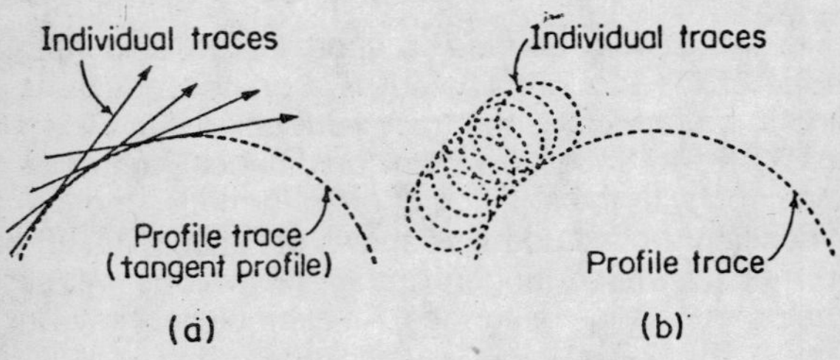

Fig. 3.15: Continuous change of disection

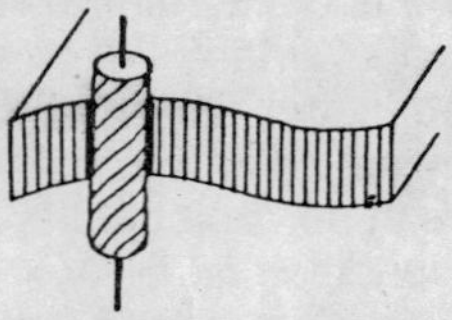
Fig. 3.16: An example

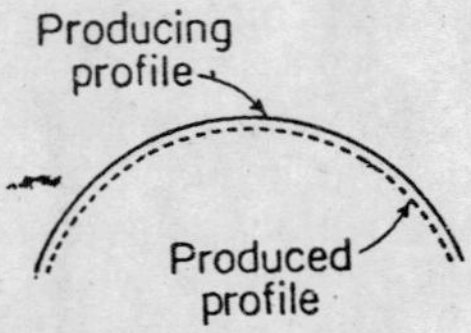

Fig. 3.17: Replication of the tool form: Forming principle

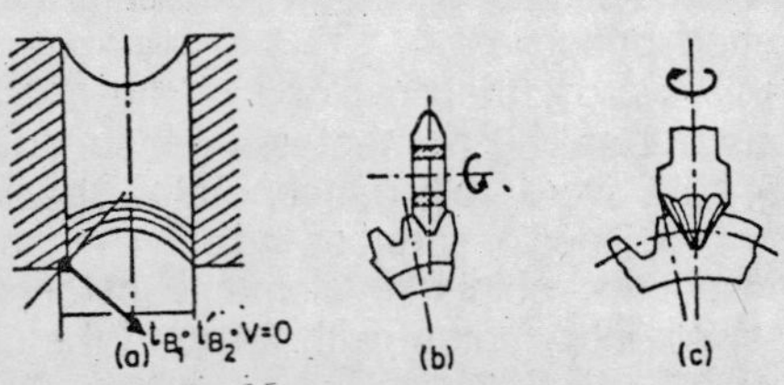

Fig. 3.18: Profiles generated by forming principle

machining is carried out by a rotating cutter which rolls on the cylindrical workpiece and thus *generates* the form. In the second example, the generatrix is a rack which moves in such a manner as if a straight line is rolling on a circle thus producing a gear profile which is of *'Involute Geometry'* shown in Fig. 3.19 (b)

These four basic principles of cutting action form the basis of devising formative motions for producing a generatrix or a directrix.

It has been noted earlier that all geometrically definable surfaces are combinations of a generatrix motion and a directrix motion.

The theoretical number of combinations between generatrix types and directrix types is 16. However, some of these combinations are not feasible which are indicated in the chart.

Methods for Producing Generatrix	Methods for Producing Directrix			
	Tracing	Enveloping	Forming	Generation
Tracing	√	×	×	√
Enveloping	√	√	×	√
Forming	√	√	×	√
Generation	√	√	×	√

A few examples from machining are given below to illustrate how kinematics of work and tool is co-ordinated:

Group	Operation	Type of	
		Generatrix	Directrix
Thread profile	Thread turning (with single point tool)	Forming	Tracing
	Thread milling (with single thread cutter)	Forming	Enveloping (Tangent tracing)
Flat surface	Shaping (with single point tool)	Tracing	Tracing
	Slab milling	Tracing	Enveloping
Gear	Gear shaping	Tracing	Generation
	Gear hobbing	Generation	Tracing
Helical relief surface	Relieving	Tracing	Tracing
Spur gear	Generation grinding	Enveloping	Generation

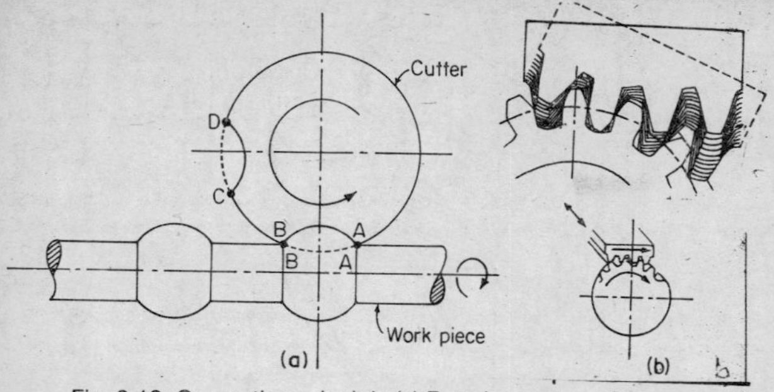

Fig. 3.19: Generation principle (a) Rotating cutter – form generation (b) Rack – involute profiling

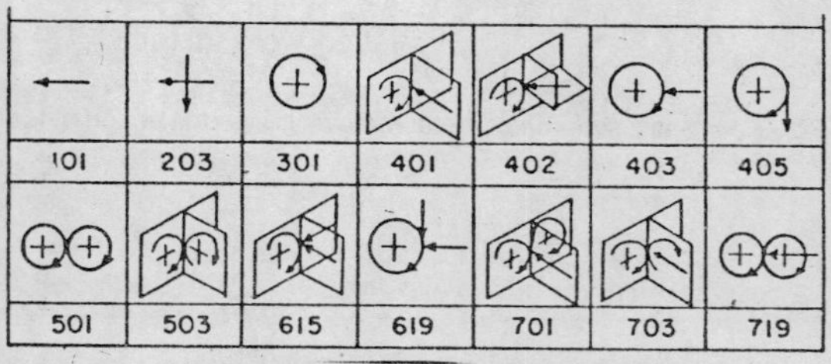

Fig. 3.20: Granovski's classification

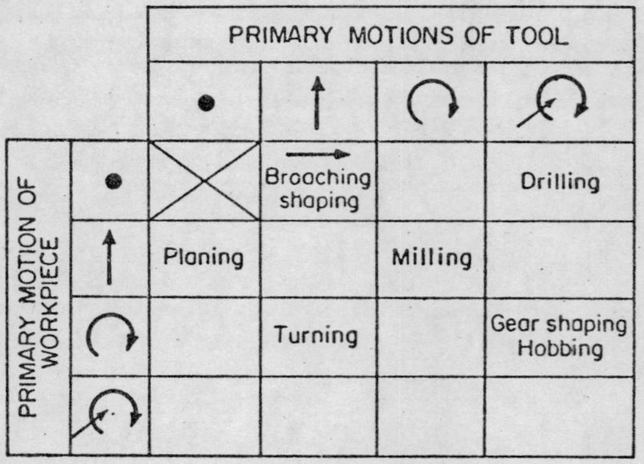

Fig. 3.21: Kaczmareck's method

There are alternative methods of describing work-tool interactions involved in various metal cutting operations. These methods are based on fundamental motion elements like translation or rotation or combination thereof such as rolling, spiralling, helical, etc. As some of the directrix and generatrix motions are invertible, primary motions are aggregated to depict the resultant metal cutting method. A kinematic classification has been suggested by Granovski as shown in Fig. 3.20, the type 101 indicates shaping or planing strokes or broaching. The other types are: 203 (hack-saw cutting), 301 (broaching of circular segments), 401 (drilling or turning), 402 (taper turning), 403 (end milling), 405 (slab milling), 501 (rotary milling), 503 (rotary turning), 615 (contour turning), 619 (contour milling), 701 (gear shaping), 703 (hobbing or drum cam milling), 719 (disc cam milling or relieving).

If tool and work piece motions are seggregated, another methodology can be developed. Kaczmareck has developed such a method indicated in Fig. 3.21.

Based on general process of shaping, sizing or finishing any workpiece to given size or dimension, the basic machining is classified into several groups depending on the nature of chip formation as shown in Fig. 3.22.

3.4. MACHINING

Basically, machining is chip-forming metal cutting process where the excess material is removed away. The initial work material consists of *P* and *E* schematically indicated in Fig. 3.23.

In order to make *P*, compatible with desired size or shape, an amount *E* is removed by an organized kinematic scheme in the form of chips. However, in almost all machining operations involving chip formation, a special wedge-shape element called cutting tool is used. Fig. 3.24 shows the basic scheme by which the wedge-shaped cutting tool

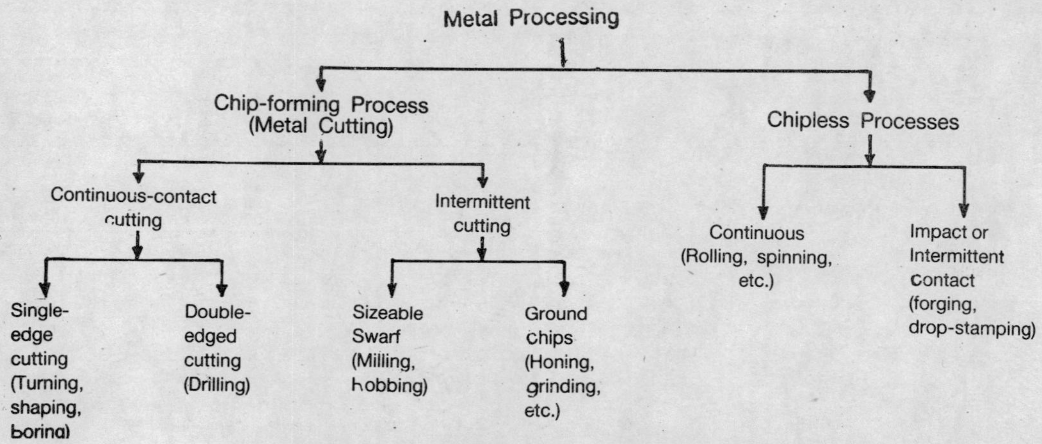

Fig. 3.22: Basic classification of machining on the nature of chip formation

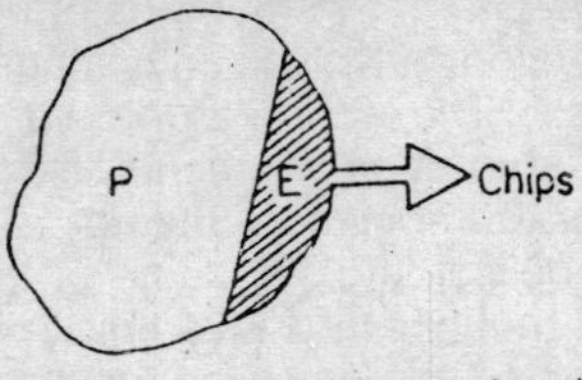

Fig. 3.23: Removal of excess metal as chips

removes excess material in the form of chips. The material in the shaded zone is pressed by the wedge and made to flow through plastic deformation over the face of the wedge as chips.

To define, machining is a metal cutting in which a finished surface of desired size, shape, accuracy or finish is obtained by separating the excess layers of material in the form of chips with the aid of an wedge-shaped device called *cutting tool.*

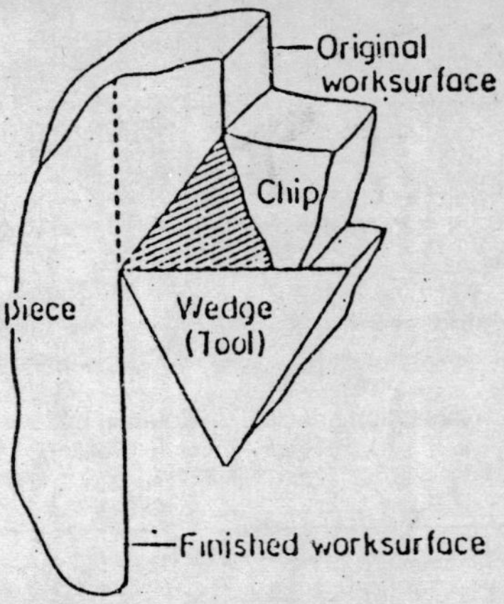

Fig. 3.24: Basic scheme of metal removal by wedge action of tool

CHAPTER 4

Process Tolerancing and Dimensional Chains

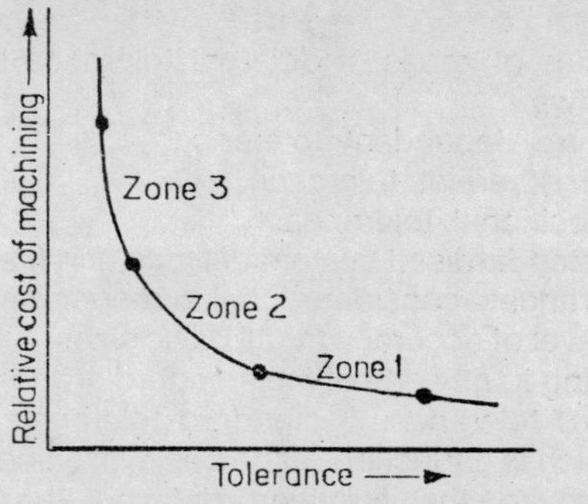

Fig. 4.1. Cost of machining as a function of tolerance.

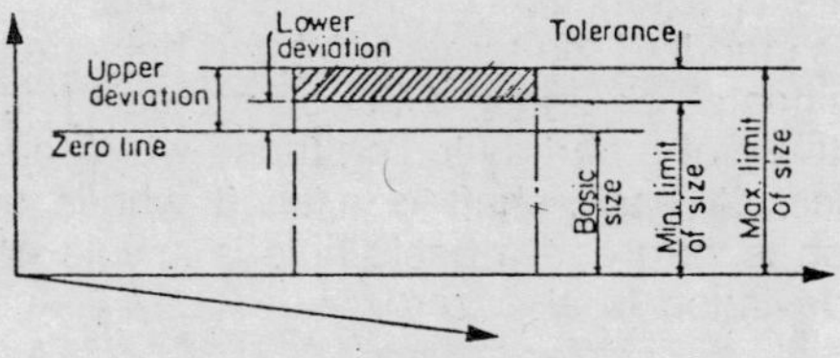

Fig. 4.2. Fundamental dimensions.

4.1. TOLERANCE

As it is impossible to produce a feature to an exact size, certain errors are to be allowed in the workmanship. The amount of error which is permissible will depend on the accuracy required and is known as tolerance.

In the design specifications of tolerance information is often provided. The ultimate goal of providing tolerance information is to show what variations in the shape of a part are acceptable. Tolerance is chosen so that it is possible to manufacture the part and to guarantee that the part will function correctly despite its possible variation in shape. It is useful to analyze a tolerance specification to check that these requirements are met.

One has to be careful in choosing the grade of accuracy. Although a higher grade of accuracy ensures a stable quality in assemblies and closer tolerances of the fits on mating part, it also increases the cost of machining as shown in Fig. 4.1. The machining cost can be divided in three zones. In zone 1 the machining cost is low and change in tolerance will not have much effect. In zone 2 the decrease in tolerance will increase the machining cost markedly and in zone 3 there will be a sharp increase in cost. A balance has to be obtained between the cost of machining and the quality of tolerance. The designer has to be more careful when working in zones 2 and 3. If the quality of fit is not impaired a little increase in tolerance can give a lot of saving in zone 3.

4.2. DIMENSIONAL SIZE

A number expressing in a particular unit the numerical value of a length is called *size.* The size of a part as may be found by measurement is called *actual* size.

The two extreme permissible sizes between which the actual size is contained are known as limits of size. The greater of the two limits of size is called maximum limit of size and the smaller of the two limits of size is called minimum limit of size. The size with reference to which the limits of size are fixed is called *basic* size (Fig. 4.2).

4.3. DIMENSIONAL DEVIATION

The algebraic difference between a size (actual, maximum, etc.) and the corresponding basic size is called *deviation.*

The algebraic difference between the actual size and the corresponding basic size is called *actual* deviation, between the maximum limit of size and the corresponding basic size the *upper* deviation, and between the minimum limit of size and the corresponding basic size the *lower* deviation. The symbol used for upper deviations is *'es'* for shaft and *'ES'* for hole. For lower deviations the corresponding symbols are *'ei'* and *'EI'*.

4.4. ZERO LINE

In a graphical representation of limits and fits, a straight line to which the deviations are referred is called the *zero line.* It is the line of zero deviation and represents the basic size. *Positive* deviations are shown above the horizontal zero line and *negative* deviations below it. One of the two deviations conventionally chosen to define the position of the tolerance zone in relation to the zero line is known as *fundamental* deviation.

4.5. TOLERANCE ZONE

The tolerance is equal to the algebraic difference between the upper and lower deviations and has an absolute value without sign. In a graphical representation of tolerance the zone bounded by the two limits of size of the part is defined by its magnitude (i.e. tolerance) and by its position in relation to the zero line, Fig 4.3.

There are various types of tolerances but for simplicity tolerances of position, size and form are only dealt with here. A tolerance zone for a cylindrical feature is an infinitely long cylindrical shell. It has three characters, position, size and thickness depending on the type of tolerance. In the case of a cylindrical zone its 'size' is its diameter. The properties of zones of different tolerance types are as follows:

(a) form dependent tolerance,
(b) size variant tolerance,
(c) positional tolerance.

In a standardised system of limits and fits, a group of tolerances considered is corresponding to the same level of accuracy for all basic sizes. A tolerance belonging to any one of the standard grades is called *standard* tolerance. A standard tolerance unit is a unit which is a function of the basic size and which is common to the formulae defining the different grades of tolerances. The tolerance is designated by a number symbol called the *grade.* The value of the tolerance is a function of the basic size.

In this standard the terms shaft and hole are used to designate all external and internal features respectively of a part including those which are not cylindrical. A basic shaft is a shaft whose upper deviation is zero and a basic hole is a hole whose lower deviation is also zero.

4.6. GRADES OF TOLERANCE

Eighteen grades of tolerances are provided with designations IT 01, IT 0 and IT 1 to IT 16. These are

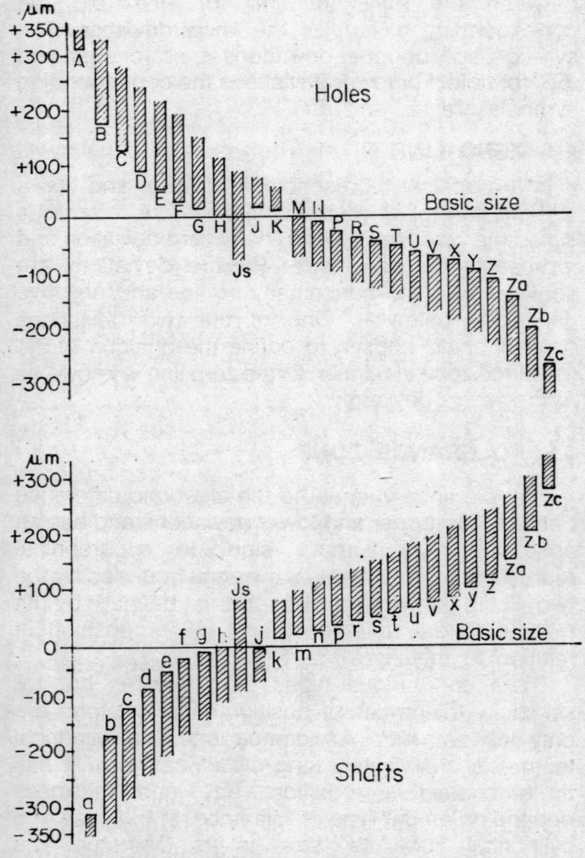

Fig.4.3. Alphabatic code for tolerancing system.

known as the *standard* tolerances. For grades 5 to 16 the values have been determined in terms of the standard tolerance unit *i*, where

$$i\ (\mu m) = 0.45\ \sqrt[3]{(D)} + 0.001\ D\ (D \text{ in mm}).$$

The relative magnitude of each grade is as given below:

Grade	IT 5	IT 6	IT 7	IT 8	IT 9	IT 10	IT 11
Value	7 *i*	10 *i*	16 *i*	25 *i*	40 *i*	64 *i*	100 *i*

IT 12	IT 13	IT 14	IT 15	IT 16
160 *i*	250 *i*	400 *i*	640 *i*	1000 *i*

The values of standard tolerances corresponding to grade 01, 0 and 1 are as given below (*D* in mm, IT in µm):

$$\text{IT } 01 = 0.3 + 0.008\ D$$
$$\text{IT } 0 = 0.5 + 0.012\ D$$
$$\text{IT } 1 = 0.8 + 0.020\ D$$

The values of IT 2 to IT 4 have been regularly scaled approximately geometrically between the values of IT 1 and IT 5.

4.7. FUNDAMENTAL DEVIATIONS

The fundamental deviations are calculated by means of formulae provided in the standard. The formulae are different for different shafts. The other deviation may be derived directly using the absolute value of the standard tolerance IT by means of the following algebraic relationship:

$$ei = es - IT$$

or

$$es = ei + IT$$

Hole limits are identical with the shaft limits of the same symbol but disposed on the other side of the zero line.

4.8. FITS

Fit is the relationships existing between two parts, shafts and holes, which are to be assembled, with respect to the difference in their sizes before assembly. The basic size which is common to the two parts of a fit is known as *basic size of a fit.* The arithmetical sum of the tolerances of the two mating parts of fit, when the difference between the sizes of the hole and shaft in a fit is positive, is known as *clearance* and when negative *interference.* A fit which provides always a clearance is called clearance fit, which provides always an interference is called interference fit and which may produce either a clearance or an interference fit is called *transition* fit.

The minimum clearance is the difference between the minimum size of the hole and the maximum size of the shaft. The maximum clearance is the difference between the maximum size of the hole and the minimum size of the shaft. The arithmetical difference between the maximum size of the hole and the minimum size of the shaft before assembly, in an interference fit is the minimum interference and between the minimum size of the hole and the maximum size of the shaft before assembly is the maximum interference.

4.8.1. Dimensioning of Cylindrical Fits

The joint formed by a shaft in a hole, as shown in Fig. 4.4, can result in either a clearance or interference fit, depending on the relative dimensions of the two parts. In order to secure a better mental picture of the condition of the fit between the two parts, a bar diagram has been formed by plotting the diametral tolerance *h* for the hole, and the diametral tolerance *s* for the shaft to large scale, as shown in the figure. A clearance fit results when there is a clearance between the smallest hole and largest shaft. The value of the clearance *a* for the tightest condition is technically called the *allowance*. An interference fit results when the shaft is larger than the hole, as illustrated in Fig. 4.4 (b).

All the parts for the tightest condition in Fig. 4.4 are at the limit containing the maximum amount of metal. Likewise, the parts for the loosest condition contain the minimum metal. For a clearance fit, the most critical situation, or the condition of least possibility of assembly, occurs for maximum metal. For an interference fit, the most critical situation, or the condition of least possibility of securing a tight joint, occurs for minimum metal.

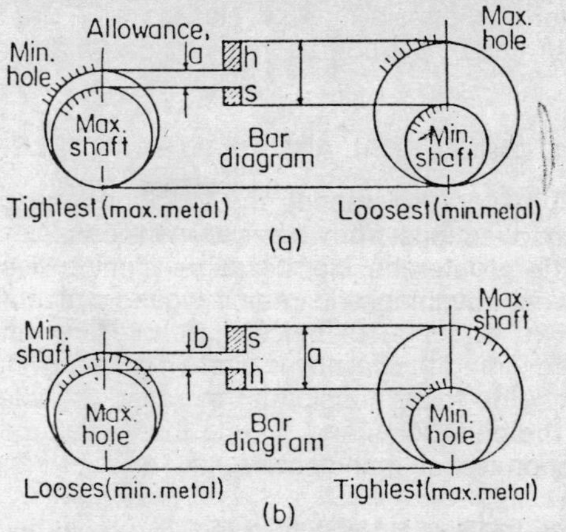

Fig. 4.4. Cylindrical fits in (a) clearance, (b) interference.

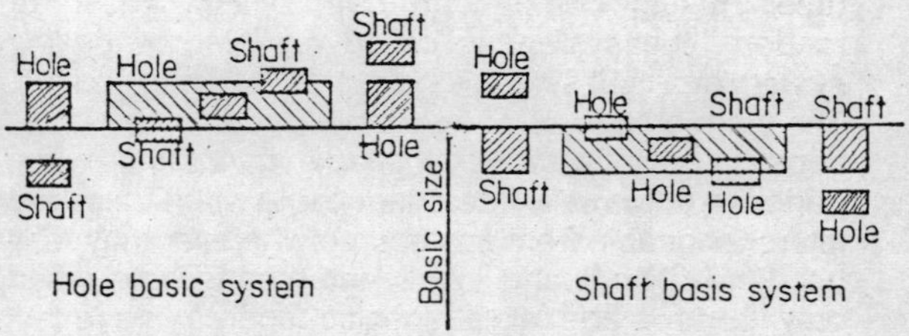

Fig. 4.5. Hole basis and shaft basis system.

4.9. SYSTEM OF FITS AND LIMITS

Limit system. It is a system of standard tolerances and deviations.

Fit system. It is a system of fits comprising shafts and holes belonging to a limit system.

Hole basis system. It is a system in which the different clearances and interferences are obtained by associating various shafts with a single hole whose lower deviation is zero (symbol *H*) (Fig. 4.5).

Shaft basis system. It is a system in which different clearances and interferences are obtained by associating various holes with a single shaft, whose upper deviation is zero (symbol *h*).

4.10. SELECTION OF FITS

IS : 2709–1964 gives recommendations for the selection of fits in engineering problems concerned with the mating of a shaft and a hole. The application of either hole basis or shaft basis system depends on many conditions, such as the nature of the product, the manufacturing methods, the condition of the raw material, etc. The hole basis system is the most extensive in use. This is because a hole is more difficult to produce than a shaft due to the fixed character of hole producing tools. The shaft basis system, however, is used mostly in industries using semi-finished shafting as raw materials. Given below are some hole producing processes to achieve different grades.

1. *H* 5—normally by precision boring, fine internal grinding or honing
2. *H* 6—by precision boring and honing or possibly by hand reaming
3. *H* 7—by grinding, broaching or careful reaming
4. *H* 8—boring or machine reaming
5. *H* 9—boring and reaming
6. *H* 10—drilled holes
7. *H* 11—coarse drilled punched or pressed holes

With the basic hole *H* 7 shafts *a* to *h* produce a clearance fit, shafts *j* to *n* produce transition fit and shafts *p* to *z* produce interference fit.

Fits between mating parts are designated by the terms 'class' and 'grade'. The class of the fit refers to the allowance *a*; the grade refers to the maximum clearance *b* for an assembly at the loosest condition.

Various possibilities of classes and grades are illustrated by the bar diagrams for shafts and holes in Fig. 4.6.

4.11. UNILATERAL AND BILATERAL TOLERANCES

In bilateral tolerances the variations are given in both directions from basic dimensions.

The unilateral system of dimensioning possesses distinct advantages in interchangeable manufacture. In Fig. 4.7, the tolerances for hole and shaft are represented to scale by h_1 and s_1, respectively. Later it may be found feasible to increase the tolerances on the parts to h_2 and s_2, and thus effect additional economies of manufacture.

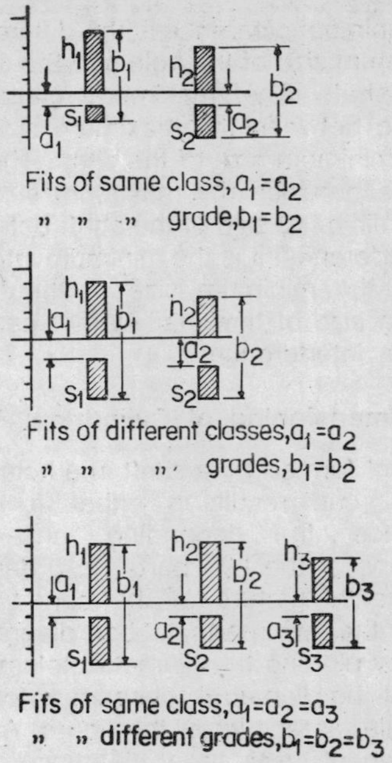

Fig. 4.6. Classes and grades for clearance fits.

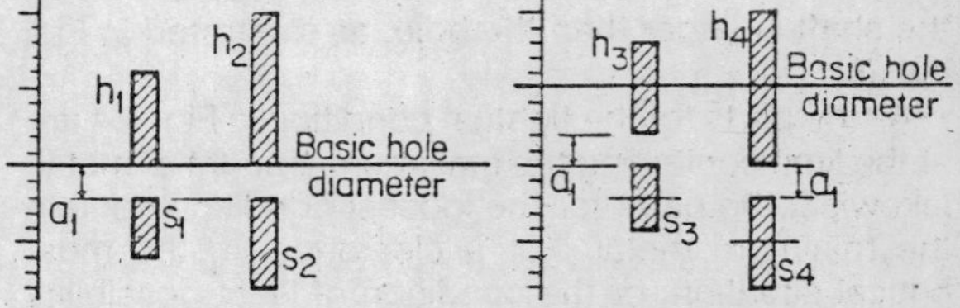

Fig. 4.7. Tolerancing of holes and shafts to scale.

4.12. INTERCHANGEABILITY

When a system of this kind is being used so that any one component will assemble correctly with any other mating component, both being chosen at random, the system is called an *interchangeable system* or a limit system or a system of limits and fits. The manufacture of components under such conditions is known as *interchangeable manufacture.* The parts are universally interchangeable when these are interchangeable from any source wherever they may be derived from and locally interchangeable when only parts made at a specific source are interchangeable. Production on an interchangeable basis results in increased output with a corresponding reduction in manufacturing cost.

The *advantages* of interchangeability are:

1. It lessens human skills to fit components and this reduces assembly time.
2. It facilitates the production of mating components at different places.
3. It facilitates repair or replacement of worn or defective components.
4. It lowers cost of maintenance when replacing parts.

4.13. SELECTIVE ASSEMBLY

BS: 2517–1954 defines selective assembly as a procedure in which parts of any one type are classified into several groups according to size (Fig. 4.8). The parts which are intended to be mated with these are also classified according to size in the same number of groups. Corresponding groups are then expected to assemble and function properly.

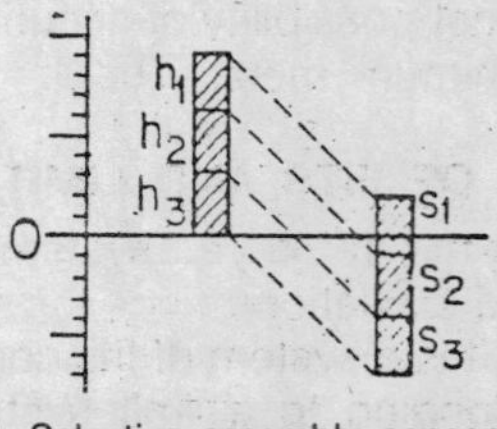

Fig. 4.8. Selective assembly concept.

No difficulty arises in maintaining suitable tolerances if the fit is very loose. The same is true for very tight fits or fits having considerable interference. However, when the difference between allowance and maximum clearance must be small, the tolerances for hole and shaft may become excessively small, with consequent increase in the cost of production.

These costs may be minimized by selective assembly, which consists of first making each of the mating parts with large tolerances and then sorting them into groups having small tolerances.

4. 14. DIMENSIONAL ANALYSIS

4.14.1. Dimension Chain

A dimension chain is a consecutive series of interrelated linear and angular dimensions which form a closed outline and refer to a single part or a group of parts. A compulsory requirement to all dimension chains is that the interrelated dimensions form a closed contour. Dimension chains may be classified as follows:

1. Dimension chains comprising linear dimensions

and parallel links. The majority of dimension chains are of this type (Fig. 4.9a).

2. Dimension chains comprising linear dimensions and non-parallel links (Fig. 4.9b). In general, any polygon can be regarded as this type if each side is specified by a linear dimension. If all the dimensions are projected on a single axis, this kind of dimension chain can be reduced to the first type.

3. Dimension chains with angular dimensions as for instance a number of holes equally spaced around a circle if the centre distances of the holes are specified in angular measures (Fig. 4.9c). This type also includes all cases in which a circle is divided into equal divisions, such as gears, worm wheels, milling cutters, etc. Because of the difficulties encountered in making angular measurements, angular dimension chains are frequently replaced by dimension chains with non-parallel links. Such dimensions can be checked by means of gauges for centre-to-centre distances.

4. Dimension chains in space are chains in which a series of dimensions do not lie in a single plane. Such chains are rarely found in practice and are solved by projecting all the dimensions onto a single plane.

The dimensions of a part or unit that make up a dimension chain are called the links of this chain. The minimum number of links required to form a dimension chain are three.

A member of a dimension chain is a dimension determining the relative distance or angular displacement of the surfaces of parts or their axes (the elements of bodies of revolution). The members of various dimension chains shown in Fig. 4.10 specify the distances between the surfaces of parts or their axes. The members shown in Fig. 4.11 specify the relative angular displacement of surfaces and axes of parts.

Deviations from squareness are apprehended by means of deviations from parallelism. This is done in machine-building practice by erecting a perpendicular in some manner (for instance, with a square or frame level) to the surface in reference to which the deviation from squareness of the other surface is to be checked. The deviation of the surface being checked from parallelism to this perpendicular is the basis for judging its deviation from squareness to the relevant surface.

The foregoing is schematically represented by drawing a dash line perpendicular to the surface chosen as the datum, and connecting the surface of the part that should be square to this datum with the perpendicular by means of a line having an arrowhead at one end, as shown in Fig. 4.12.

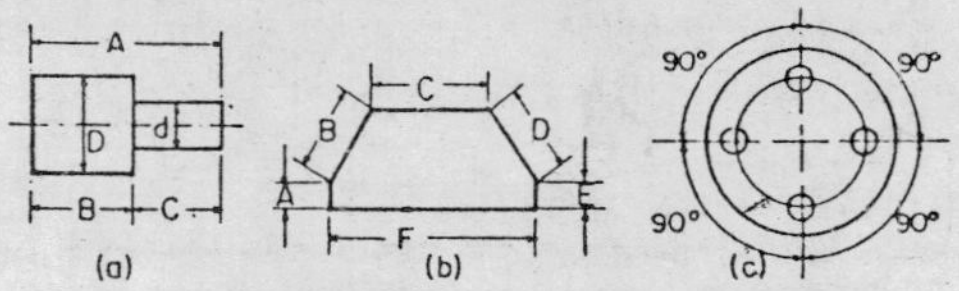

Fig. 4.9: Types of dimension chains.

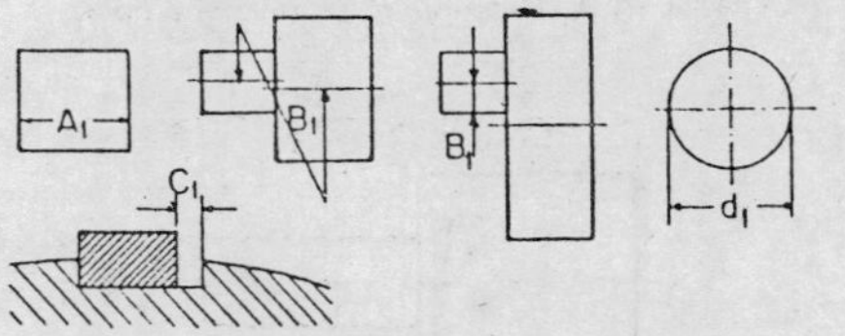

Fig. 4.10: Examples of dimension chain members that specify the distances (dimensions) between surfaces of the part.

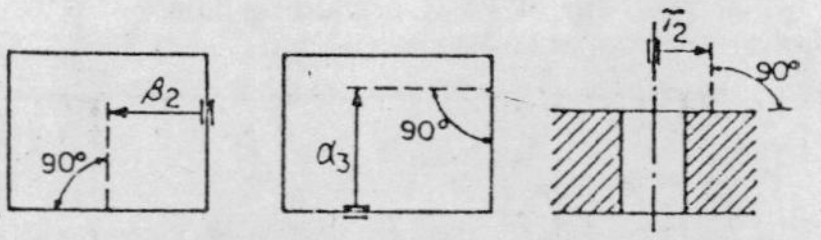

Fig. 4.11: Examples of dimension chain members that specify the relative angular displacement between surfaces of the part.

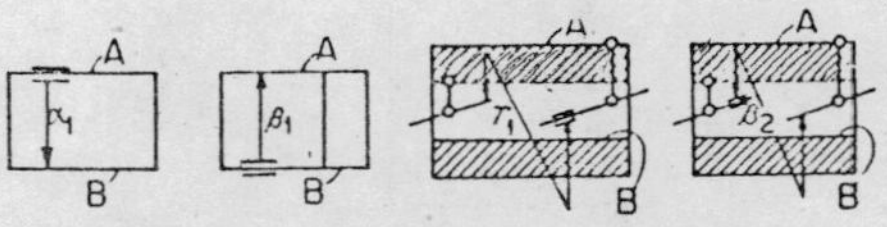

Fig. 4.12: Symbols showing the deviation from squareness of one surface of a part to another surface.

A *concluding link* is one that is obtained as a final element in manufacturing or assembling parts. Any link may be the concluding one in accordance with the sequence of operations in machining or assembly.

The initial or concluding member of a dimension chain is one that directly interconnects part surfaces or axes whose relative distances and angular displacements are to be specified or measured. An example is the distance between the axes of the centres of a lathe measured in one of the planes (Fig. 4.13).

It is called the *initial* member if the construction of the dimension chain by means of which the given problem is solved begins with this member.

It is called the *concluding* member when, being last in constructing the dimension chain, it interconnects the surfaces of axes of the part whose position is to be specified or measured.

To distinguish it from the other members, it is convenient to denote the initial or concluding member by a capital letter with the subindex Δ, for instance A_Δ, B_Δ, C_Δ, α_Δ, β_Δ, etc.

4.14.2. Initial or Concluding Member of Dimensional Chain

All the links of a dimension chain may be classified as component links and the concluding link.

4.14.3. Increasing and Decreasing Member

An *increasing link* is one that increases the concluding link when it itself is increased. A

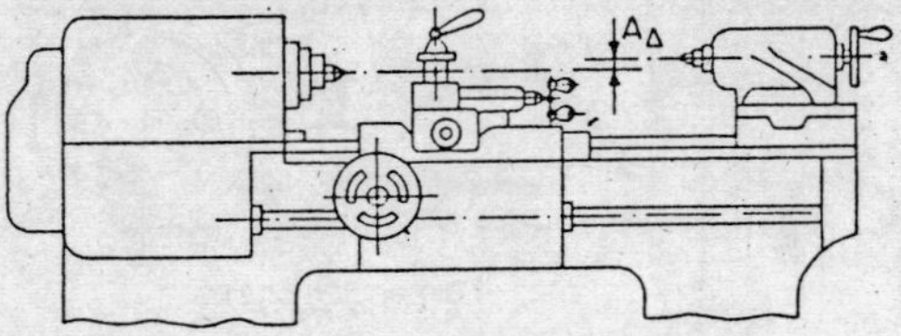

Fig. 4.13: An example of an initial member.

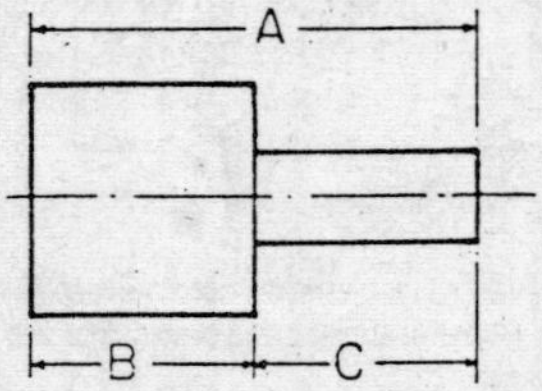

Fig. 4.14. Concluding link.

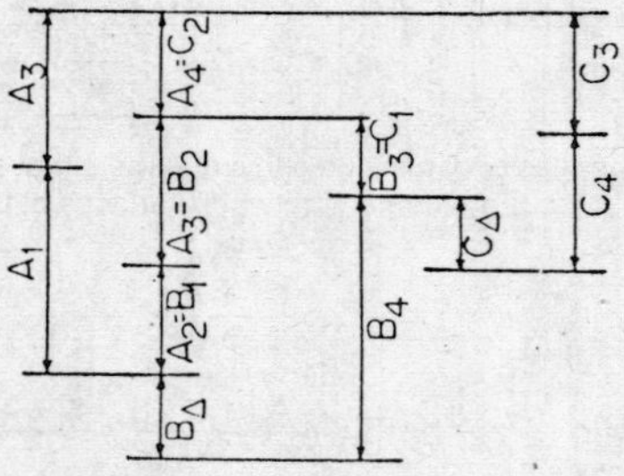

Fig. 4.15: Parallel-interlinked dimension chains.

decreasing link, on the contrary, decreases the concluding link when being increased. For example, if link C is the concluding link (Fig. 4.14), then A will be an increasing link and B a decreasing link.

Proper dimensional analysis depends upon whether the increasing and decreasing links have been correctly determined.

A general rule for determining increasing and decreasing links is to set up an equation of the nominal dimensions, interrelating all the links of a dimension chain, and to solve this equation for the nominal size of the concluding link. Then, on the right-hand side of the equation, all the terms having a plus sign are increasing links and all those with minus signs are decreasing links.

4.14.4. Linking of Chains

Parallel-interlinked dimension chains have one or several common members. Such chains are shown in Fig. 4.15.

Series-interlinked dimension chains form a system in which each consecutive chain has a single common datum with the preceding chain. In Fig. 4.16 the common datum planes are denoted by the letters *a* and *b*.

Compound-interlinked dimension chains consist of several chains having both parallel and series interlinkages. Chains A and B in Fig. 4.17 are

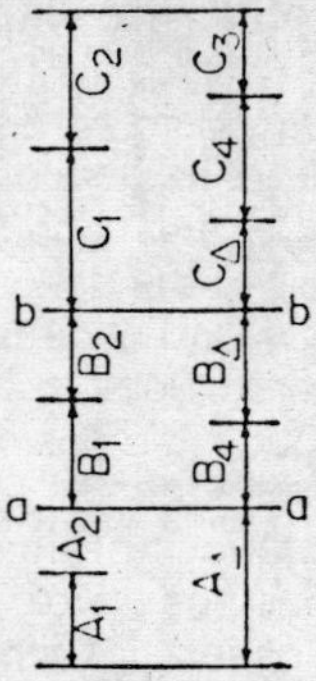

Fig. 4.16: Diagram of series-interlinked dimension chains.

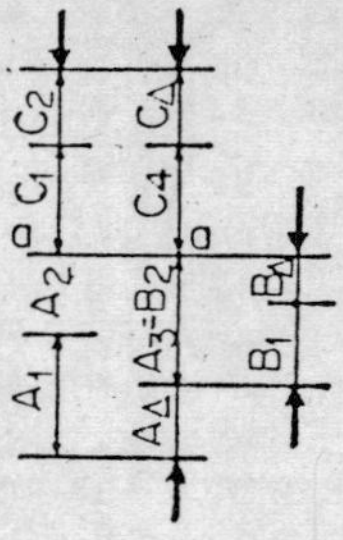

Fig. 4.17: Diagram of compound-interlinked dimension chains.

parallel-interlinked, while chains A and C are series-interlinked.

4.14.5. Plane and Spatial Dimension Chains

A *plane* dimension chain is one whose members lie in a single plane or a several parallel planes and can be projected onto a single plane without their magnitudes being changed.

A *spatial* dimension chain is one having at least one member that does not comply with the conditions defining a plane dimension chain. In calculations and analysis, all the members of such a chain can be projected onto the three co-ordinate planes, thereby reducing the spatial chain to three plane dimensional chains.

PLANE DIMENSION CHAINS

Any machine tool may be considered as a group of links connected in sequence by joints which constitute a closed dimensional circuit. The relative motion between the performance surfaces is affected by the kinematic connections.

In a turret lathe, shown in Fig. 4.18, the performance surface is governed by the co-axiality of the spindle and turret. Under static conditions, the equation for the dimensional circuit referred to the bed as datum is given by

$$\Delta A+A_1-A_2-A_3-A_4=0$$

where ΔA is the closing link signifying the error in the co-axiality.

Besides co-axiality, there are other forms of closing links to be investigated, viz. rotation, perpendicularity, etc. In Fig. 4.19, the accuracy of distance of the tool post from the workpiece is governed by

$$A_\Delta+A_1+A_2-A_3+A_4-A_5+A_6=0$$

while the parallelism of surface with respect to tool post is given by

$$a_\delta+a_1-a_2+a_3=0$$

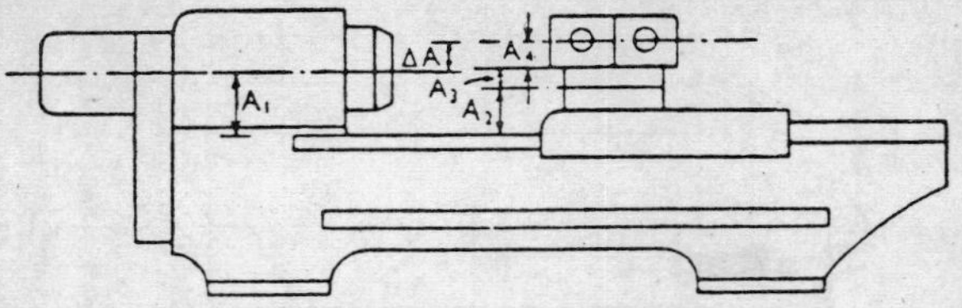

Fig. 4.18: Dimensional circuit of a turret lathe.

In the general form,

$$A_\Delta = A_1 + A_2 + \ldots + A_{m-1} = \sum_{i=1}^{m-1} A_i$$

4.15. ERROR DYNAMICS

The dimension in the direction of a given force P_y, forms the dimensional linkage system which governs the error in the final dimension of the product. Hence:

$$\varepsilon_\Delta = f(D_1, D_2, D_3, \ldots, D_n)$$

where ε_Δ= dynamical error on the workpiece, n= number of component links in the dimensional circuit, and D= values of the component links in the dimensional circuit in the deformed state.

If it is assumed that one component link D_i undergoes deformation, then

$$\varepsilon_\Delta = \Phi(D_i)$$

Further, if the value of the field of deviation of the deformed link D_i is ΔD_i, then

$$\varepsilon_{\Delta i} + \Delta\varepsilon_{\Delta i} = \Phi(D_i + \Delta D_i)$$

Expanding the right-hand side of this equation by Taylor's theorem:

$$\Phi(D_i + \Delta D_i) = \Phi(D_i) + \frac{\partial \varepsilon_{\Delta i}}{\partial D_i}\Delta D_i + \frac{1}{2!}\frac{\partial^2 \varepsilon_{\Delta i}}{\partial D_i^2}\Delta D_i^2$$

As ΔD_i is very small in comparison with the dimension, the higher order terms can be neglected. Hence,

$$\Phi(D_i + \Delta D_i) = \Phi(D_i) + \frac{\partial \varepsilon_{\Delta i}}{\partial D_i}\Delta D_i$$

$$= \varepsilon_{\Delta i} + \Delta\varepsilon_{\Delta i}$$

from which, for small values of deformations,

$$\Delta\varepsilon_{\Delta i} = \frac{\partial \varepsilon_{\Delta i}}{\partial D_i}\Delta D_i$$

Hence, the deviation of the closing link for all the links is given by

$$\Delta\varepsilon_\Delta = \sum_{i=1}^{m}\left[\frac{\partial \varepsilon_\Delta}{\partial D_i}\right]\Delta D_i$$

4.16. PROCESS CAPABILITY

The dimensional circuits are composed of linkages under no-load conditions. The dimensions of these linkages may be different due to

(i) workmanship (tolerances provided),
(ii) deformation under forces,
(iii) thermal distortions.

From, load-rigidity characteristics,

$$\Delta D_i = \frac{\Delta P_i}{K_i}$$

where, ΔP_i= value of the change of cutting force on the system, kg, K_i= rigidity of the link kg/mm.

$$\text{Hence, } \Delta\varepsilon_\Delta = \sum_{i=1}^{m}\Delta P_i \frac{1}{K_i}\left[\frac{\partial \varepsilon_\Delta}{\partial D_i}\right]$$

If the cutting tool and workpiece are separated out as distinctive links, then

$$\Delta\varepsilon_\Delta = \Delta P\left(\frac{1}{K_w}\frac{\partial \varepsilon_\Delta}{\partial D_w} + \frac{1}{K_T}\frac{\partial \varepsilon_\Delta}{\partial D_T}\right) + \sum_{i=1}^{m-2}\Delta P_i \frac{1}{K_i}\frac{\partial \varepsilon_\Delta}{\partial D_i}$$

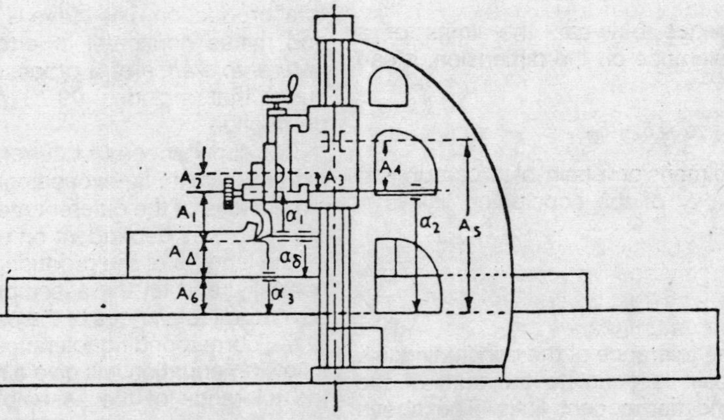

Fig. 4.19: Dimensional circuit of a planer.

Hence, the process capability can be improved by either of the following techniques:

(a) by decreasing the number of links,
(b) by reducing the range of cutting force,
(c) by changing the rigidity of the links,
(d) by changing the 'transfer coefficient' $(\partial\varepsilon_\Delta/\partial D_i)$ of the links.

4.17. ALGEBRA OF DIMENSIONS

4.17.1. Rule of Tolerancing

In the general case, for any number of links of a dimensional chain, the nominal dimensional equation will be

$$C = \sum A - \sum B$$

Thus, the nominal size of the concluding link equals the sum of the nominal sizes of all the increasing links minus the sum of the nominal sizes of all the decreasing links.

To illustrate the solution of direct problems for the maximum and minimum values, the simplest kind of dimension chain (Fig. 4.20) can be considered with the concluding link C. Here, as mentioned above, A is an increasing, and B, a decreasing link. It is evident from the drawing that $A = B + C$, from which $C = A - B$.

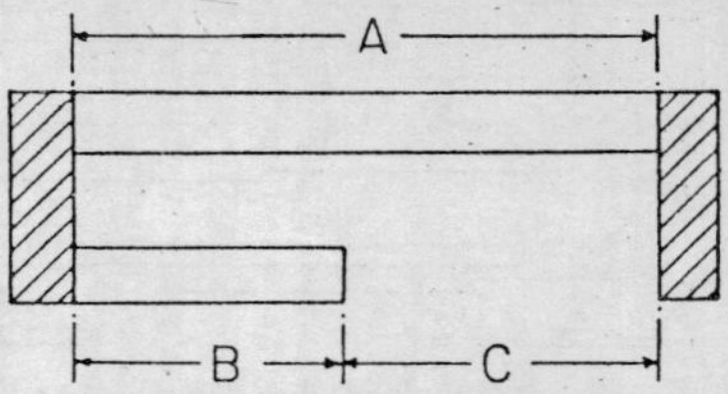

Fig. 4.20. Simplest tolerancing–direct.

It is evident from Fig. 4.20 that the limiting values of the concluding link C depend on the limiting values of the component links A and B. Thus

$$C_{max} = A_{max} - B_{min} \text{ and } C_{min} = A_{min} - B_{max}$$

where A_{max}, B_{max}, and C_{max} = maximum limits of the links, A_{min}, B_{min}, and C_{min} = minimum limits of the links.

Subtracting the second equation from the first we obtain

$$C_{max} - C_{min} = (A_{max} - B_{min}) - (A_{min} - B_{max})$$

and, after transposing the terms,

$$C_{max} - C_{min} = (A_{max} - A_{min}) + (B_{max} - B_{min})$$

Since the difference between the limits of a dimension is the tolerance on the dimension, it can be written as,

$$\delta_C = \delta_A + \delta_B.$$

Hence, for any dimension chain of n component links, the tolerance δ_c of the concluding link is

$$\delta_c = \sum^{n} \delta_i$$

Consequently, the tolerance of the concluding link of a dimension chain is equal to the sum of the tolerances of all the component links. The above equation is the fundamental relationship in dimensional analysis, and from it the two main rules are deduced.

4.17.2. Bilateral Tolerances on Dimensional Assembly

The assembly of a number of parts into a final product is schematically shown in Fig. 4.21. If parts 1, 2, 3 and 4 have the dimensions shown as mean value of nominal dimensions,

$$B + C - A - D = G$$

With bilateral tolerancing, and min-max strategy,

$$B + \begin{bmatrix} +b \\ -b \end{bmatrix} + C + \begin{bmatrix} +c \\ -c \end{bmatrix} - A - \begin{bmatrix} -a \\ +a \end{bmatrix} - D - \begin{bmatrix} -d \\ +d \end{bmatrix}$$

$$= G + \begin{bmatrix} -g \\ +g \end{bmatrix}$$

$$\begin{bmatrix} b \\ -b \end{bmatrix} + \begin{bmatrix} c \\ -c \end{bmatrix} + \begin{bmatrix} a \\ -a \end{bmatrix} + \begin{bmatrix} d \\ -d \end{bmatrix} = \begin{bmatrix} g \\ -g \end{bmatrix}$$

Hence, $\pm\ (a+b+c+d) = \pm g$

This equation assumes that the tolerances for an assembly are arithmetic or additive.

4.17.3. Pythagorean Tolerancing

Application of the normal frequency curve to the variations in the length of a part is illustrated with Fig. 4.22. The area under the curve represents the total production. The curve is asymptotic to the plus and minus horizontal or error axis. The so-called *natural tolerance* of a process is that spread of the curve that includes 99.73 per cent of the total production.

The simultaneous occurrence of all parts at the extreme limits is exceedingly rare. Suppose the dimensions of the different members of an assembly are in no way dependent on each other. In addition, let the lengths of the production parts be distributed normally, and let the assembly be random. Then, if the natural tolerances of the parts are all proportional to the corresponding tolerances of the drawings, the following equation will give a more realistic value for the tolerance of the assembly,

$$\pm g = \sqrt{a^2 + b^2 + c^2 + d^2}$$

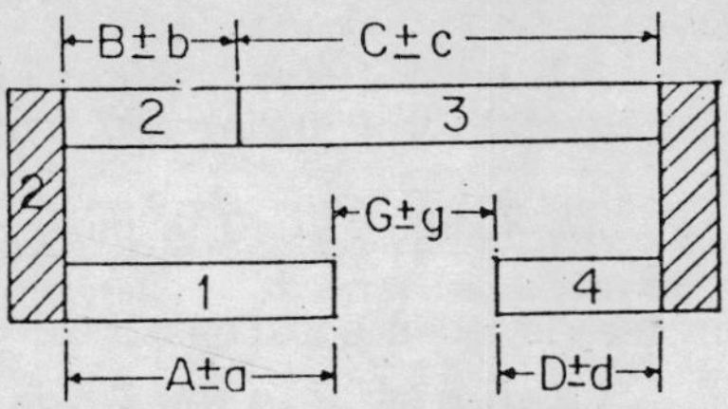

Fig. 4.21. Tolerancing on assembly.

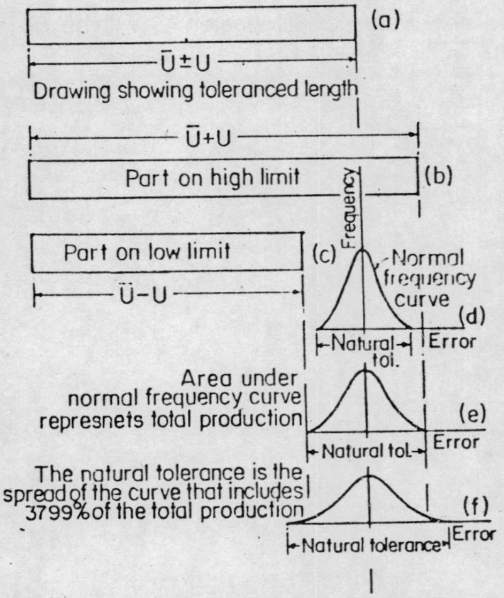

Fig. 4.22. Varied frequency distributions.

4.17.4 Shortest Dimension Chain Rule

First rule. It is always necessary to select the link requiring the least accuracy (from the viewpoint of operation of the part or unit) as the concluding link of the dimension chain, so that the sum of the tolerances of the whole chain can be assigned to this link.

This rule ensues from the fact that since the concluding link is obtained last in the manufacturing sequence, its size will contain the accumulated errors of all the previous operations for obtaining the component links.

Second rule. Dimension chains should be designed with the least possible number of links. This will facilitate dimension chain calculations.

This rule is sometimes called the *shortest dimension chain rule.* It is based on the fact that with a large number of dimensions (10 to 15), the tolerance on the concluding link will become so large that it cannot be assigned to any link of the chain.

It will prove insufficient to know the nominal size and tolerance of the concluding link. Its deviations must also be determined.

4.17.5. Deviation Analysis

Let, the upper deviation is the difference between the maximum limit A_{max} and the nominal size A, and the lower deviation is the difference between the minimum limit A_{min} and the nominal size A, then

$$\Delta_{UA}=A_{max}-A \text{ and } \Delta_{LA}=A_{min}-A$$

where

Δ_{UA} = upper deviation of dimension A,
Δ_{LA} = lower deviation of dimension A.

It follows that

$$A_{max}=A+\Delta_{UA} \text{ and } A_{min}=A+\Delta_{LA}$$

Similar equations can obviously be written for the remaining links of a dimension chain. This will come to six auxiliary equations:

The maximum value equation for the concluding link will be of the form

$$G_{max}=A_{max}-B_{min}$$

After substituting one auxiliary equation for each term of the preceding equation,

$$C+\Delta_{UC}=A+\Delta_{UA}-B-\Delta_{LB}$$

Subtracting from it the nominal size equation, $C=A-B$,

$$\Delta_{UC}=\Delta_{UA}-\Delta_{LB}$$

In the general case, for any dimension chain made up of any number of links, the equation for the upper deviation of the concluding link will be of the form.

$$\Delta_{UC}=\sum \Delta_{UA}-\sum \Delta_{LB}$$

In this manner, the upper deviation of the concluding link is equal to the sum of the upper deviations of all the increasing links *minus* the sum of the lower deviations of all the decreasing links.

The minimum value equation for the concluding link will be of the form

$$C_{min}=A_{min}-B_{max}$$

After substituting one auxiliary equation for each term, the final equation for the lower deviation of the concluding link will be

$$\Delta_{LC}=\sum \Delta_{LA}-\sum \Delta_{UB}$$

Thus, the lower deviation of the concluding link is equal to the sum of the lower deviations of all the increasing links *minus* the sum of the upper deviations of all the decreasing links.

Problem

Determine the final value of dimension C in nominal size equation on $A-B=C$. Given,

$A = 120^{+0.05}_{-0.03}$ and $B = 70^{-0.13}_{-0.20}$ mm

Solution.

1. From the nominal size equation
$C = A - B = 120 - 70 = 50$ mm
2. From the tolerance equation
$\delta_C = \delta_A + \delta_B$
3. $$\begin{bmatrix} p \\ q \end{bmatrix} - \begin{bmatrix} m \\ n \end{bmatrix} = \begin{bmatrix} s \\ t \end{bmatrix}$$

$$[p-q]-[m-n]=\delta_c$$

in which

$$\delta_A=(+0.05)-(-0.03)=0.08 \text{ mm}$$
$$\delta_B=(-0.13)-(-0.20)=0.07 \text{ mm}$$

then

$$\delta_c=0.08+0.07=0.15 \text{ mm}$$

4. From the deviation equations

$$\Delta_{UC}=\Delta_{UA}-\Delta_{LB}=(+0.05)-(-0.20)=+0.25 \text{ mm}$$
$$\Delta_{LC}=\Delta_{LA}-\Delta_{UB}=(-0.03)-(-0.13)=+0.10 \text{ mm}$$

The final value of dimension C will be $50^{+0.25}_{+0.10}$ mm.

CHAPTER 5

Machining Accuracy

5.1. ACCURACY

By comparing two parts—a real manufactured part and the one specified by the designer and shown in the part drawing—it can be established that they differ. The degree with which the real part differs from the specified one is determined by the quality of manufacturing and characterizes the accuracy, Fig. 5.0.(a).

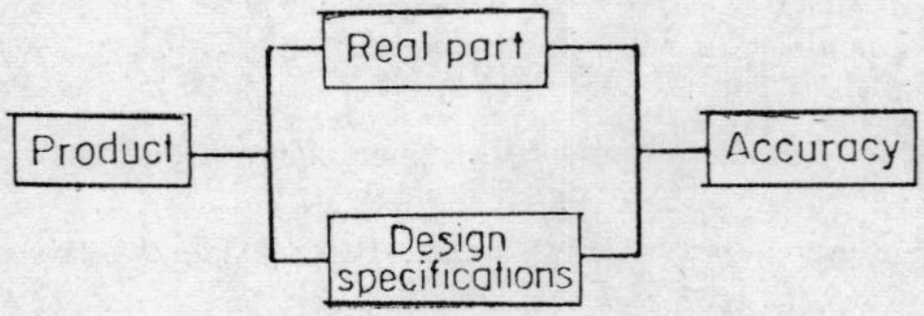

Fig. 5.0.(a) Accuracy–product quality

5.2. Product Attributes

In most cases, a part is a combination of geometrical bodies bounded by surfaces of the simplest shapes, such as flat, cylindrical, and conical surfaces. Therefore, the following principal factors can be established to indicate how close a real part approaches the one specified by the drawing, Fig. 5.0.(b):

(1) accuracy of form, or geometrical features, i.e. the degree, with which various portions (surfaces) of the part approach the geometrical bodies that identify them;
(2) shape of contour, path and surfaces;
(3) dimensional accuracy of portions (surfaces) of the part;
(4) coordinational accuracy of the same portions (surfaces); and
(5) surface finish, i.e. the degree with which the real rough surface approaches a true geometrical surface.

5.3. ERRORS OF GEOMETRICAL FEATURES

5.3.1. Deviations

Errors of geometrical features can be defined as that amount by which the actual surfaces vary from the perfect geometry implied in drawings, and are classified as follows:

(1) Deviations from true cylindrical shape: out-of-roundness and lobed shape in the cross-section, and barrel shape, bow shape, lack of straightness and taper in the longitudinal section.

(2) Deviations from a true surface: lack of straightness, lack of flatness, concavity and convexity.

(3) Deviations from the correct co-ordination of surfaces: misalignment of axes, radial runout, axial runout, incorrect location of intersecting axes, non-parallelism and lack of squareness of flat surfaces.

5.3.2. Deviations from True Cylindrical Shape

(a) *Out-of-roundness* (also called *ovality*) is the difference between the maximum and minimum diameters in the same cross section (Fig. 5.1a). Thus

$$\Delta = D - d$$

In some cases the roundness tolerance may exceed the diameter tolerance, for example in machining thin-walled bushings or sleeves that may be distorted in machining but are forced into the correct shape in assembly. In such cases, the diameter error is taken to be the deviation of one half of the sum of the maximum and minimum diameters, determined by measurement, from the nominal diameter.

(b) The *amount of lobedness* is the difference between the diameter of the circle into which the lobed cross-section can be inscribed and the distance between parallel planes tangent to the surface of the part (Fig. 5.1b). Thus,

$$\Delta = D - l$$

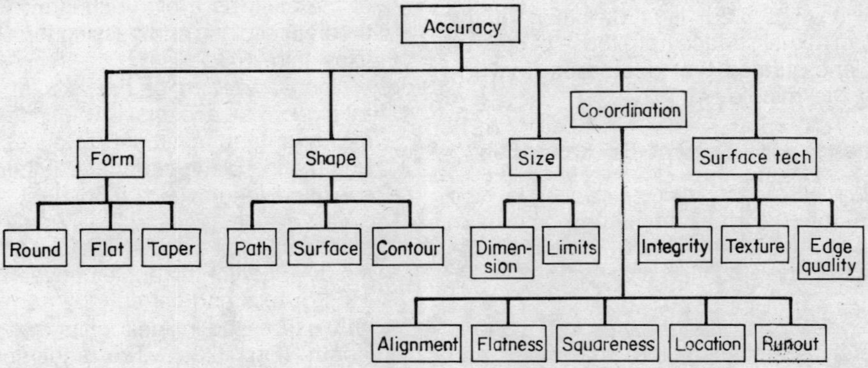

Fig. 5.0.(b) Different attributes of accuracy

Lobed shape is one in which the cross-section is composed of interconnected arcs drawn from different centres.

(c) The *amount of barrel shape* (Fig. 5.1c) *or bow shape* (Fig. 5.1d) is the difference between the diameters of the middle and end cross-sections of the workpiece. Thus

$$\Delta = D_1 - d_1 \text{ and } \Delta = D_2 - d_2$$

(d) The *lack of straightness* of the line passing through the loci of the centres of the cross-sections of a cylindrical surface is determined by the amount of camber of the axis of the cylinder or of its elements (Fig. 5.1e). Thus,

$$\Delta = D_3 - d_3$$

(e) *Taper* is the lack of parallelism of opposing elements (or generating lines) and is determined as the ratio of the difference in diameter of two cross-sections to the distance between these sections (Fig. 5.1f). Thus,

$$\Delta = \frac{D_1 - d_1}{l}$$

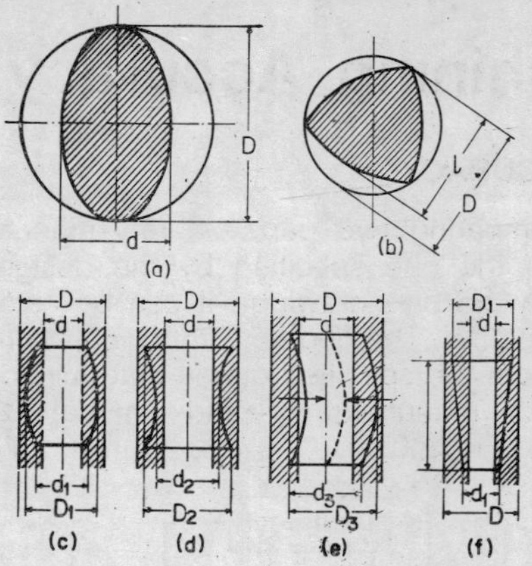

Fig 5.1. Deviations from true cylindrical shape in the cross and longitudinal sections:

a–out-of-roundness; *b*–lobed shape; *c*–barrel shape; *d*–bow shape; *e*–lack of straightness; *f*–taper.

5.3.3. Deviations from True Shape of Flat Surfaces

(a) The *lack of straightness* of a *flat surface* is the distance Δ between two parallel lines located in a plane A square to the tested surface. Plane A intersects the surface in the required direction and the two parallel lines enclose the section profile of the surface (Fig. 5.2a).

(b) The *lack of flatness* is the distance Δ between two parallel planes A and B which enclose the tested surface between them. The distance Δ is measured in plane C which is square to planes A and B (Fig. 5.2b).

(c) Elementary kinds of *straightness and flatness errors* are concavity and convexity.

Concavity (Fig. 5.2c) is a deviation of the points of an actual surface from those of the true mating plane surface that increases from the edges to the middle.

(d) *Convexity* (Fig. 5.2d) is a deviation of the points of an actual surface from those of the true mating plane surface that decreases from the edges to the middle.

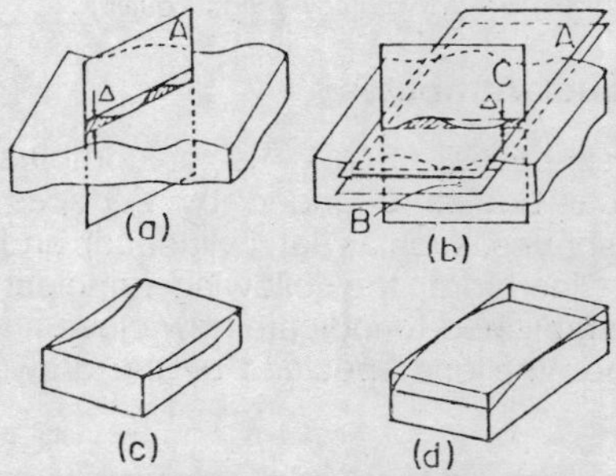

Fig. 5.2. Deviations from true shape of flat surfaces:

a–lack of straightness; *b*–lack of flatness (the profile of the cross-section is shown hatched); *c*–concavity; *d*–convexity.

5.3.4. Deviations from Correct Co-ordination

(a) *Non-parallelism of flat surfaces* is the difference of the maximum and minimum distances between the true mating plane surfaces over the specified area or length (Fig. 5.3a).

(b) *Non-parallelism of straight lines* in a plane is the difference between the maximum and minimum distances between the lines over the specified length (Fig. 5.3b).

(c) *Non-parallelism of the axes of surfaces of revolution* (or straight lines in space) is equal to the non-parallelism of the projections of the axes on their common theoretical plane passing through one axis and through a point of the other axis (Fig. 5.3c).

Misalignment of axes (or straight lines in space) is the non-parallelism of the projections of the axes on a plane square to the common theoretical plane and passing through one of the axes (Fig. 5.3*c*).

(d) *Non-parallelism of the axis of a surface of revolution to a plane* is the difference of the maximum and minimum distances between the true mating plane surface and the axis over the specified length (Fig. 5.3d).

(e) *Out-of-squareness* (squareness error) *of flat surfaces, axes* or *of an axis to a flat surface* is the deviation of the angle between the surfaces, axes or axis and flat surface from a right angle (90°) expressed in linear units over the specified length (Fig. 5.3e). The squareness error is measured from true mating flat surfaces or straight lines.

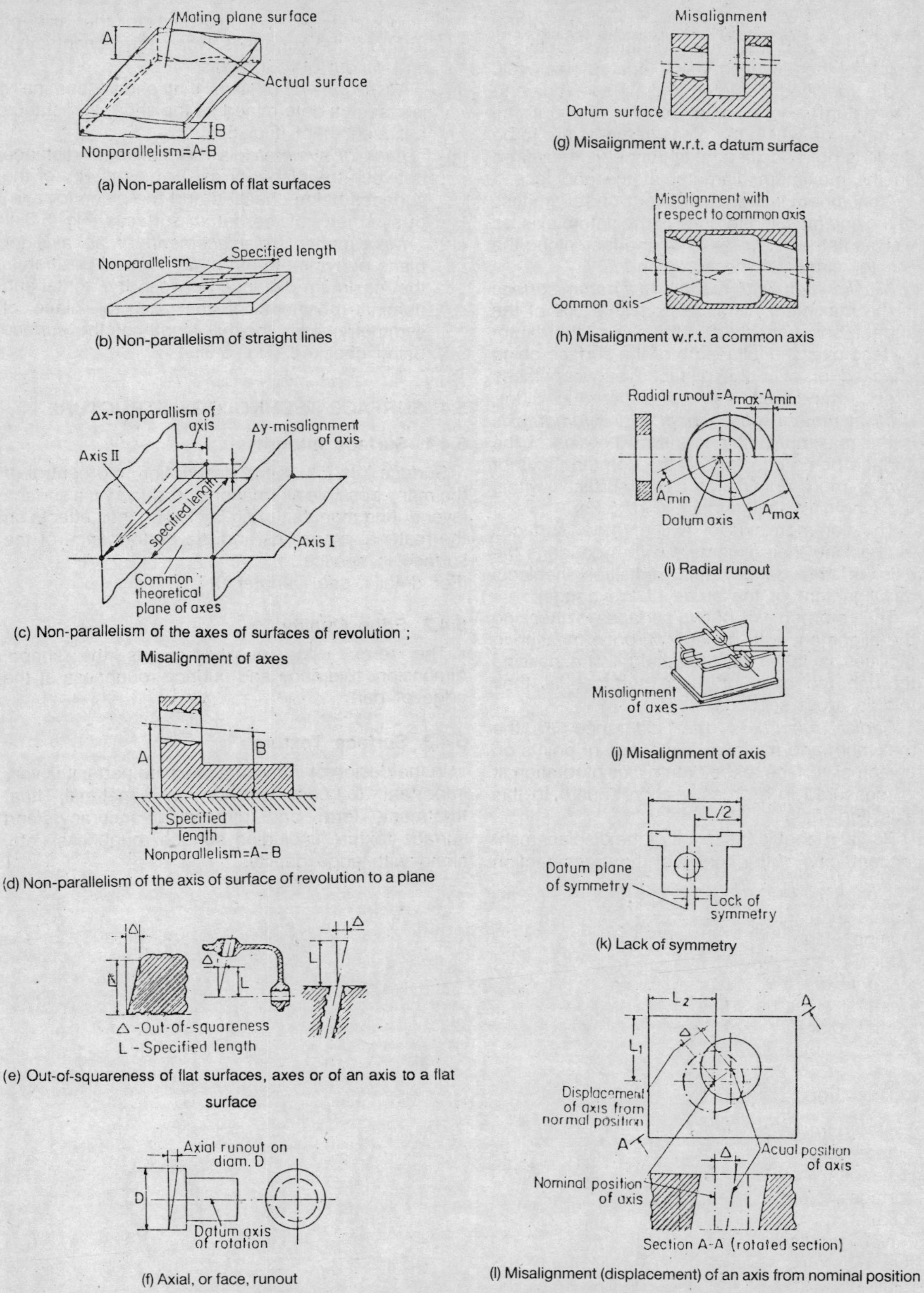

(a) Non-parallelism of flat surfaces

(b) Non-parallelism of straight lines

(c) Non-parallelism of the axes of surfaces of revolution ; Misalignment of axes

(d) Non-parallelism of the axis of surface of revolution to a plane

(e) Out-of-squareness of flat surfaces, axes or of an axis to a flat surface

(f) Axial, or face, runout

(g) Misalignment w.r.t. a datum surface

(h) Misalignment w.r.t. a common axis

(i) Radial runout

(j) Misalignment of axis

(k) Lack of symmetry

(l) Misalignment (displacement) of an axis from nominal position

Fig. 5.3: Deviations from correct coordination of surfaces

(f) *Axial*, or *face runout* is the difference of the maximum and minimum distances between points of the actual end face, located on a circle of the specified diameter, and a plane square to the datum axis of rotation (Fig. 5.3f). If the diameter on which the measurements are to be made is not specified, axial runout is measured on the maximum diameter of the end face.

Axial runout is the result of out-of-squareness of the end face in reference to the datum axis, as well as flatness errors of the end face along the line (or circle) of measurement.

(g) *Misalignment with respect to a datum surface* is the maximum distance from the axis of the surface being checked to the axis of the datum surface over the full length of the surface being checked or over the distance between these axes in the specified cross-section (Fig. 5.3g).

(h) *Misalignment with respect to a common axis* is the maximum distance from the axis of the surface being checked to the common axis of two or more nominally coaxial surfaces within the length of that surface (Fig. 5.3h).

The common axis of two or more surfaces in checking their alignment with a gauge is the axis of the gauge (this definition neglects misalignment of the steps of the gauge).

The common axis of two surfaces in checking the alignment with general purpose measuring facilities is taken as the straight line passing through the axes of these surfaces at their middle cross-sections.

(i) *Radial runout* is the difference in the maximum and minimum distances of points on an actual surface to the datum axis of rotation. It is measured in a cross section square to this axis (Fig. 5.3i).

Radial runout is the result of the displacement (eccentricity) of the centre of the cross-section being checked with respect to the axis of rotation (the runout is twice the eccentricity), and of out-of-roundness.

(j) *Misalignment of axes* that should nominally intersect is determined as the shortest distance between them (Fig. 5.3j).

(k) *Lack of symmetry* is the maximum distance between the plane (or axis) of symmetry of the surfaces being checked and the plane (or axis) of symmetry of the datum surfaces (Fig. 5.3k).

(l) *Misalignment (displacement) of an axis* (or plane of symmetry) from its nominal position is the maximum distance between the actual and nominal positions of the axis (or plane of symmetry) over the full length of the surface being checked (Fig. 5.3l).

5.4. SURFACE TECHNOLOGY STRUCTURE

5.4.1. Surface Integrity

Surface integrity is the description and control of the many possible alternations produced in a surface layer during manufacturing, including their effects on the material properties and the performance of the surface in service.
(For details, see Chapter 6)

5.4.2. Edge Quality

The term 'edge quality' means the shape, dimension, tolerance and surface roughness at the edge of part.

5.4.3. Surface Texture

In the design of individual machine parts, it is very important to consider not only material, heat treatment, form and dimension accuracy, and surface texture (including surface roughness), etc. along with edge quality.

CHAPTER 6

Surface Integrity

6.1. THE SURFACE

Surface integrity is the sum of all the elements that describes all the conditions existing on or at the surface of a workpiece. Surface integrity has two aspects. The first is *surface topography* which describes the roughness, 'lay' or texture of this outermost layer of the workpiece, i.e., its interface with the environment. The second is *surface metallurgy* which describes the nature of the altered layers below the surface with respect to the base of matrix material. This term assesses the effect of manufacturing processes on the properties of the work piece material. Fig. 6.1 depicts a simulated section showing the various layers between the base material and the environment. In between lies the altered material zone or AMZ. Kahles defines surface integrity as 'the unimpaired or enhanced surface condition or properties of a material resulting from a controlled manufacturing process'.

The ever-increasing capabilities of new aerospace materials have been accompanied by an increase in sensitivity to processing variables. The concern for surface integrity is due to the following facts:

(i) more sensitive and difficult alloys have been introduced;
(ii) higher stress levels are encountered;
(iii) reliability demands are stringent;
(iv) limit design concepts are often used;
(v) thinner sections are becoming prevalent.

The principal causes for the surface alterations that have been found in material removal operations are:

(i) high temperature or high thermal gradients;
(ii) mechanical working beyond the limit of plastic deformation;
(iii) chemical reactions and subsequent absorption into the nascent machined surfaces.

To obtain good process understanding and thereby achieve best quality control, it is essential to evaluate the effects produced by various processing techniques. A surface integrity (SI) domain is constituted from the assessment of the material properties and for a specific level of process. Fig. 6.2 shows the (SI) domain in which the state of surface integrity is determined from

(a) particular state of a specific *material;*
(b) operating *process* and its *energy level;*
(c) assessment of individual *property.*

Definitive and complete sets of such data are rare. The General Electric Co., of USA has been reported to compile an encyclopaedia of such surface integrity data.

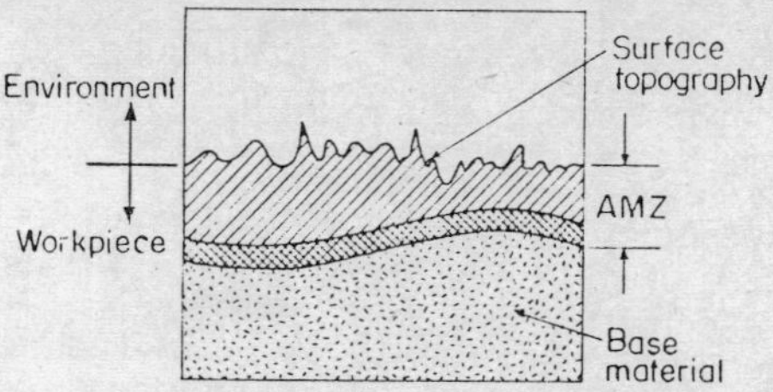

Fig. 6.1. A simulated section of various layers

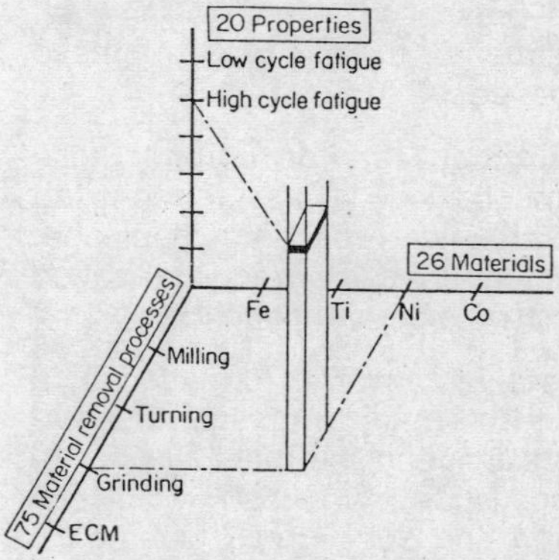

Fig. 6.2. The SI domain

6.2. Surface Metallurgy

Surface metallurgy is concerned mainly with a number of effects a process has below the outermost layer. The subsurface characteristics occur in various layers or zones. These are usually termed as AMZs, altered material zones. These alterations range from the simple residual stressed state to a more complex grain structure changes interlaced with intergranular attack (IGA). Such alternations are caused by chemical, thermal or mechanical energy and affect the physical and metallurgical properties of the material.

The *mechanical* modes of change are: plastic deformations, tears, laps, macroscopic and microscopic cracks, residual stress, endurance limit changes, hardness alterations, etc. The *metallurgical* modes are: transformation of phases, grain size and shape, twinning, recrystallization, precipitate size and distribution, etc. The *chemical* ones include intergranular attack (IGA), intergranular corrosion (IGC), intergranular oxidation (IGO), embrittlement, pits or selective etch, corrosion, etc. The *thermal* modes include heat-affected zone (HAZ), redeposited material, resolidified material etc. The *electrical* changes include alterations in conductivity or magnetic properties.

A few AMZs commonly occurring are defined below:

1. **Cracks:** These are fissures in materials observed with the unaided eye or with × 10 or less magnification. The microcracks are seen only with greater magnifications.

2. **Plastic Deformation:** Microstructural changes caused by exceeding the yield point of the material.

3. **Hardness Alteration:** Changes in hardness of surface layers as a result of heat, mechanical stressing or chemical change during material processing.

4. **Residual Stresses:** Those stresses which are present in a material after all external forces or thermal gradients have been removed.

5. **Metallurgical Transformations:** These include resolidified layers, redeposited material, chemical reaction, depletion, grain structure change or recrystallization as a result of external influences.

6. **Recrystallization:** The formation of a new strain-free grain or crystal structure from that existing in the material prior to processing usually as a result of plastic deformation and subsequent heating.

7. **Intergranular Attack:** A form of in-process corrosion or attack in which preferential reactions are concentrated at the network of grain boundaries usually in the form of sharp notches or discontinuities.

8. **Selective Etch:** A form of in-process corrosion or attack in which preferential reactions are concentrated within and through the grains or concentrated on the certain constituents of the base material.

9. **Heat-Affected Zone (HAZ):** That portion of the material not melted, yet subjected to sufficient thermal energy to contain microstructure alterations.

6.3. SURFACE TOPOGRAPHY

6.3.1. General Characteristics of a Surface

Outermost layers of all machined surfaces display a great number of both macrogeometrical and microgeometrical deviations from the ideal geometrical surface. Deviations of the first order include errors of form and are macrogeometrical. Deviations of the second order are waves while the deviations of third and higher orders refer to surface roughness. The principal elements of surfaces are defined below:

SURFACE: The surface of an object is the boundary which separates that object from another substance. Its shape and extent are usually defined by a drawing or descriptive specifications.

PROFILE: The contour of any specified section through a surface.

ROUGHNESS (Primary texture): Relatively fine-spaced surface irregularities. On surfaces produced by machining and abrasive operations, the irregularities produced by the cutting action of tool edges and abrasive grains and by the feed of the machine tool are roughness. It may be considered as being superposed on a wavy surface.

WAVINESS (Secondary texture): The surface irregularities which are of greater spacing than the roughness. On machined surfaces such irregularities may result from machine and work deflections, vibrations, etc.

FLAWS: Irregularities which occur at one place or at relatively infrequent intervals in the surface, e.g. a scratch, ridge, hole, crack, etc.

LAY: The direction of predominant surface pattern.

A typical surface is shown in Fig. 6.3. The identification of the macrogeometrical or microgeometrical errors is based on (l/h) ratio as shown in Fig. 6.4.

When, $(l_1/h_1) > 1000$, the deviation is macrogeometrical denoting out-of-roundness, taper or barrel form. When $500 > (l_2/h_2) \geq 150$ the deviations denote waviness. When $(l_3/h_3) \leq 50$, the microgeometrical deviations are characteristic of surface roughness.

6.3.2. Factors affecting Surface Quality

There are seven major factors influencing surface quality in machining operations:

(i) machine tool rigidity and bearing accuracies,
(ii) finishability of work material,
(iii) type and condition of cutting tool,
(iv) application of cutting fluid,
(v) method of chip removal,
(vi) geometry of cutting tool,
(vii) cutting variables: feed, depth of cut and cutting speed.

Waviness of the surface may be due to vibrations which are in turn due to

(i) external sources of shock and vibrations,
(ii) internal unbalancing of machine elements,
(iii) defects in the drive in machine tool,
(iv) self-excited vibrations.

The 'inherent' surface roughness is dependent on feed and plane geometry of the cutting tool. However, this inherent roughness can only be approached when continuous chips are produced without built-up edges under ideal machine tool conditions.

The maximum height H_{max} of the roughness form produced by a single point radius turning tool, as shown in Fig. 6.5 is given by

$$H_{max} = \frac{s^2}{8R} \quad \text{... Eqn. 6.1}$$

where s = feed in mm/revolution and R=nose radius, mm.

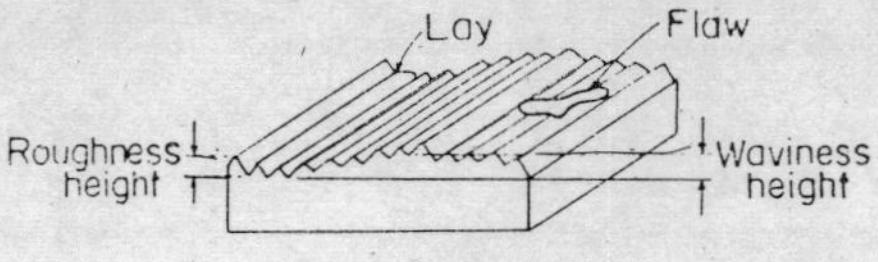

Fig. 6.3 : A typical surface highly magnified

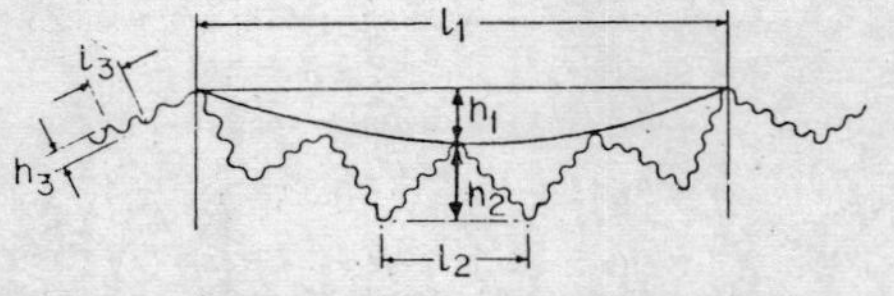

Fig. 6.4 : Macrogeometrical and microgeometrical errors

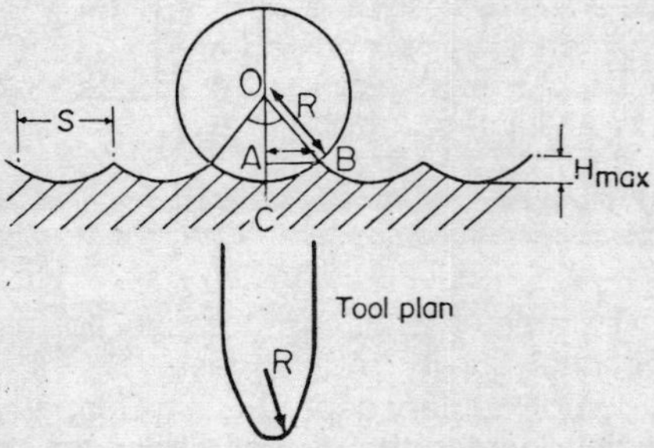

Fig. 6.5 : Roughness form produced by simple radius turning tool

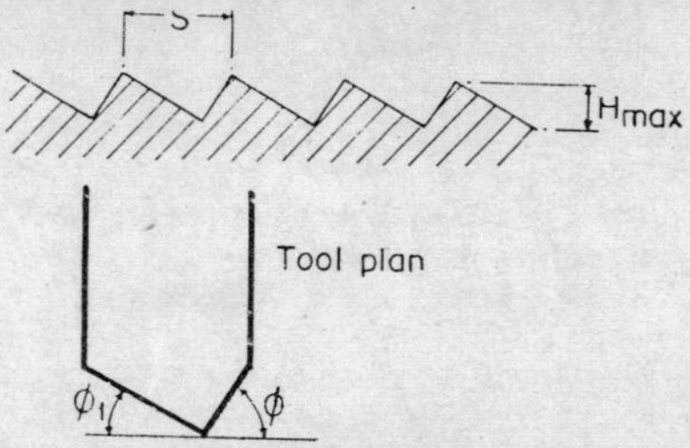

Fig. 6.6 : Surface roughness of a single point turning tool

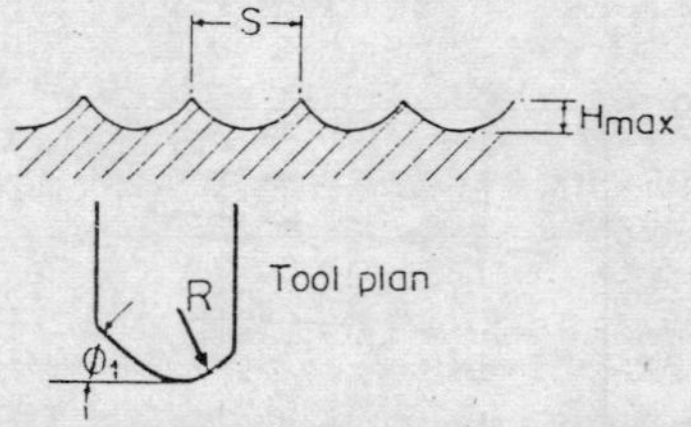

Fig. 6.7 : Surface roughness of a complex radius turning tool

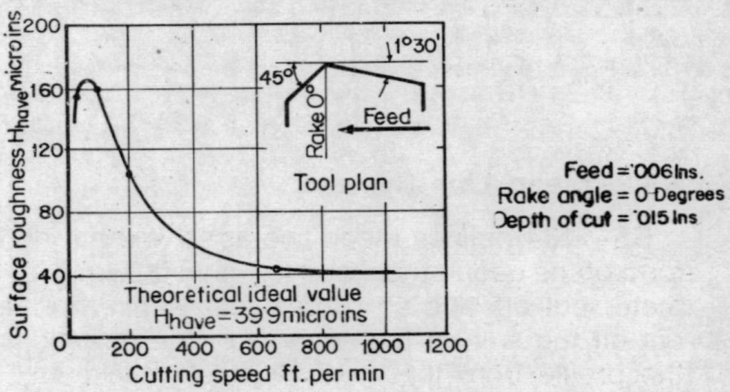

Fig. 6.8 : Effect of cutting speed on surface roughness

For cylindrical milling:

$$H_{max} = \frac{s_t^2}{4D} \quad \text{... Eqn. 6.2}$$

when D=diameter of cutter, mm, and s_t=feed/tooth, (mm/tooth).

For a single point turning tool as shown in Fig. 6.6.

$$H_{max} = \frac{s}{\cot\phi + \cot\phi_1} \quad \text{... Eqn. 6.3}$$

While for a complex turning tool as shown in Fig. 6.7.

$$H_{max} = s \tan\phi_1 + \frac{R}{2}\tan^2\phi_1 - \sqrt{2\, s\, R \tan^3\phi_1} \quad \text{... Eqn. 6.4}$$

The influence of cutting speed on surface roughness is illustrated in Fig. 6.8 where Chisholm has demonstrated that H_{max} approaches the ideal value obtainable by Eqn. 6.3 when cutting speed is increased.

The effect of feed and nose radius is shown in Fig. 6.9. (Opitz and Moll), The depth of cut does not appreciably incluence surface finish as shown in Fig. 6.10. An extensive survey of surface finish research has been made by Spears, where also the effect of other machine tool variables besides cutting variables, have been discussed.

6.4. NUMERICAL ASSESSMENT OF THE SURFACE

A number of methods have been developed for numerical assessment of the surfaces. Usually a reference line is selected relative to which the variations are determined. In one case, E-system, the shape of the reference line is generated by the surface irregularities themselves.

The basic measuring systems are:

(i) peak-to-valley measures,
(ii) mean line measures (M-system),
(iii) crest line measures (Nicolau's method),
(iv) envelope method (E-system, Von Weingraber).

6.4.1. Peak-to-valley Measures

In this system the height of the separation of two reference lines touching the highest peak and the lowest valley occurring within a given length of surface is assessed. This overall height is called *rauhtiefe*.

In the Swedish system, the peaks are systematically cut off whereupon the depth R is determined. This is shown in Fig. 6.11.

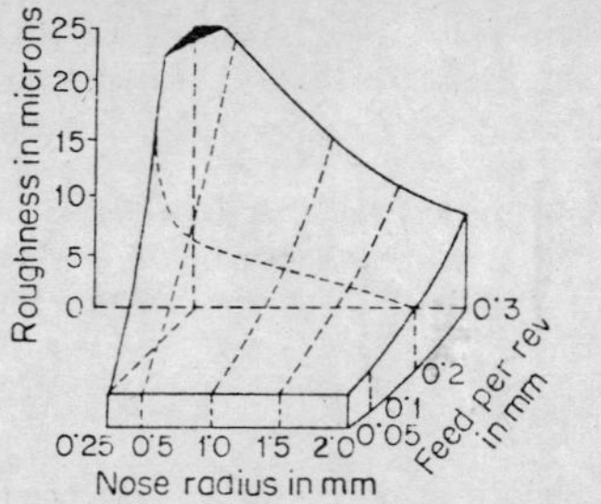

Fig. 6.9 : Effect of feed and nose radius on surface roughness

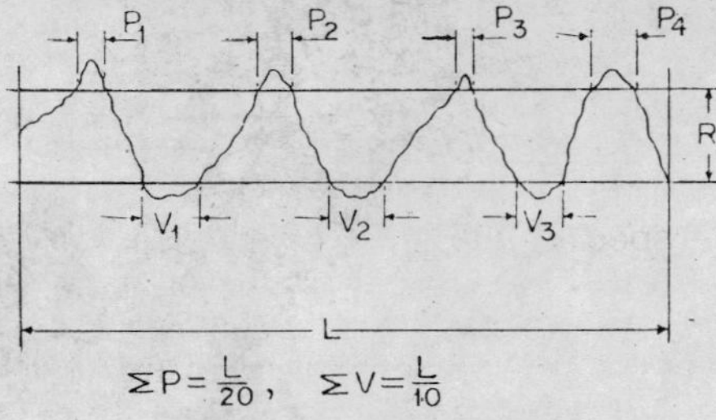

Fig. 6.11 : Swedish system

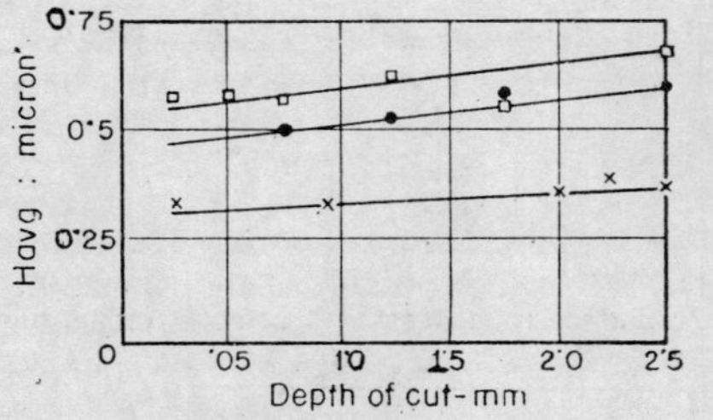

Fig. 6.10 : Effect of depth of cut on surface finish

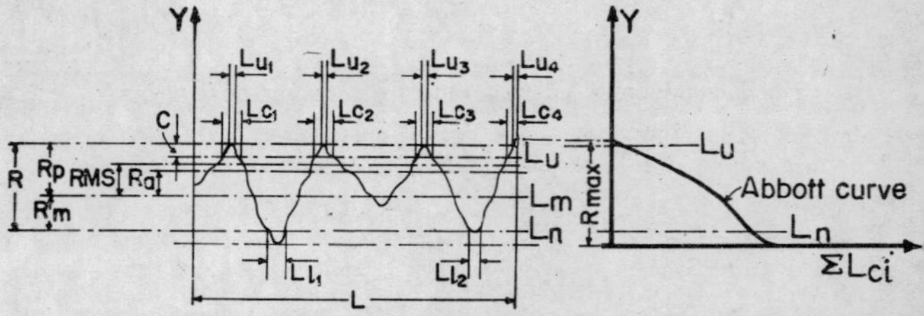

Fig. 6.12: Section of a surface profile depicting depth and length measures in mean line system and to the right, the Abbott curve shown

6.4.2. Mean Line System

BS–1134 defines mean line as "a line having the form of the geometrical profile within the limits of the meter cut-off, and so placed that within the meter cut-off the sum of the squares of the deviations of the profile from the mean line is a minimum."

Fig. 6.12 shows a profile. The figure shows the mean line as L_m while the upper and lower reference lines are shown by L_u and L_n. The total sampling length is L.

In some standards the upper and lower reference lines indicate the lines touching the highest and the lowest peaks. In Danish and Swedish standards, exceptions are systematically excluded by determining the lines in such a way that

$$\Sigma l_u = 0.05\, l \text{ and } \Sigma l_n = 0.10\, l$$

The definitions of the different roughness values are based on mean line. There are various such parameters required for *functional* assessment of surface texture. Some of these are given, below:

(i) Total depth, R: It is the peak-to-valley value which is perceived by looking at and touching the surface and is generally influenced by scratch and tool marks. [In German: this depth is called *Rauhtiefe*.]

(ii) Levelling depth or depth of smoothness, R_p [in German: *Glättungstiefe* and in French: *Aplanation*] is the distance between the mean line L_m and upper surface line L_u. This is approximately obtained from,

$$R_p = \frac{1}{l}\int y\, dx \qquad \text{... Eqn. 6.5}$$

This parameter is often necessary for assessing the frictional quality of a surface.

The complimentary depth, R_m, [in German: *mittlere Rauhtiefe*] denotes the distance between the mean line and bottom surface line, i.e.

$$R_m = R - R_p.$$

(iii) The centre-line-average value is defined according to BS–1134 as "the arithmetical average (AA) value of the departure of the whole of the profile both above and below its centre line throughout the prescribed meter cut-off in a plane substantially normal to the surface." Mathematically,

$$R_a = \frac{1}{l}\int_0^l |y|\, dx \qquad \text{... Eqn. 6.6}$$

The centre-line-average value is denoted by 'CLA' in the UK, 'AA' in the USA and 'R_a' in ISO-TC-57.

The CLA parameter does not describe any of the known functions in particular but provides a sufficiently useful datum for many functional purposes.

The *advantages* of the CLA are:

- easy and reliable instrumentation,
- small amount of scatter,
- considerable sorting power and discriminative quality on irregular and aperiodic profiles.

There are some *disadvantages* for the CLA index:

- it has no relationship to actual depth of the irregularities,
- it does not reveal maximal values,
- on regular and periodic profiles, the CLA has only limited discriminative quality and sorting power.

The above disadvantages of the CLA point out that R_p is a better index than CLA. However, present status of instrumentation prevents adoption of R_p and is considered as a 'parameter of future'. Some industrial undertakings in France and Germany insist on the specifications of R_p on the drawings.

(iv) The RMS or root-mean-square value is "the geometrical average value of the departure of the whole of the profile both above and below its centre line throughout the prescribed meter cut-off in a plane substantially normal to the surface." Mathematically,

$$\text{RMS} = \sqrt{\left(\frac{1}{l}\int_0^l y^2\,dx\right)} \qquad \text{... Eqn. 6.7}$$

The RMS value was used in the USA until 1955 when it has been replaced by R_a in the standard of 1955.

(v) Filling-out factor *(K)* is a proportion measure to denote to which extent the area between the two reference lines L_u and L_n is filled in with metal and is given by

$$K = \frac{R_m}{R} \qquad \text{... Eqn. 6.8}$$

In German: this is called *Völligkeitsgrad.*

(vi) Horizontal measures include the wavelength and the spacing of the surface irregularities. Though at present time these measures are seldom used for specification, yet in many cases these have importance.

If $B_b = \sum_1^n L_{ci}$ be the bearing length for a very important measure as it gives data on the size of the bearing proportion, then the factor is determined as

$$K_b = \frac{B_b}{l} \times 100\% \qquad \text{... Eqn. 6.9}$$

where K_b = the bearing area, B_b = the bearing length.

When the profile (surface) extends uniformly across the surface in the direction normal to the intersecting plane, K_b will indicate the bearing area which is to be expected after a given amount of lapping or wear.

The bearing area is determined by the horizontal extension of the peaks and by measuring B_b in different distances, *C*, from the upper reference line L_u. it is possible to draw important conclusions concerning the properties of the surface.

The position of the bearing line is determined by the distance, *C*, below the upper reference line L_u according to the requirements, and this distance is frequently expressed as a fraction of the peak-to-valley value *R*.

When the measurement of the bearing length belongs to a number of lines at various distances, the bearing length may be used for obtaining the so-called Abbott curve (Fig. 6.12) from which it is possible to judge different conditions such as the resistance to wear, bearing and sliding properties.

6.4.3. Crest Line Measure

Nicolau has developed methods, shown in Fig. 6.13. in which a datum is fixed either contacting the highest peaks or the lowest valleys. The average distance of the profile from the datum is given by

$$G_m = \frac{1}{L}\int_0^L y\,dl \qquad \text{... Eqn. 6.10}$$

and is the crest line measure of the surface roughness.

6.4.4. Envelope Method

In this method, a crest line is generated as shown in Fig. 6.14 from the locus of the centre of a rolling circle moving across the surface, the locus being displaced towards the surface until it contacts the crests. The average depth of the profile below the contacting envelope is the *Glättungstiefe* or levelling depth, G_E, given by

$$G_E = \frac{1}{L}\int_0^L y^2\,dl \qquad \text{... Eqn. 6.11}$$

German standard DIN–4762 of 1960 defines envelope profile of a surface as a line touching the effective profile, being congruent with the centre line path of a circle of radius *r*, rolling along the effective profile and touching the highest peaks of that profile as shown in Fig. 6.15.

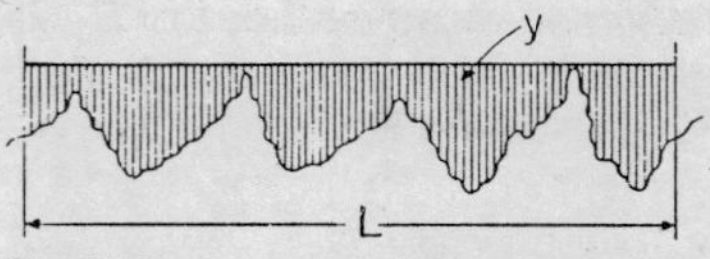

Fig. 6.13 : Crest iine measure (Ni-colau)

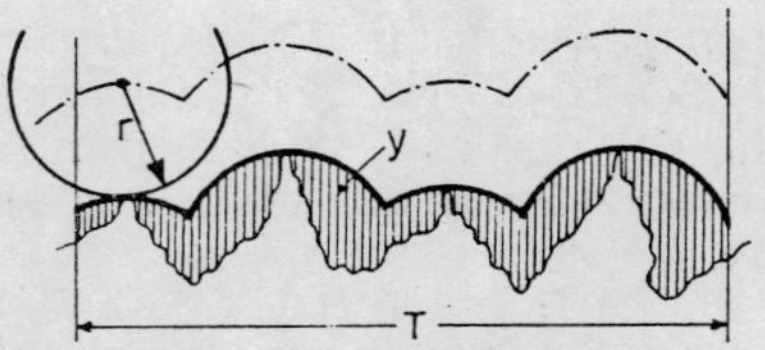

Fig. 6.14 : Envelope method

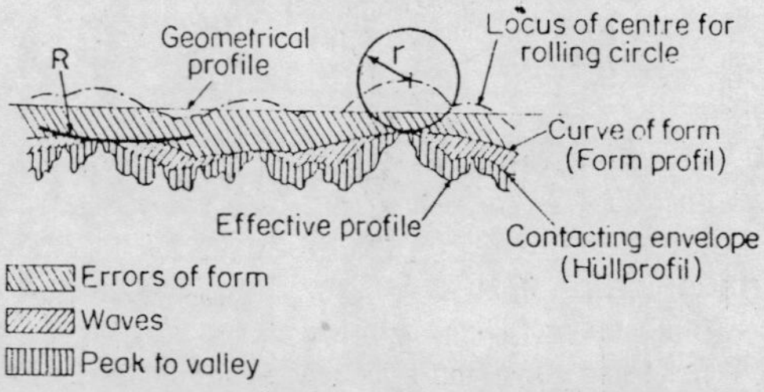

Fig. 6.15 : Principles of assessment by E-system

Generally circles of two different radii, R and r ($R > r$) are rolled for determining 'profile of form' and 'profile of surface'. When a large radius R=250 mm is rolled across the surface, it will describe a curve which is displaced in a direction perpendicular to the geometrical profile to a position where it is contacting some of the highest peaks in the effective profile. The locus of centre of the 'rolling circle' determines the 'curve of form' (in German *Formprofil*). In a similar manner, rolling a smaller radius of 25 mm would generate 'contacting envelope' (in German *Hüllprofil*) or 'curve of surface'.

Now, the area between the geometrical profile and the 'curve of form' represents the errors of form while the area between the 'curve of form' and the 'curve of surface' represents the waviness or secondary texture. The area between the 'curve of surface' and the 'effective profile' represents the 'surface roughness' or primary texture.

In M-system, the selection of sampling length decides whether the measurement involves form or surface. In E-system, the selection of roller radius decides whether form or surface texture is evaluated.

The contacting envelope of 'profile of surface' when displaced to a position where the areas enclosed by the effective profile are equal is shown in Fig. 6.16. This displaced curve is called 'mean curve' (in German: *mittleres Profil*) which corresponds to the mean line of M-system. If the contacting envelope is further displaced to pass through the deepest valleys, the 'lower contacting envelope' is obtained (in German: *Gründprofil*). Thus the E-system becomes similar to M-system except for mean line which is the 'mean profile' in E-system.

The outstanding disadvantages of E-system is the fact that there is uptil now no cheap instrument available for evaluating surface texture according to E-system.

6.4.5. ISO Recommendation for Assessment

Based on M-system, the ISO has recommended a standard (ISO–TG–57–No. 221 of September 1958) where the CLA value R_a of M-system has been chosen. In the ISO recommendation three items are necessary:

(i) R_a (ii) R_{max} (iii) R_z

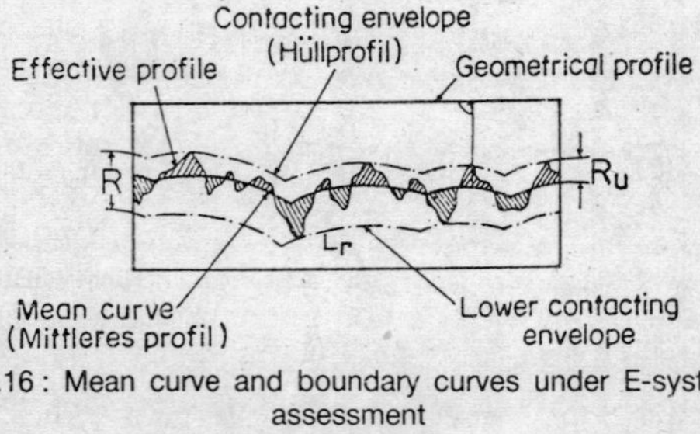

Fig. 6.16 : Mean curve and boundary curves under E-system of assessment

R_{max} is the maximum distance between two reference lines parallel to the mean line touching the highest and the lowest points within the sampling length (l) as shown in Fig. 6.17. R_z is a ten-point averaged height of irregularities. It is given by the difference between the average height of five highest peaks and five lowest peaks within the sampling length l measured with respect to a datum parallel to mean line as shown in Fig. 6.18. Mathematically,

$$R_z = \frac{(R_1+R_3+R_5+R_7+R_9)-(R_2+R_4+R_6+R_8+R_{10})}{5}$$

Eqn. 6.12

6.5. NATIONAL STANDARDS IN VARIOUS COUNTRIES FOR SURFACE ASSESSMENT

The following countries have adopted assessment standards based on M-system:

The USA, the USSR, the UK, Czechoslovakia, Japan, Poland, Netherlands, Denmark, Finland, etc.

The following countries either have adopted or are considering adoption of assessment standards based on E-system:

France, Germany, Italy, Switzerland.

BS–1134 of 1961 recommends preferred values for CLA following Renard series with a factor of 2 ranging from 1 to 1000 micro-inch along with standard meter cut offs for various finishing processes.

American ASA–B–46.1 of 1955 recommends R_a (CLA) values ranging from 0.25 to 1000 micro-inch according to Renard 10/3 series of factor 2. The sampling lengths are 0.003″, 0.010″, 0.3″, 0.100″, 0.300″ and 1.00″.

The Russian GOST–2789 of 1959 is similar to ISO–221 also following R–10/3 series.

Indian standard—EDC-20 of 1953 requires roughness height (CLA), waviness, height, direction of lay and roughness width to be stated on drawings.

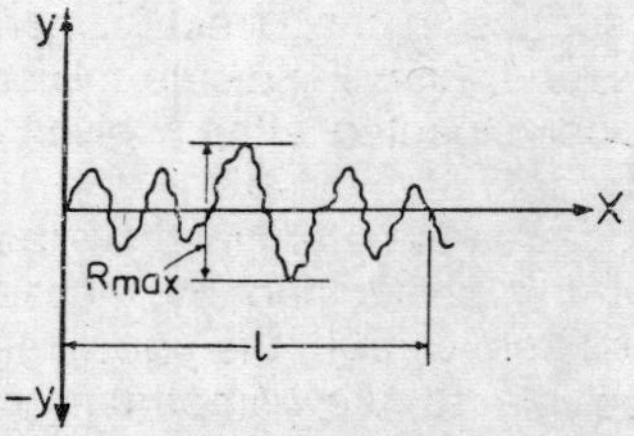

Fig. 6.17 : R_{max} within a sampling length

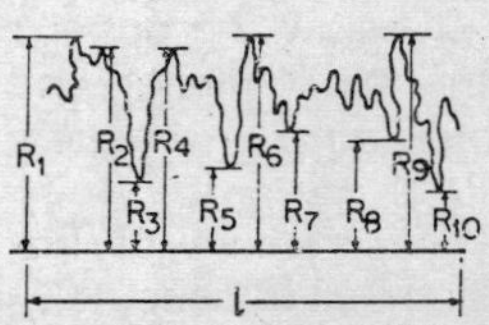

Fig. 6.18 : Determination of ten-point averaged R_z

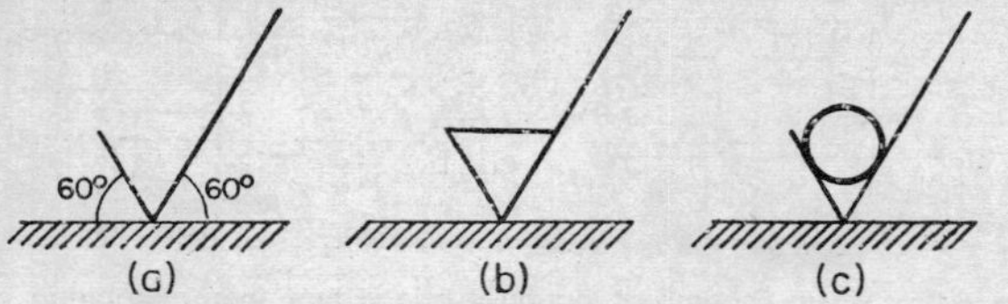

Fig. 6.19 : Indian standard roughness symbols : IS 696 : 1972

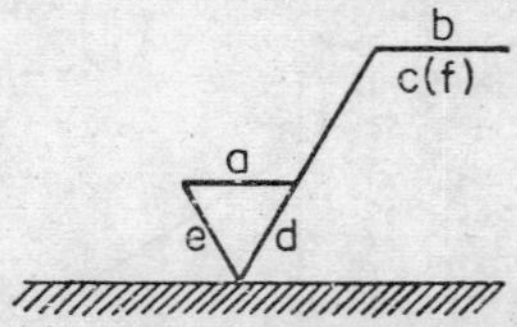

Fig. 6.20 : General roughness symbol in metal cutting

6.5.1. Roughness Symbols

Indian Standard IS–696 of 1972 provides roughness symbols for use on drawings. The most common technique is to indicate surface roughness by a basic symbol consisting of two legs of unequal length at approximately 60° to the line representing the surface under consideration. Fig. 6.19 shows three types, the first one to be used for any production method, the second one for metal cutting operations and the third one for metal forming operations.

The general symbol for surface roughness in metal cutting operations is shown in Fig. 6.20. The various specifications are as follows:

a=roughness value, R_a,
b=production method, treatment or coating,
c=sampling length,
d=direction of lay,
e=machining allowance,
f=other roughness values (in bracket).

The symbols for direction of lay are:

(i) = parallel to the plane of projection of the view in which the symbol is used.
(ii) ⊥ perpendicular to the plane of projection of the view in which the symbol is used.
(iii) × crossed in two slant directions relative to the plane of projection.
(iv) M multidirectional.
(v) C approximately circular relative to the centre of the surface.
(vi) R approximately radial relative to the centre of the surface.

Simplified symbols for machining surfaces are often used. Fig. 6.21 shows one such type used in Hindusthan Machine Tools.

6.6. ON THE PROPERTIES AND COMPARISON OF VARIOUS METHODS FOR SURFACE ROUGHNESS MEASUREMENTS

All methods like peak-to-valley measure (H), CLA value (R_a) or crest-line measure (G) are in one way or other are arbitrary so far as to characterize surfaces

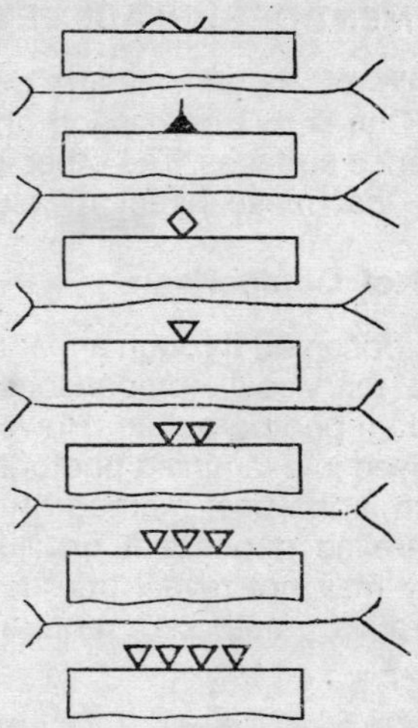

Fig. 6.21 : Simplified symbols used in HMT

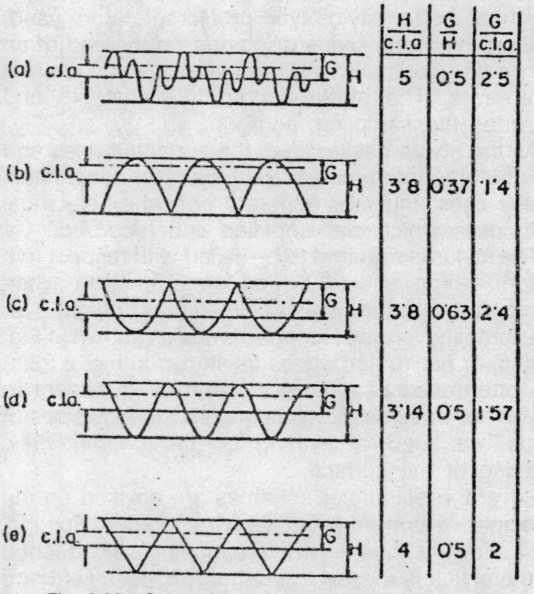

Fig. 6.22 : Comparison of shape factors using ratios

with respect to physical performance. At best, these are used to compare one piece with the other made by the same process. Peak-to-valley measure may be significant in connection with fatigue strength. Centre-line method is useful, as simple direct instruments can be devised using this technique. The G value can be determined from a profile graph but direct reading instrumentation is difficult.

Many a time a ratio such as (H/R_a) or (G/H) or (G/R_a) is used in order that degree of rounding or shape of surface profiles can be discriminated.

Fig. 6.22 shows the comparison of shape factors using such ratios. It seems that (G/R_a) is a more useful factor when various shapes are compared for equal value of H.

6.7 MEASUREMENT OF SURFACE TEXTURE

Measurements of roughness are made in two general ways. One is by comparison of surfaces to standard reference surfaces. The other is by the use of instruments that make direct measurements.

6.7.1. Method of Comparison

Surfaces are observed through a magnifying glass or microscope for visual comparisons. Magnified surfaces are often photographed. However, caution must be exercised in examining photographs. Often sight-and-touch is the best method of comparison. Photographs are so dependent on illumination or focus that they may not reveal true depth pattern. Most comparative methods require skill and judgement.

6.7.2. Stylus Type Profilometers

For actual quantitative measurements of surface roughness often stylus type profilometers are used. A sharp diamond stylus or pointer on the end of an arm is dragged back and forth across a surface at a given rate. The stroke length is adjustable and provides the sampling length.

As the stylus passes over the surface it rises and falls by amounts proportional to surface roughness. These rises and falls generate optical or electrical responses which are amplified and recorded.

The stylus is required to be slided with respect to a datum which should be accurate enough when compared with the measurable smallest height over the sampling length. Rounded *skids* as shown in Fig. 6.23(a) or flat-footed *shoes* as shown in Fig. 6.23(b) are often used as reference datum with respect to which the traverse of stylus occurs. Various sizes of skids are used depending on the characteristic spacing of the surface.

Several profilometer schemes are devised on the principle of current or voltage generation. For Fig. 6.24 shows a scheme where a moving coil attached to the stylus is moved through a magnetic field and generates current. An *R–C* circuit is incorporated for compensating the effect of frequency change. Fig. 6.25 shows another scheme where a carrier wave is modulated through the movement of the stylus. This

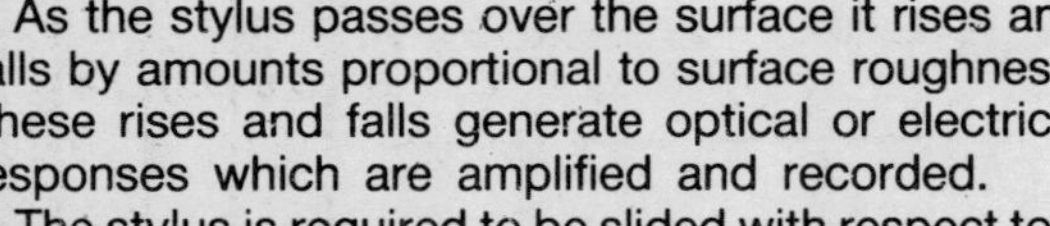

Fig. 6.23 : Skid and shoe as datum

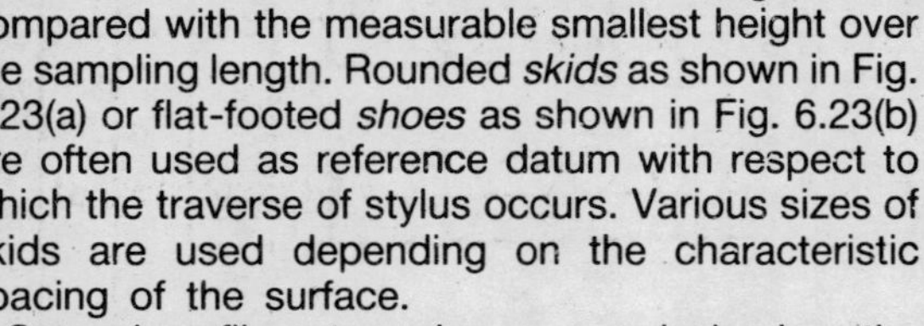

Fig. 6.24 : Current generating type instrument

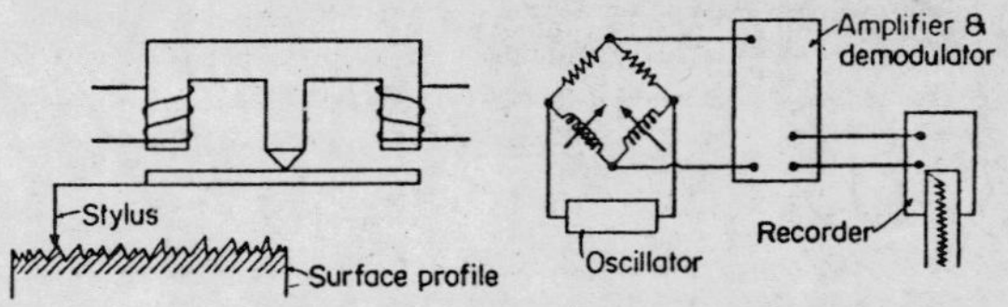

Fig. 6.25 : Basic scheme of inductive-bridge type instrumentation for surface roughness measurement

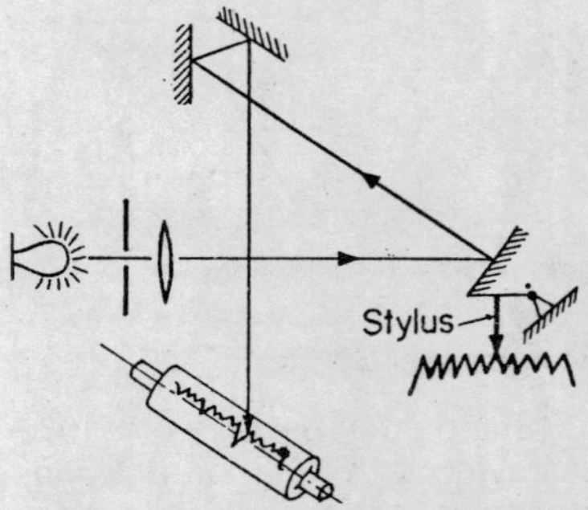

Fig. 6.26 : Scheme of optical profilograph

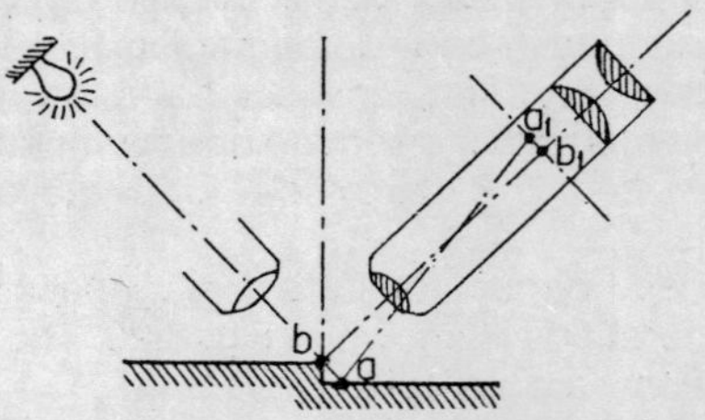

Fig. 6.27 : Optical diagram of Linnick: Double microscope

is subsequently amplified and de-modulated before recording.

Fig. 6.26 shows the basic principle of an optical profilograph where a light beam falling on a tilting mirror, when deflected through the stylus movement, is reflected by a mirror system on to a recording film system.

6.7.3 Method of Interferometry

Fig. 6.27 shows the scheme of a Linnick micro-interferometer based on the principle of light interference. A beam of light strikes the surface being inspected at angle of 45° creating a light cross-section through the surface under observation. The image when observed through a secondary lens system is displaced from b_1 to a_1 at the reticule due to the fact that some light is reflected from point *a*. This displacement is proportional to the height of step. Thus the image would be distorted according to the pattern of the micro-irregularities.

6.7.4 Method of Cast Replica

Several methods have been proposed for obtaining cast replicas in plastic material which is in a soft state (either through the application of heat or some solvent) is held on to the surface. In faxfilm process a cellulose acetate strip softened with a solvent is pressed to obtain surface texture.

CHAPTER 7

Machines for Machining

7.1. MACHINING MACHINES

The generalized concept of any form of 'Machine Tools' is related to 'Machining' with the M-W-T-F system, Fig. 7.0(a).

The process of machining is affected by a relative motion between the workpieces held against the hard edge of a cutting tool. Such relative motion is produced by a combination of rotary and translatory movements either of the workpiece or of the cutting tool or of both. The nature of this relative motion varies for different processes like turning, shaping, planing, boring, etc, (Table 7.1).

TABLE 7.1
NATURE OF RELATIVE MOTION OF VARIOUS PROCESSES

Operation Processes	Motion of Job	Motion of Cutting Tool
Shaping	Intermittent translation	Translation
Planing	Translatory	Intermittent translation
Milling	Translatory	Rotation
Turning	Rotary	Translation
Boring	Forward translation	Rotation
Drilling	Fixed	Rotation as well as Translatory feed
Hobbing	Rotation	Rotation and Translation
Surface grinding	Translation	Rotation

7.2. CLASSES OF MACHINE TOOLS

The common features of various machine tools can be grouped depending on,

(a) the types of surfaces, (b) kinematic interlinks, (c) multiple tooling setups, (d) extended capabilities, (e) concentrations, compounding, and integrations, (f) modified forms, etc.

The classifications can be subdivided as shown in Fig. 7.0(b) and Table 7.2.

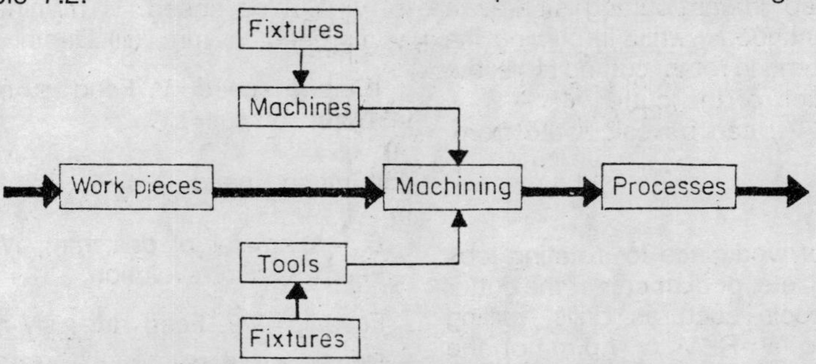

Fig. 7.0.(a) The generalised concept

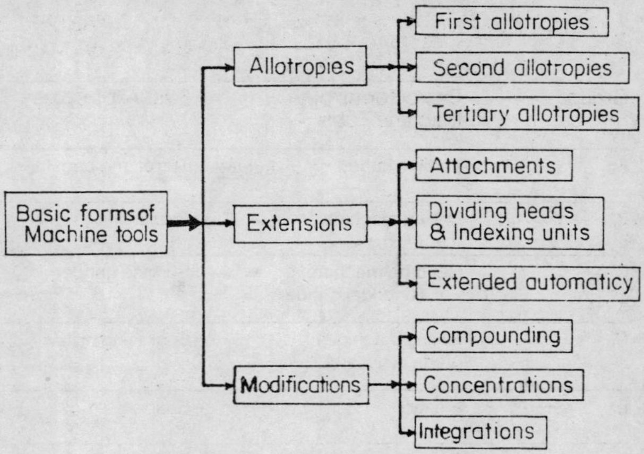

Fig. 7.0.(b) Classification of machine tools

7.3. A FEW BASIC MACHINE TOOLS AND PRIMARY CUTTING MOTIONS

The most commonly used types of primary cutting motions are rotation and straight-line reciprocation.

The primary cutting motion of certain machine tools may be of a more complex nature but it can also be described as a combination of rotary and reciprocal motions.

Rotary motion may be transmitted either to the work, as in the lathe group of machine tools (Fig. 7.1a), or to the tool, as in the milling (Fig. 7.1c), drilling (Fig. 7.1d), grinding (Fig. 7.1b) and other machines, or to both simultaneously, as in drilling small-diameter holes.

A straight-line reciprocating primary cutting motion is employed in planers, shapers, slotters, broaching machines, power hacksaws, etc. This motion can be transmitted either to the tool, as in shapers (Fig. 7.1f), slotters (Fig. 7.1e) and certain other machines, or to the work, as in planers (Fig. 7.1g).

7.4. KINEMATIC ELEMENTS OF MACHINING PROCESSES

7.4.1. Speed-Feed-Depth of Cut

For carrying out the process of cutting, the workpiece and the cutting tool must be moved

TABLE 7.2
GROUPS OF MACHINE TOOLS

Group	Basic Forms of Machine Tools	First Allotropies	Second Allotropies	Tertiary Allotropies
A	Turning lathes	Turret and capstans	Auto-screw lathes, etc.	Facing lathes
B	Drilling machine	Radial drill	Boring machine	Jig boring
C	Grinding machine (Cylindrical grinder)	Surface grinder	Tool and cutter grinder	Centreless grinders
D	Milling machines (Vertical millers)	Engraving mills	Horizontal millers	Plano-millers
E	Planers	Shapers	Slotters	Broaching machine
F	Gear shapers (Generators)	Hobbers	Bevel gear generators	Worm-wheel cutting machine
G	Thread cutting machine	Thread grinders	Thread millers	Worm grinder
H	Micro-finishing machine	Honing machine	Lapping machine	Superfinish machine
I	Cutting-off machine (Lathe)	Circular saw cutting machine	Abrasive cutting machine	Hack-sawing machine
J	Miscellaneous machines	Tool-testing machine	Dividing machine	Balancing machine

relative to each other for setting the depth of cut. This is an auxiliary motion by which the cutting tool is moved before and at the end of the cut.

Conventionally, the translatory displacement of the cutting edge of the tool along the work surface during a given period of time is called 'feed' *(s)*, while the rate of traverse of work surface past the cutting edge is designated as 'cutting velocity' or simply 'speed' (V_c). The presence of these principal motions, feed and cutting velocity permits the carrying out of the process of cutting continuously. The feed motion enables the machining to be performed all over the work surface.

Depth of cut is the perpendicular distance between the atmospheric surface and the generated surface on the workpiece, being the projection of the cutting edge measured on the reference plane, measured in mm or inches. In turning, the feed is measured in mm/revolution of the workpiece and in planing, it is measured in mm/cutting stroke. In turning, the feed is continuous, while in planing the feed motion is intermittent. In rotary cutting tools, the feed is in mm/revolution of the cutter.

The cutting speed (V_c) can be calculated from

$$V_c = \frac{\pi DN}{1000} \text{ m/minute}$$

where D = diameter of workpiece for rotating jobs such as turning, boring, etc, or diameter of the cutter for rotating cutting tools such as drills, milling cutters, etc., mm and N=RPM or r.p.m. of the workpiece or cutter as the case may be.

Hence, in case of single point turning, it is the peripheral velocity of the rotating workpiece in feet per minute or metres per minute. In case of slab milling, it is the peripheral velocity of the milling cutter in m/minute.

7.4.2. Kinematic Elements of a Few Processes

The various relative motions between workpiece and cutting tool during several conventional operations are shown in Figs. 7.2-7.7.

(a) Cutting speed = $V = \frac{\pi DN}{1000}$ m/min (Fig. 7.2.)

where N = r.p.m. and D = diameter, mm

Feed = s, mm/revolution
Feed rate = sN = mm/minute
Depth of cut = t.

(b) (i) Cutting speed, V, m/minute (Fig. 7.3.)
(ii) Feed, s, mm, (iii) Depth of cut, t, mm

Cutting speed=V, Feed=s, mm/stroke
Depth of cut=t.

(c) Cutting speed, $V = \frac{\pi DN}{1000}$ m/min (Fig. 7.4.)

D = diameter of drill, mm, N = r.p.m.
Feed=s, mm/revolution.

Feed/lip=s/2, Feed rate=sN mm/minute
Depth of cut per lip = $\frac{D}{2}$ mm

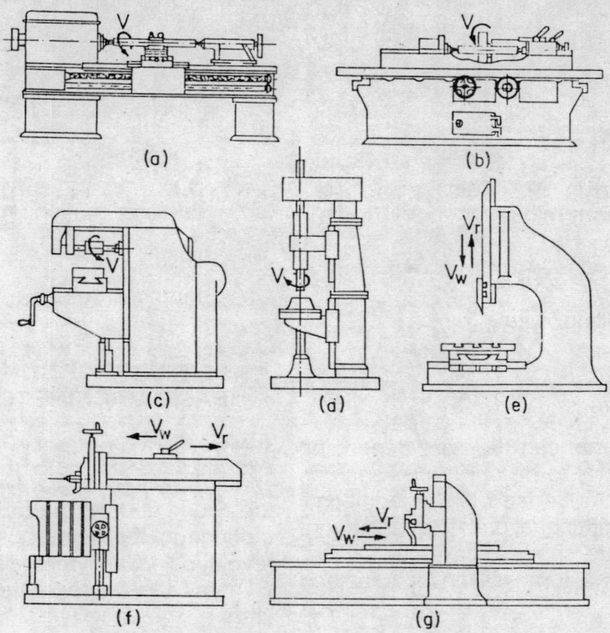

Fig. 7.1: Primary cutting motion in vartious types of machine tools.

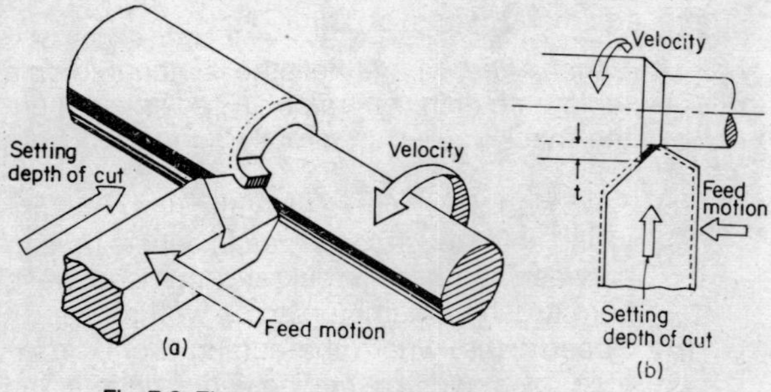

Fig. 7.2: Elements of cutting in turning operations.

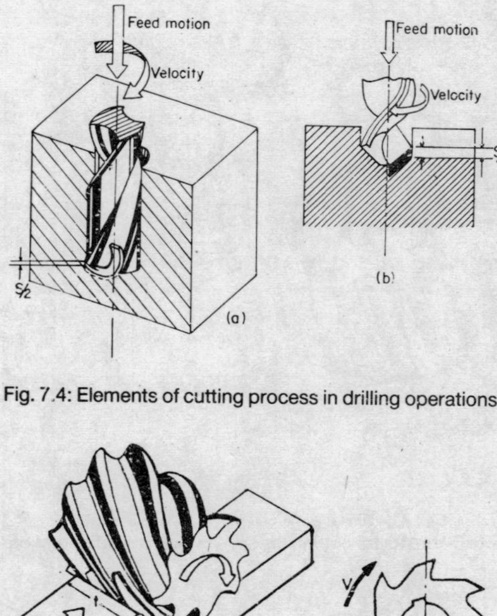

Fig. 7.4: Elements of cutting process in drilling operations.

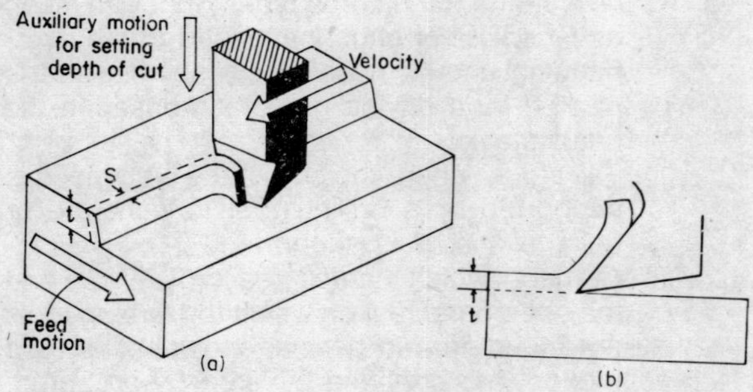

Fig. 7.3: Elements of cutting in shaping operations.

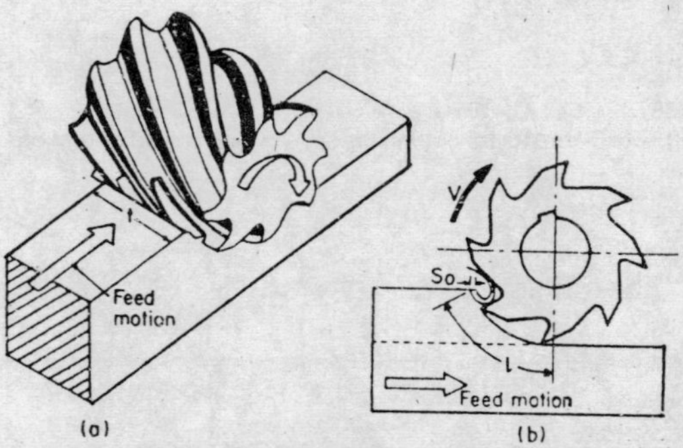

Fig. 7.5: Elements of cutting process in peripheral milling operations.

(d) Cutting speed, $V = \frac{\pi DN}{1000}$ m/min (Fig. 7.5.)

D = diameter of milling cutter, mm
N = r.p.m.
Feed/tooth = s_0, mm
Feed/revolution = s_0Z
Z = number of teeth in the cutter
Feed rate = s_0ZN, mm/minute

(e) Cutting speed = $V = \frac{\pi DN}{1000}$ m/min (Fig. 7.6.)

D = diameter of milling cutter, mm

N = r.p.m.

Feed/tooth = s_0, mm.

(f) The kinematic elements can be more depending on the various operations to be performed such as indexing motion (i) or radial feed (s_r) or additional rotation (Δ_v) in order to get spiral path in gear cutting, etc. A few examples are shown from gear cutting, etc. Fig. 7.7 (a,b) shows form milling of spur and helical gears.

$V = \frac{\pi DN}{1000}$ m/minute
D = diameter of form milling cutter, mm
N = r.p.m.

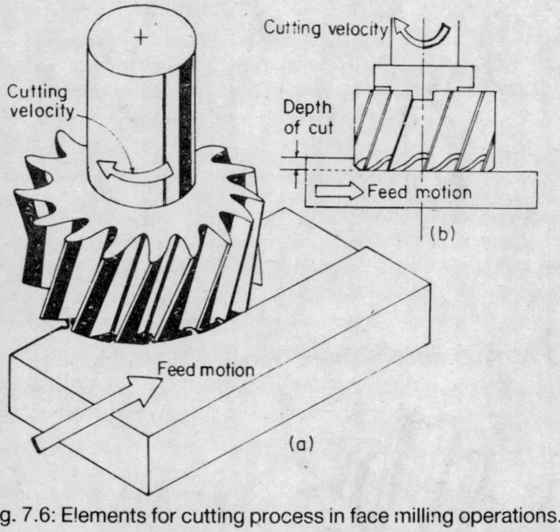

Fig. 7.6: Elements for cutting process in face milling operations.

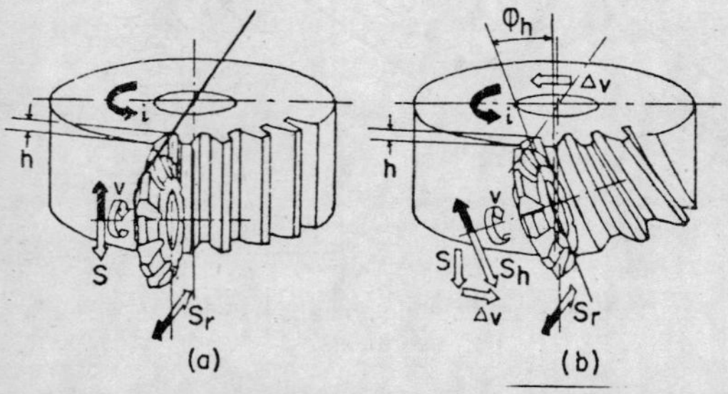

Fig. 7.7: Elements of cutting process in form milling of gears. (a) spur (b) helical

s = feed, axial

s_r = feed, radial

Δ_v = additional workpiece rotation for helical traverse

s_h = resultant feed

i = indexing motion, mm

ϕ_h = helical angle

h = depth of cut, mm.

7.5. FEATURES OF CONSTRUCTION OF A FEW BASIC MACHINE TOOLS

7.5.1. Lathe

A *lathe* is a particular type of machine tool in which the work is held and rotated against a suitable cutting tool for the purpose of producing surfaces of revolution in any material.

A typical power operated floor-type modern centre lathe is shown in Fig. 7.8. The principal features of construction are :

(a) *Types*

There are many types of lathe, viz. :

(i) *Bench lathe,* where the lathe is mounted on a table or bench, or a *floor-type lathe*, where the machine bed is directly supported from floor ;
(ii) *Solid bed lathe,* where the lathe has constant swing, or the *gap-bed lathe,* where a gap between the head stock and carriage ways permits larger diameters to be worked upon ;
(iii) *Speed lathe,* where the cutting tool is actuated by hand, or *engine lathe,* where the cutting tool is fed by power.

(b) *Features :*

(i) *Head stock,* where the workpiece is held in a 'live centre' or in some types of device, such as faceplate or chuck, attached to the overhanging end or nose of the spindle and is rotated by a driving gearbox, housed in the head stock.
(ii) *Tail stock*, where the end of the workpiece is supported or a drilling tool to generate or enlarge a hole is supported.
(iii) *Carriage* which carries the cross-slide and the compound rest on which the turning tools are mounted and gear arrangement of feed motion, are provided covered by an apron. The sectional view of a common type of apron is shown in Fig. 7.9.
(iv) *Bed* is the principal supporting structure on which the head stock and tail stock are mounted with the carriage sliding on ways.

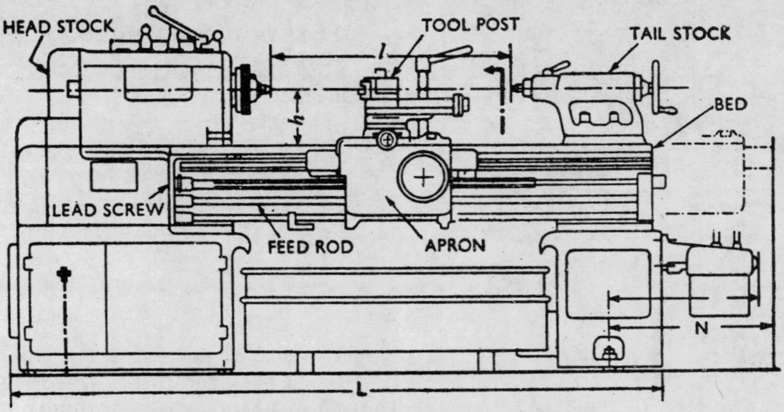

Fig. 7.8: A typical heavy duty centre lathe.

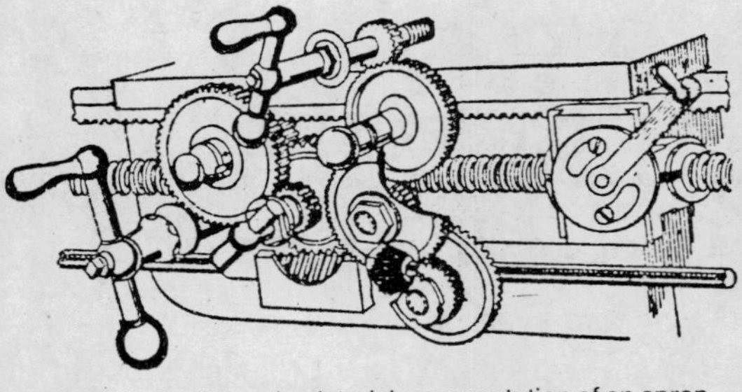

Fig. 7.9: A schematic pictorial representation of an apron.

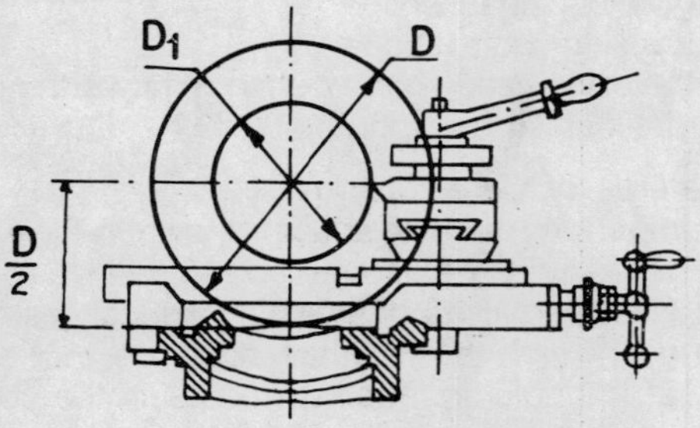

Fig. 7.10: Main dimensions of an engine lathe.

(c) *Specifications and Size*

The size (Fig. 7.10) of a lathe is usually specified by :

(i) height of centre *h* (continental system), or by the swing *(2h)* of the workpiece (American system) ;

(ii) the maximum admit length *(l)* of a job between centres (continental system) or the overall length *(L)* of the bed ;

(iii) the power ;

(iv) range and number of steps of feeds and speeds.

7.5.2. Drilling Machines

Drilling is a term which designates the particular operation of producing holes.

A drilling machine or a drill press is the type of machine where the drill is held and the means for furnishing proper rotation, speed and axial feed are provided.

(a) *Classification of Machines*

Drilling machines are classified either according to features of design or application of power.

According to features of construction :

(i) Portable or stationary,

(ii) Pillar or column drill or radial drill,

(iii) Single spindle or multiple spindle.

According to application of power:

(i) Hand or power-driven spindle.

(ii) Sensitive (hand) feed or power feed.

(iii) Line shaft driven or individual drive.

(b) *Features of Construction of Few Drills*

(i) The principal features of construction of a pillar drilling machine (Fig. 7.11).

(a) *Pillar* which supports the head stock and feed drive ;

(b) *Table* which supports the workpiece ;

(c) *Quills* that is fed up and down.

(d) *Screw* is for radial location of the head;

(e) *Elevating screw* is for vertical location of the head.

(ii) The different parts of a radial drill are shown in Fig. 7.12. The principal parts of radial drilling machine are:

(a) *Column* which is fixed at the base and is the main supporting structure ;

(b) *Arm* which along with the column rotates around the stump ;

(c) *Stump* is inner structure arouna which the column is supported ;

(d) *Head* where the main gearbox for spindle drive and feed movement is housed.

(c) *Specifications*

The principal specifications of a drilling machine are:

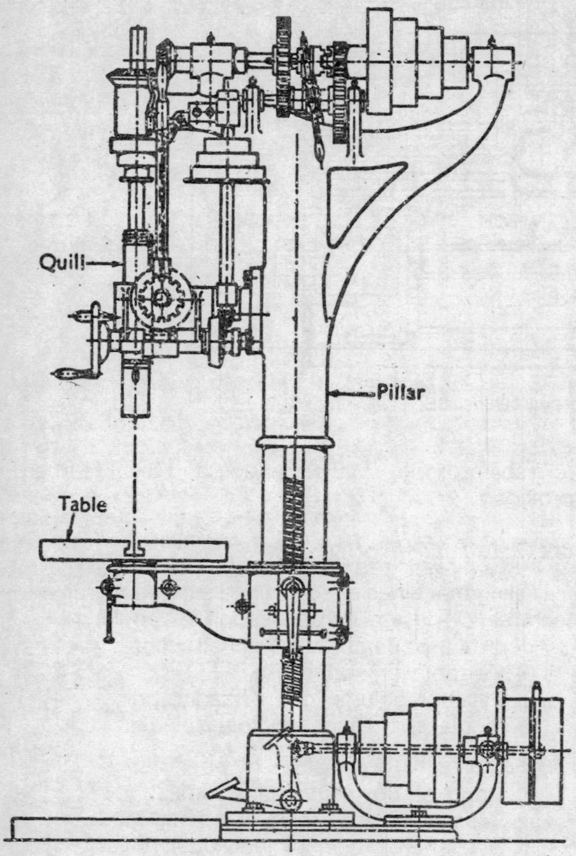

Fig. 7.11: A typical pillar drill.

(i) size of maximum drill that can be done in a given material ;
(ii) the maximum radial dimension of a workpiece that can be admitted between the upright or column and the drill ;
(iii) horsepower or killowatt of the main motor ;
(iv) range and number of steps for spindle speeds and feed drive.

7.5.3. Planing Machines

(a) Object

A planer, also used to produce flat surfaces on workpieces that are too large or otherwise impossible to be handled on a shaper, differs from shaper in that the work table reciprocates while the cutting tool is fed on a cross rail which straddles the table.

(b) Class of Machines

The planing machines are classified according to
(i) type of housing: openside type or double housing type,
(ii) method of obtaining quick return motion,
(iii) method of driving the table.

(c) Features of Construction.

The main features of Construction of a planer consist of
(i) table,
(ii) bed on whose slides the table moves,
(iii) housing, which is the main supporting structure for tool head,
(iv) the cross rail which carries the tool head,
(v) the tool head.

These features are indicated in Fig. 7.13.

(d) Size

The size of a planing machine is denoted by :
(i) stroke length (or size of table),
(ii) horizontal distance between the housings,
(iii) vertical distance between the rail and table.

7.5.4. Shaping Machines

(a) Object

A shaper, first developed to machine flat surfaces on small workpieces, is a machine where the cutting tool is held in a tool holder held on a clapper in the head at the end of a reciprocating ram.

(b) Types and Classes

There are two broad types of shapers based on the method of application of feed :
(i) travelling head shaper
(ii) travelling table shaper.

In a travelling head shaper, often, Whitworth quick return mechanism is adopted.

(c) Features of Construction

The important parts of shaper shown in Fig. 7.14 are :
(i) Ram: heavily ribbed semicylindrical cast iron part which reciprocates in the dovetail slide of the column. The ram contains the stroke positioning and downfeed mechanisms.
(ii) *Rocker arm*: actuates the ram through a linkage.
(iii) *Base and column*: this is the supporting structure for the operating members of the machine.
(iv) *Table*: on it the workpiece is clamped. The table is firmly supported by saddle which moves on a cross rail fixed to the column.
(v) *Bull gear*: it is large geared wheel, mounted inside the column carrying a radial slide that controls the stroke length.

(d) Size and Specification

A shaping machine is specified by:
(i) maximum stroke length(s);
(ii) maximum traversing distance of the table;
(iii) horsepower;
(iv) range and steps of ram speeds and feeds.

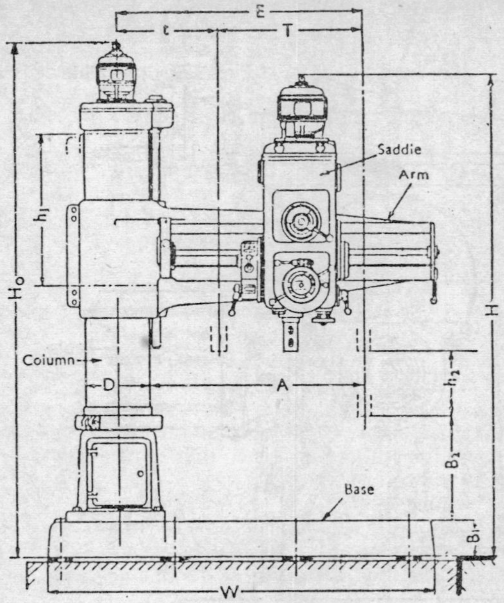

Fig. 7.12: A typical radial drill.

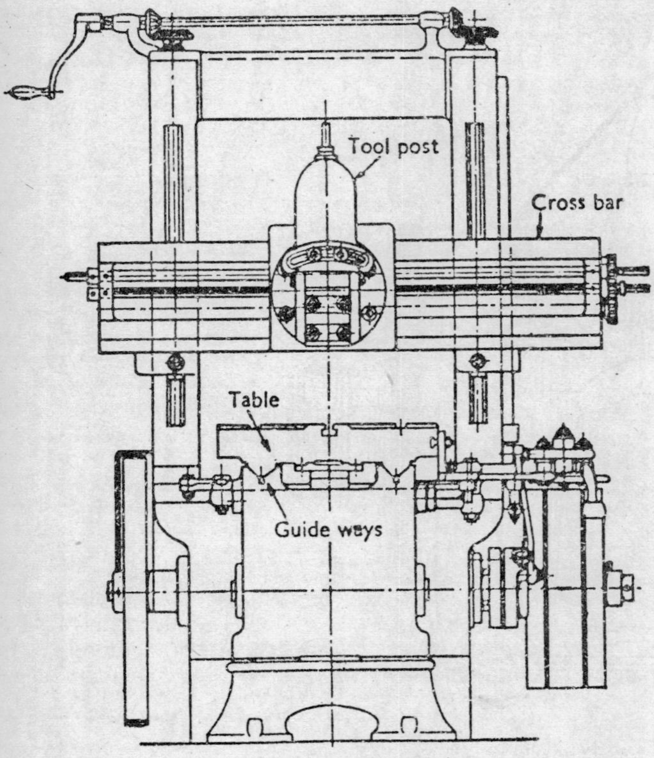

Fig. 7.13: A typical planer.

7.5.5. Slotters

(a) Object

The vertical shaper, commonly known as a slotting machine, is a machine with a vertical ram which is specially adopted to slotting holes, keyways and grooves.

(b) Features of Construction

The work is usually supported on a rotating table over a cross slide and carriage. The general features of design of a slotter are shown in Fig. 7.15.

(c) Specification

Slotting machines are mainly specified by

(i) length of stroke
(ii) diameter of table working surface,
(iii) maximum table travel,
(iv) tool speed
(v) table feed,
(vi) power of motor.

7.5.6. Broaching Machines

(a) Object

Broaching is a machining process in which a typical cutter called *broach* is drawn or pushed past the surface thus machining a hole or a groove (Fig.7.16).

(b) Classes of Machine

Broaching machines are classified as:

(i) pull, or push or continuous type,
(ii) mechanical or hydraulic drive,
(iii) horizontal or vertical.

(c) Specification

A broaching machine (Fig. 7.16) is specified by:

(i) Pulling force,
(ii) Length of slide stroke,
(iii) Speed of cutting stroke,
(iv) Power of the motor.

7.5.7. Boring Machines

(a) Object

Boring machines are used to enlarge or bore holes. However, these machines can also perform drilling, milling, facing and similar other operations.

(b) Features of Construction

The features of construction include:

(i) rigid box bed on which the work is supported;
(ii) elevating spindle at the headstock for holding boring bars;
(iii) supporting column at the end;
(iv) *table* is supported on the guide ways of the bed and traverses longitudinally and transversely.

(c) Size and specification

A boring machine (Fig. 7.17) is specified by:

(i) diameter of movable spindle;
(ii) dimension of work table;

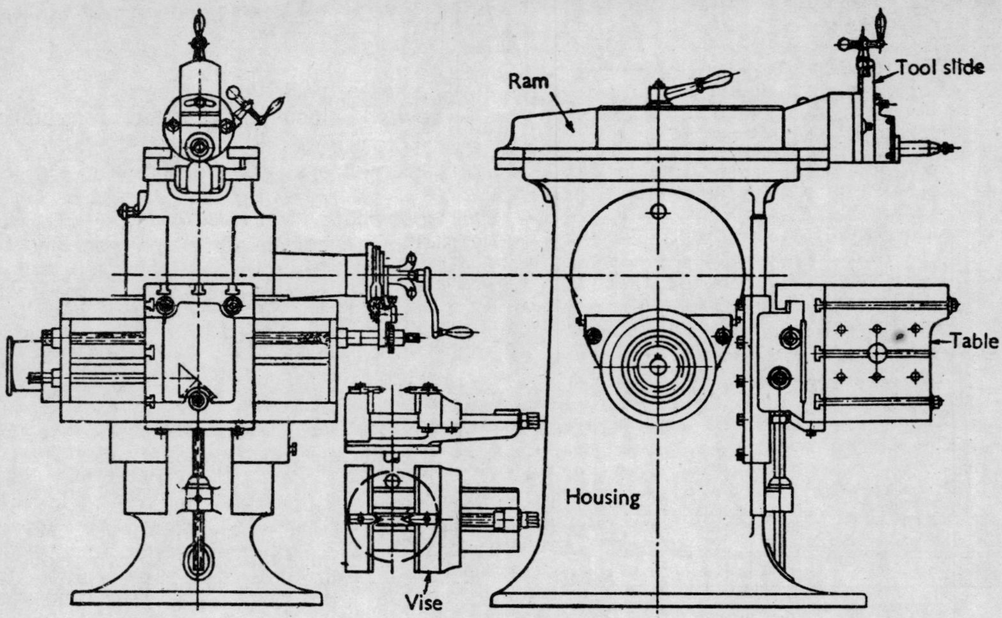

Fig. 7.14: A typical shaping machine.

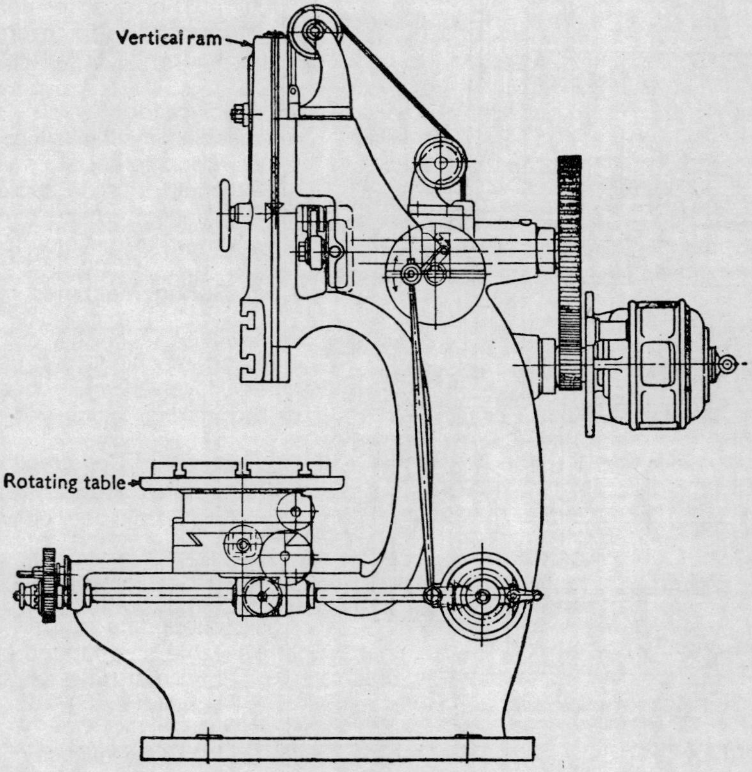

Fig. 7.15: Slotting machine.

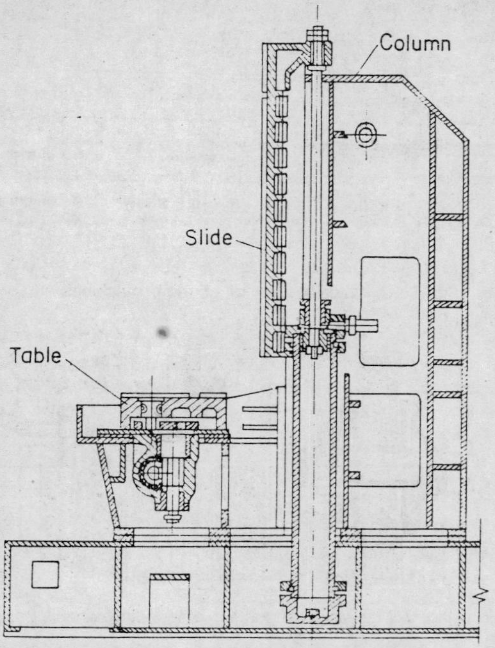

Fig. 7.16: A vertical surface broaching machine.

(iii) maximum cross travel of table;
(iv) maximum vertical travel of hand stock;
(v) maximum weight of the workpiece mounted on the table;
(vi) maximum axial travel of movable spindle;
(vii) RPM of spindle;
(viii) RPM of faceplate;
(ix) speed of axial spindle feed;
(x) power of the motor.

7.5.8. Milling Machines

(a) Object

The term, *milling*, as applied to certain kinds of machining operations is based on the action of a multiple-toothed wheel on the workpiece in the same manner as the old 'mill stone' acted on the grain in early days.

Milling machines employ such multiple rotating cutters, called milling cutters.

(b) Classes of Milling Machine

There are various kinds of milling machines. The most important types of classification are:

(i) column and knee-type or fixed bed-type,
(ii) plain or universal,
(iii) horizontal spindle (arbour) (Fig. 7.18) or vertical (Fig. 7.19) spindle,
(iv) mechanical drive or hydraulic drive,
(v) general purpose or production type.

(c) Features of Construction

The main features of construction of a general purpose column and knee-type milling are:

(i) *Knee:* the milling machine saddle and table are supported by a 'knee' which slides up and down on a column.
(ii) *Arbour*: the spindle that holds the cutter is called arbour. It is fixed to the spindle nose by self-releasing taper and draw down bolts.
(iii) *Overarm*: it supports the arbour through hangers and braces. It also dampens the vibrational effect of cutting impulses.
(iv) *Upright or Housing*: is the main supporting frame for all operating elements.

(d) Size and Specification

The size of a milling machine is usually specified by:

(i) dimension of table working surface ;
(ii) maximum table travel:
  (*a*). logitudinal,
  (*b*). cross,
  (*c*). vertical,
(iii) maximum angle of table swivel,
(iv) range of spindle speeds,
(v) range of table feeds,
(vi) power of the motor.

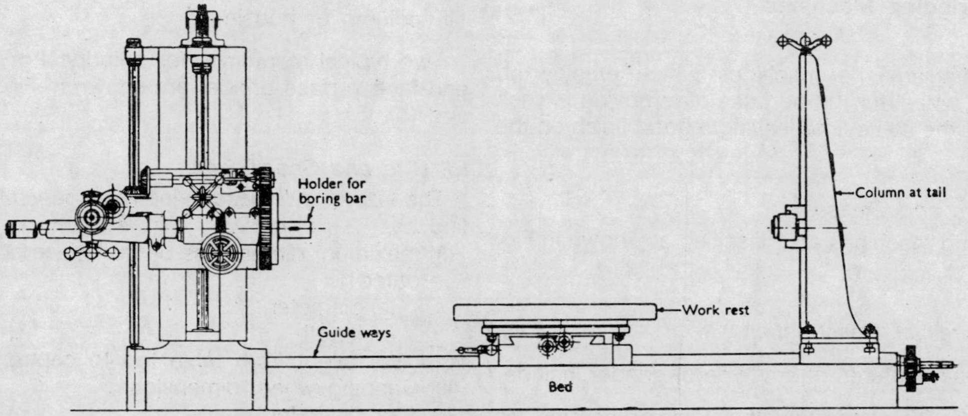

Fig. 7.17: A boring machine.

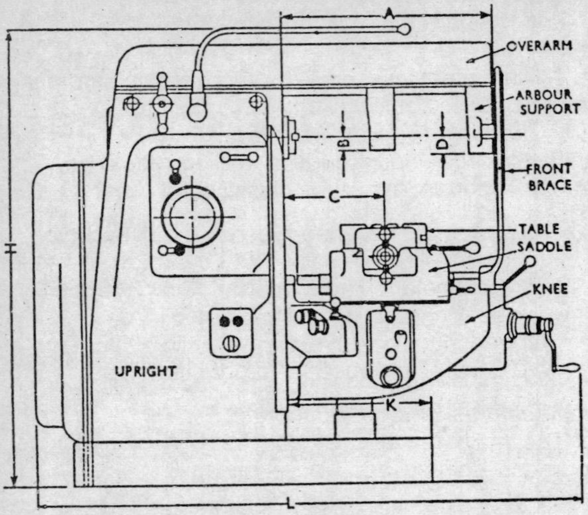

Fig. 7.18: Horizontal arbour milling machine.

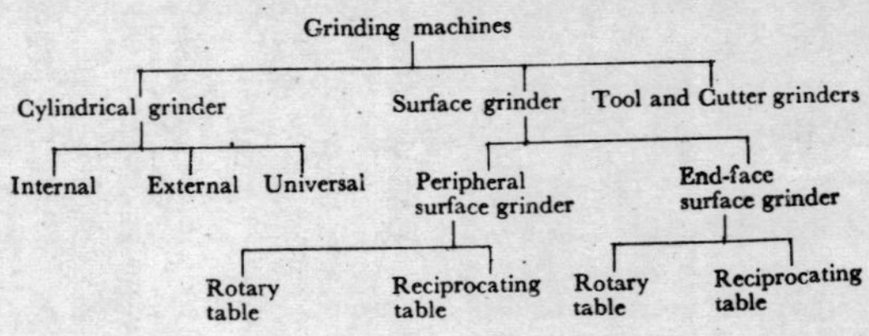

Fig. 7.20: Classification of grinding machines

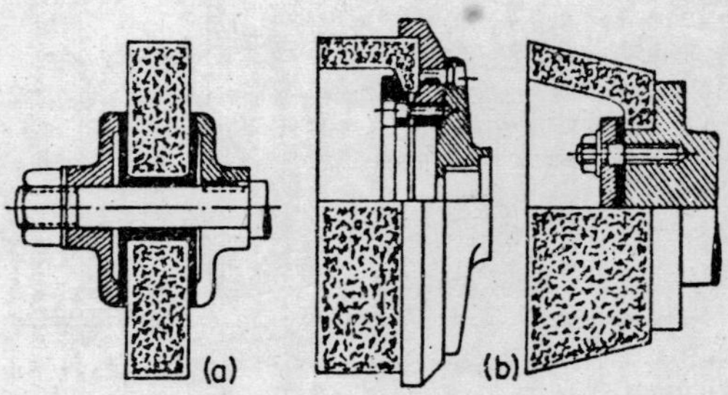

Fig. 7.21: Typical mounting of grinding wheels.

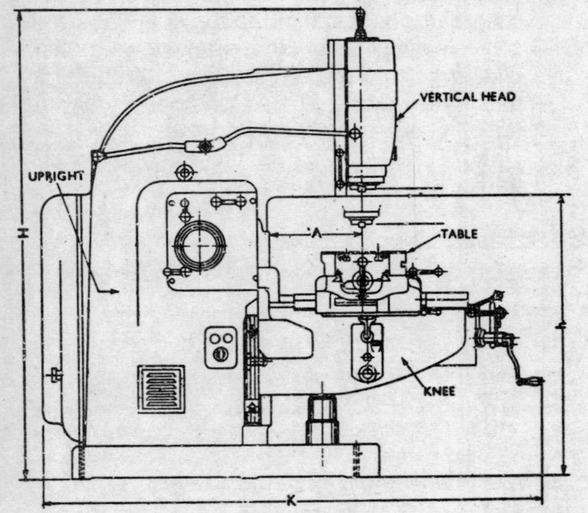

Fig. 7.19: Vertical milling machine.

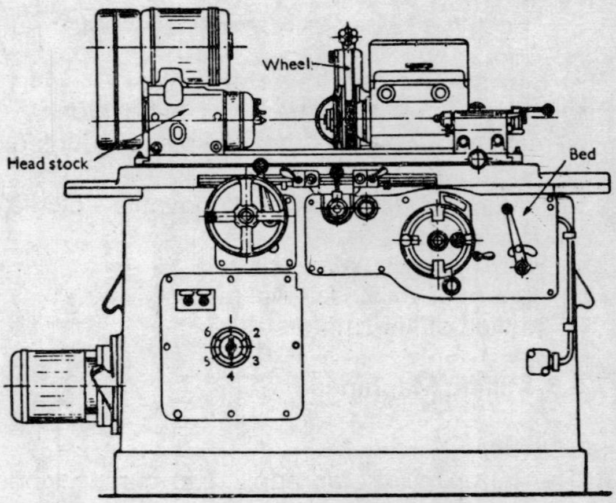

Fig. 7.22: Typical cylindrical grinding machine.

7.5.9. Grinding Machines

(a) Object

Grinders are the machines which employ an abrasive wheel for the purpose of removing excess stock, at the same time leaving a good finish on the work.

(b) Types and Classes

Grinding machines are classified as shown in Fig. 7.20.

(c) Features of Construction

The main features of any type of grinder include:

(i) *Table:* for supporting the work or centres for holding the work;

(ii) mounting device for the grinding wheel;

(iii) column or housing.

Two typical mountings for cylindrical grinder and end-face surface grinder are shown in Fig. 7.21 (a) and (b).

(d) Size and Specification

The size of *cylindrical* grinding is generally specified by:

(i) maximum dimensions of workpiece accommodated:
   (a) diameter,
   (b) length,
   (c) height from table top to centre;

(ii) Grinding wheel dimensions:
   (a) diameter,
   (b) width;

(iii) RPM of workpiece;
(iv) speed of table traverse and wheel head traverse;
(vi) angle of table swivel
cross feed of wheel head per double stroke of table;
(vii) power of the motor.

The size of surface grinding (Fig. 7.22) is usually specified by:
(i) size of the table;
(ii) maximum distance from grinding wheel spindle to table surface;
(iii) diameter of grinding wheel;
(iv) RPM of grinding wheel;
(v) cross feed rate and vertical feed rate;
(vi) speed of longitudinal feed;
(vii) power of the motor.

7.6. PROCESS TOPOLOGY

7.6.1. Lathe and Turning Operations

Sketch | *Operations*

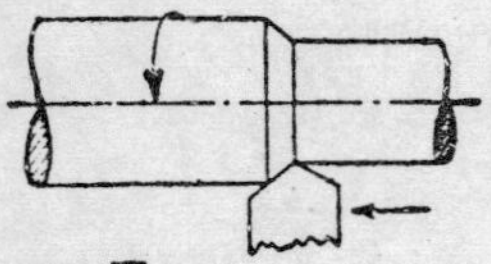

(a) Simple turning with a turning tool.

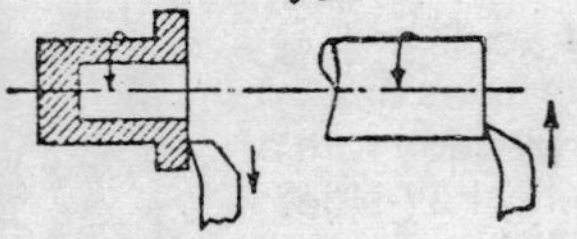

(b) Facing a job in chuck or in centres.

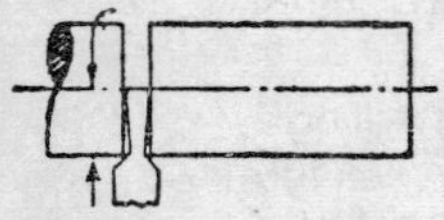

(c) Parting by means of a parting tool.

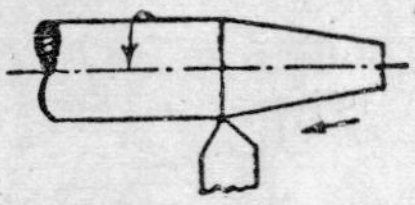

(d) Taper turning with a turning tool.

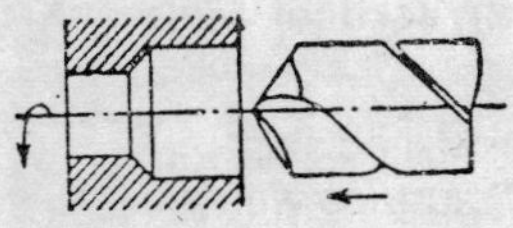

(e) Similar to drilling, enlarging a drilled hole with a twist drill.

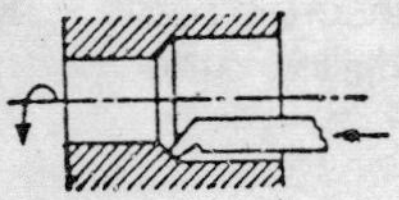

(f) Boring a cylindrical hole with a boring tool.

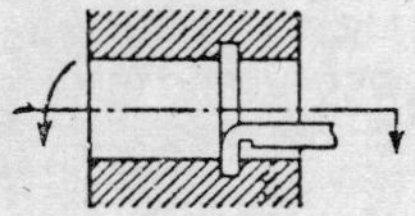

(g) Recessing groove.

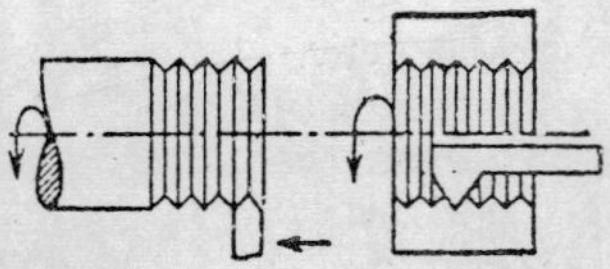

(h) External and internal thread cutting.

Sketch | *Operations*

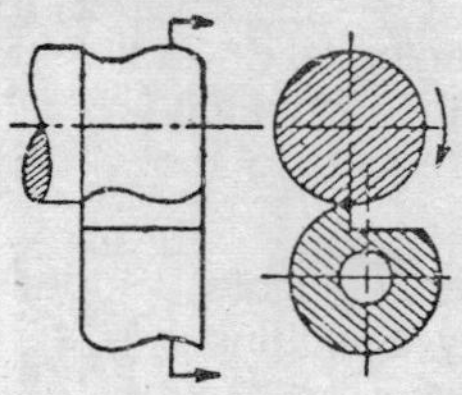

(i) Forming with a circular forming tool.

7.6.2. Drilling Operations in Drilling Machines

(a) Drilling a hole with a twist drill.

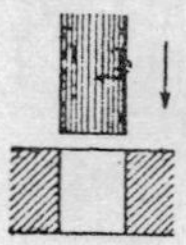

(b) Reaming a hole by a reamer.

(c) Counter sinking a conical recess by a counter sink.

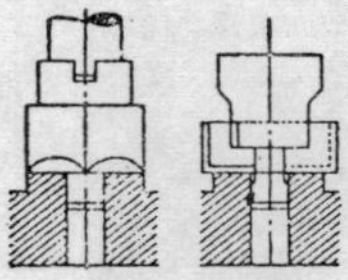

(d) Spot facing a boss with a spot facing tool:
(i) shell type,
(ii) blade type.

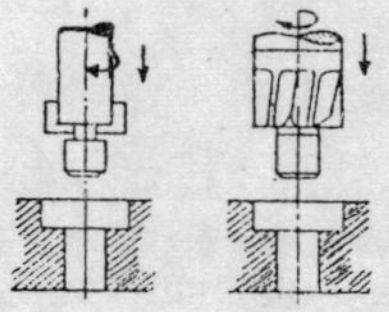

(e) Counter boring and counter sinking recesses with the corresponding tools.

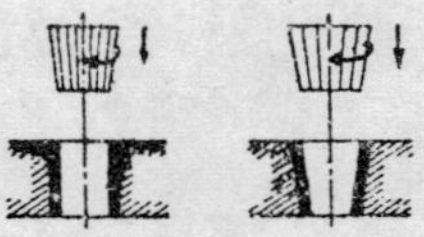

(f) Reaming a hole with taper reamers:
(i) roughing
(ii) finishing

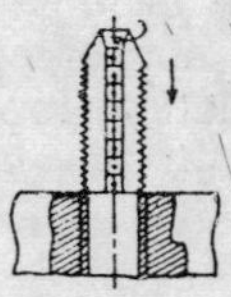

(g) Cutting thread with a tap (tapping).

7.6.3. Planer and Shaper Operations

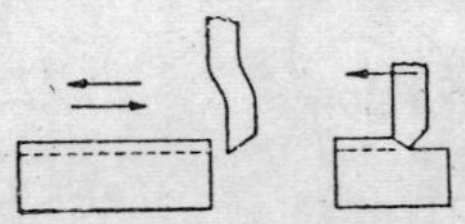

(a) Planing a flat surface with straight planing tool.

Sketch | *Operations*

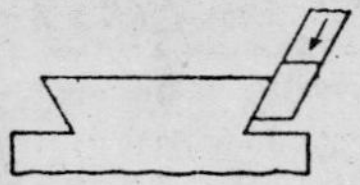

(b) Planing a dovetail with a dovetail side cutting tool.

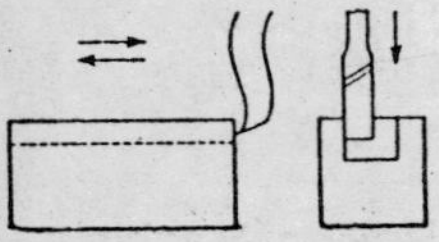

(c) Cutting slots by a slotting tool.

(d) Cutting grooves with grooving tools.

(e) Cutting a T-slot with a recessing tool.

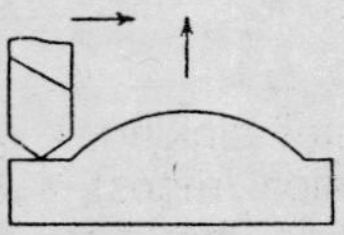

(f) Planing a formed surface with a straight planing tool.

7.6.4. Slotter Operations

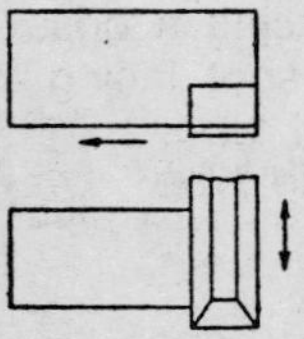

(a) Producing a flat surface with a plain slotter tool.

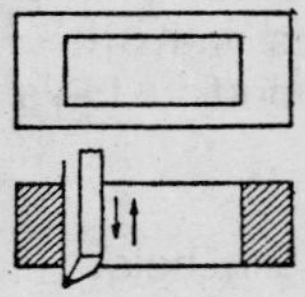

(b) Cutting a rectangular slot with a slotting tool.

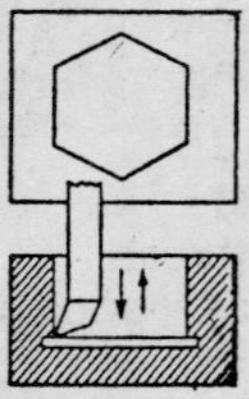

(c) Cutting a blind hexagon socket with a slotting tool.

7.6.5. Milling Operations in Horizontal Milling Machines

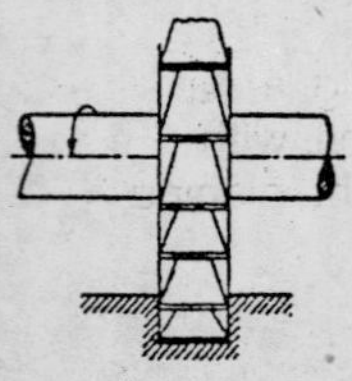

(a) Milling a slot with a milling cutter.

Sketch | *Operations*

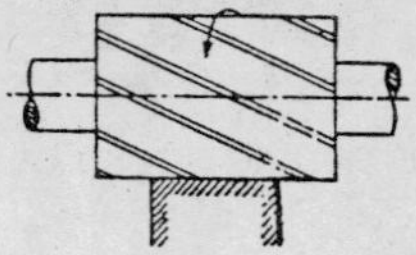

(b) Milling a flat surface with a helical tooth plain milling cutter.

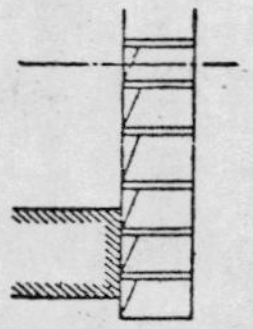

(c) Milling a flat surface with the side of a side milling cutter.

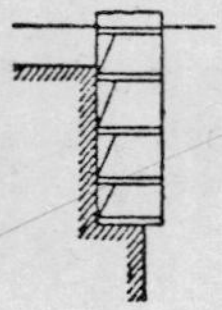

(d) Side milling with a half-side milling cutter.

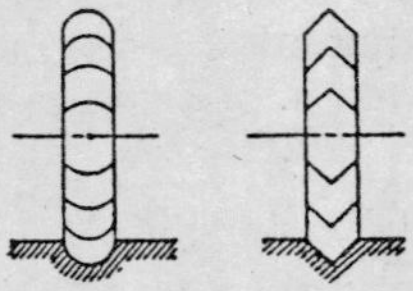

(e) Milling a semicircular and a vee-groove.

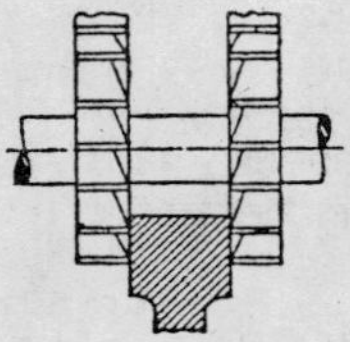

(f) Straddle milling the sides of a projection with a set of half-side milling cutters.

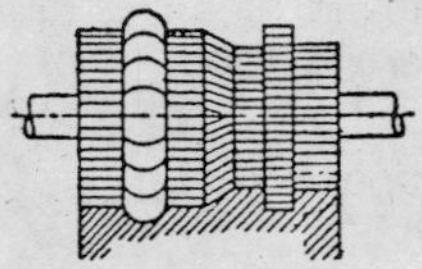

(g) Gang milling a complex form with a set of milling cutters of various types.

7.6.6. Milling Operations in Vertical Mills

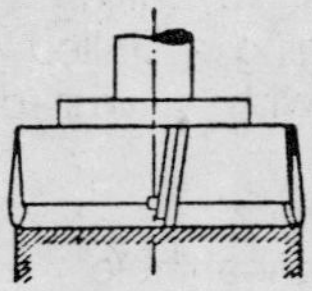

(a) Milling a flat surface with an inserted blade milling cutter.

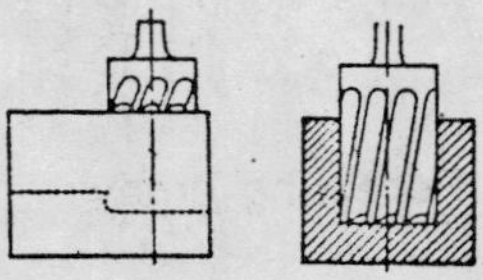

(b) Milling a slot with an end mill.

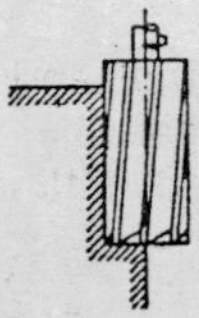

(c) Milling two surfaces of a recess by an end mill.

7.6.7 Broaching Operations

Sketch *Operations*

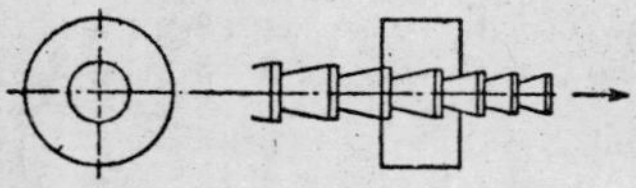

(a) Broaching a round hole.

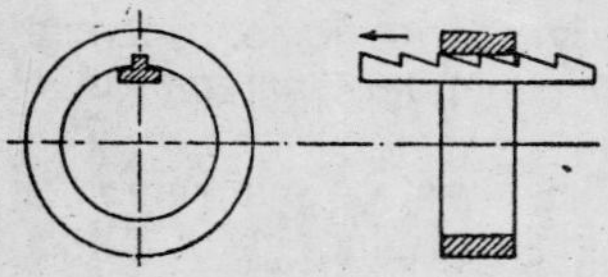

(b) Broaching a key slot with a keyway broach.

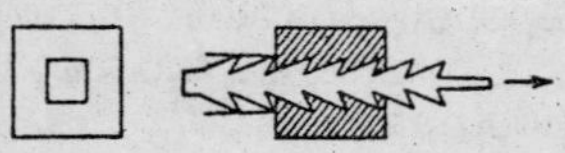

(c) Broaching a square hole by a square broach.

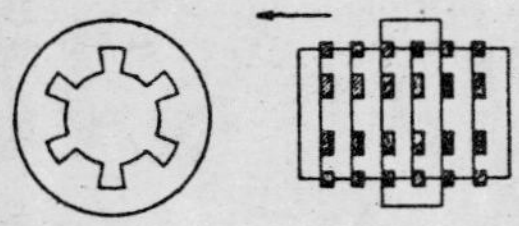

(d) Broaching spline hole with a splin broach.

7.6.8. Cutting of Gears

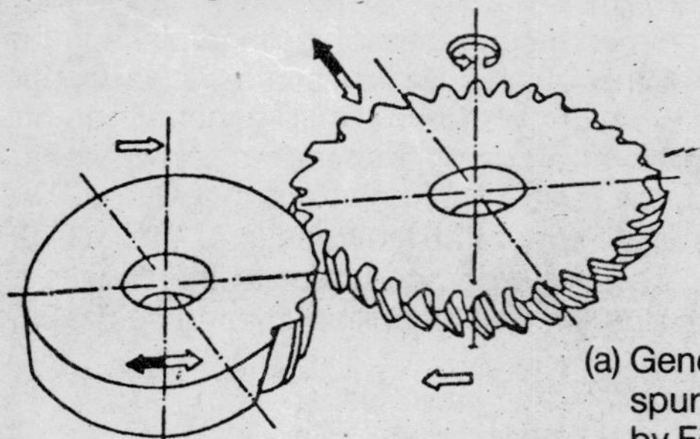

(a) Generating spur gears by Fellow's process.

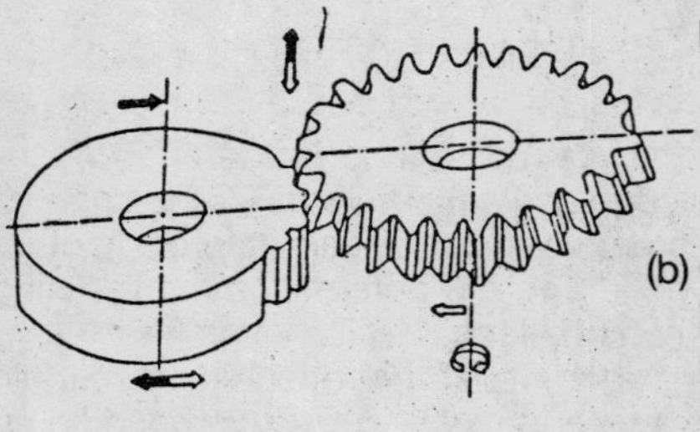

(b) Generating helical gears by Fellow's process.

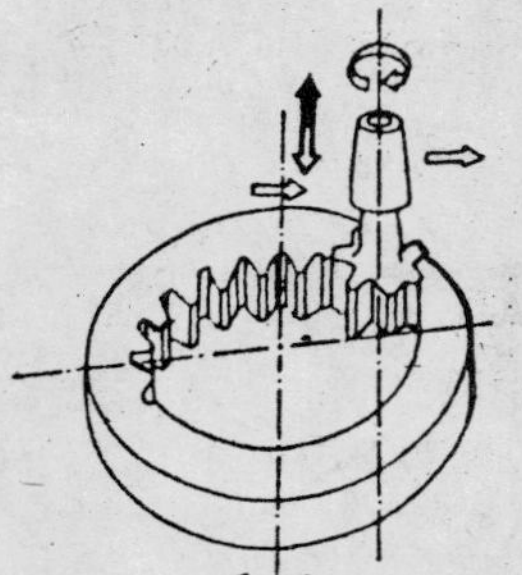

(c) Generating a spur internal gear by Fellows, process.

Sketch *Operations*

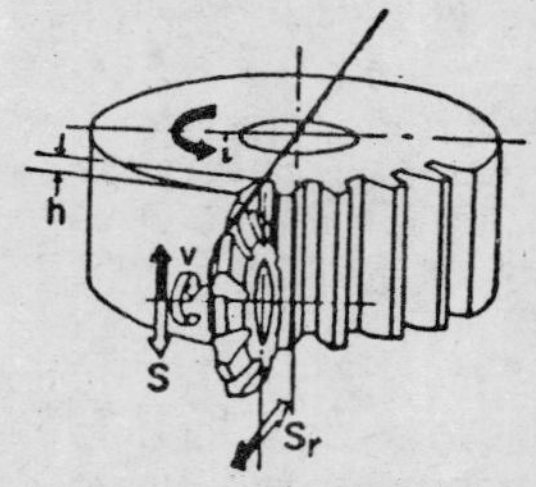

(d) Form milling of spur gears.

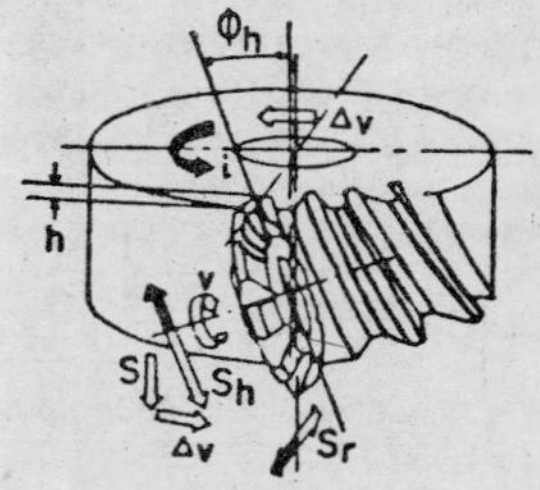

(e) Form milling of helical gears.

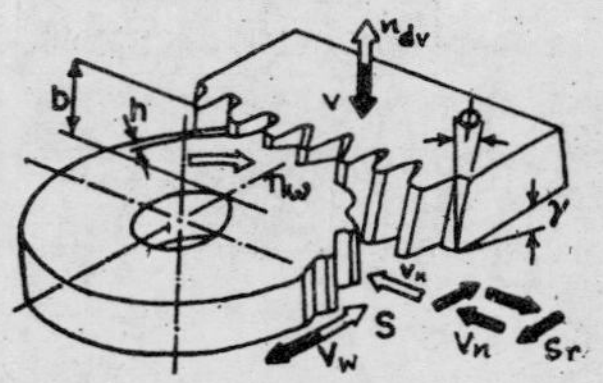

(f) Generation of spur gears by Sunderland's processes

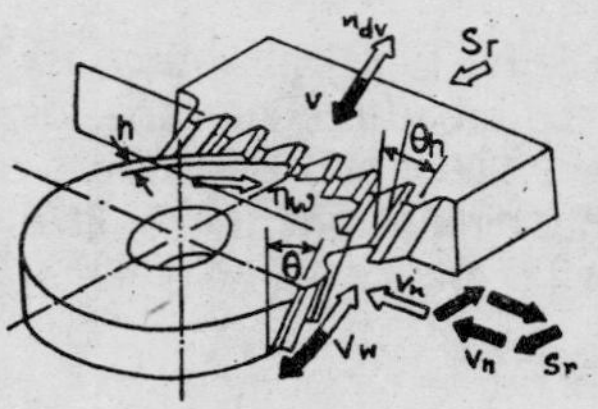

(g) Generation of helical gears by Sunderland's processes.

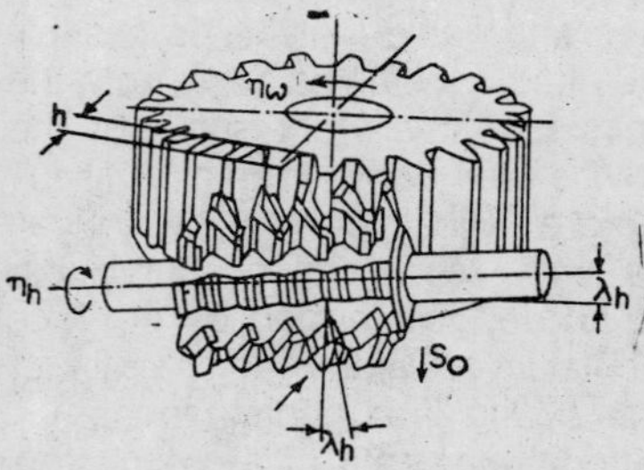

(h) Generation of hobbing of spur gears.

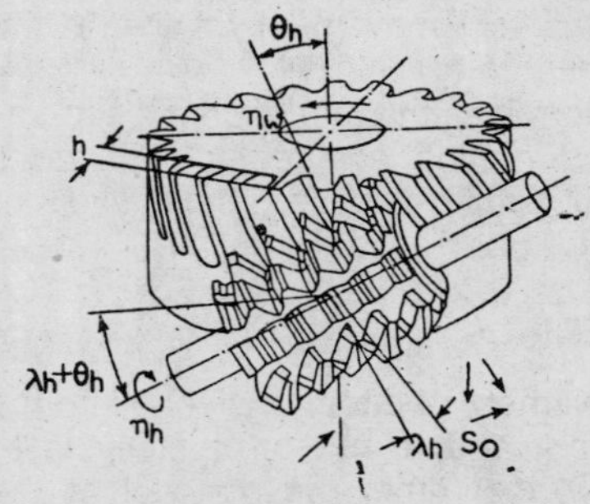

(i) Generation of hobbing of helical gears.

CHAPTER 8

Cutting Tools

8.1. GENERAL FEATURES

Depending on the type, purpose, and conditions of the machining operations to be done, cutting tools are shaped to a certain geometry and dimensions that would suit most effectively the cutting process.

The various cutting tools comprise three essential but elemental characteristics as shown in Fig. 8.0.

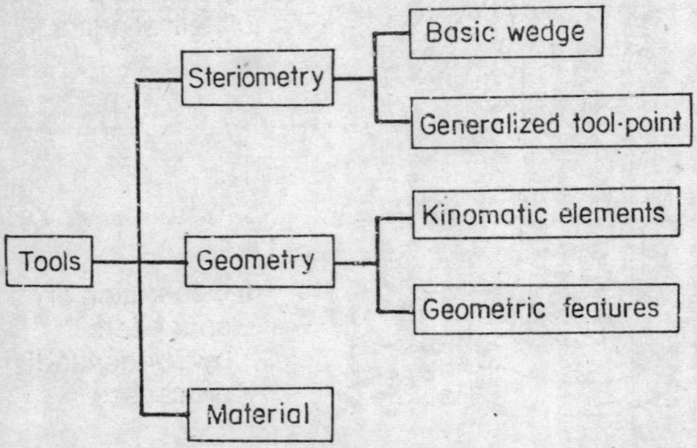

Fig. 8.0: Elemental characteristics of cutting tools.

Cutting tools include lathe tools, milling cutters, broaches, drills, reamers, thread-cutting taps, dies and chasers, gear cutters, files, etc. Regardless of their type, purpose, and dimensions, all of these tools have many geometric and design elements in common.

The working element of any cutting tool is, essentially, a wedge whose cutting edges are bounded by the face and flanks. In the process of cutting, the tool face directs the chips away from the tool, while the flanks face the work surface. The cutting elements can be bounded in form not only by planes, but also by more intricate conical, cylindrical, or helical surfaces with rectilinear or curvilinear generatrics. The cutting edges formed by the face intersection with the flanks are classified as main, or side, auxiliary, or end, and intermediate ones.

The basic geometrical elements of a cutting tool are governed by the tool movement relative to the work in the process of machining.

Main motion determines the rate of chip's separation, whereas feed motion ensures a continuous plunging of the tool cutting edge into fresh layers of metal.

8.2. TOOL MATERIALS

For the cutting element of a tool to plunge into the material of the work and to cut off a stock layer within a long period of time, the material of the cutting element must possess certain properties.

First of all, this material should display very high hardness and strength. These features must be retained by cutting tool even after the latter has been heated to rather high temperatures in the process of cutting; in other words, the tool material must show adequate thermal stability, or red hardness.

The cutting element must withstand the abrasive effect of the work material within considerable periods of time and under the conditions of high pressures and temperatures, i.e. tool materials must offer high wear-resistance.

In addition, tool materials must be machinable in order to ensure the original manufacture of cutting tools and their resharpening in operation.

All these features depend upon the chemical composition, the structure, the physical and mechanical properties of the tool materials.

Today, cutting tools are made of carbon and alloy tool steels, high-speed tool steels, cemented carbides, mineral ceramics, diamonds, and abrasive materials.

8.3. BASIC WEDGE

Almost all cutting tools used in metal cutting operations consist of the basic form of a wedge which is defined as one form of inclined plane in the shape of a triangular prism. Assume that a wedge under the action of force P is penetrating into another body at a constant speed as shown in Fig. 8.1. The body resists the motion of the wedge. The reactions N, N appear at the cheeks of the wedge. The forces N, N are perpendicular to the cheeks in absence of friction. From equilibrium of forces of Fig. 8.2.

$$\frac{N}{P} = \frac{KM}{KL} = \frac{1}{2\left(\frac{KL/2}{KM}\right)} = \frac{1}{2\sin\frac{\beta}{2}}$$

Thus the mechanical advantage in force is dependent on the wedge angle β. The smaller the angle of a wedge, the greater will be the gain in force. In other words, the wedge angle β determines the resisting force of the cutting edge.

The cutting edge must be oriented at certain required angles with the work surface depending on the nature of operation to be performed. Fig. 8.3

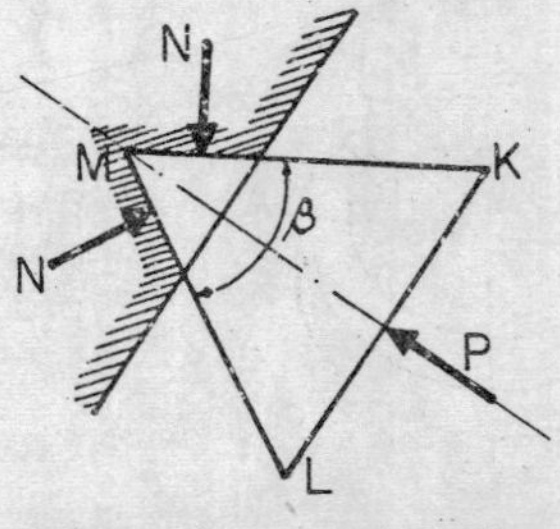

Fig. 8.1: Forces acting on an indenting wedge.

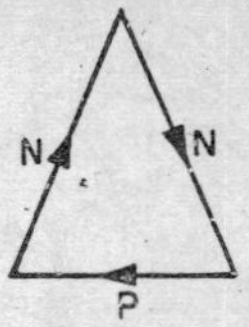

Fig. 8.2: Force triangle at the wedge cheek.

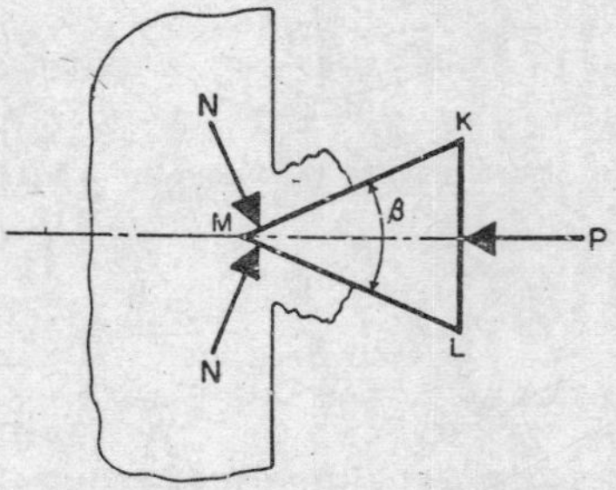

Fig. 8.3: Orientation of the wedge during the parting or cutting off by the indentation process.

shows that the wedge must be set at right angles to the work surface so that the driving force P is in the direction of parting. During chipping, the wedge must be set at an angle inclined to work surface so that separation of chips can be done (Fig. 8.4).

The only definable geometric parameter of the wedge is the wedge angle β and the axis of symmetry along which P is assumed to act, Fig. 8.5.

But as the wedge is poised for parting or chipping, the wedge stereometry becomes dependent on kinematic parameters. When the axis of symmetry of the wedge is at right angles to the work surface, the process is parting but when the axis of symmetry of the wedge is inclined to the work surface, a chipping action can be initiated.

During the process of chip removal (chipping action), the inclination of the wedge axis is contingent on several conditions with respect to Fig. 8.6.

(i) The top face, also called lip face, of the wedge should be so set that at the cutting point the condition $\bar{N}. \bar{V} = \bar{0}$ is complied with.

(ii) The axis of symmetry of the wedge is so oriented with respect to work surface as to make chipping action possible.

(iii) The lower face of the wedge must obey the condition of clearance so as to recede from the finished surface.

8.4. GENERALISED TOOL POINT : TOOL-IN-HAND NOMENCLATURE

In all cutting tools two groups of functional parts are distinguished. The first group consists of shank, disk, bar, etc., on which the cutting tool elements are established while the second group is the 'tool points' which are characterized by trihedral or polyhedral corners arising out of the interpenetration of several surfaces. As all the tool points when held

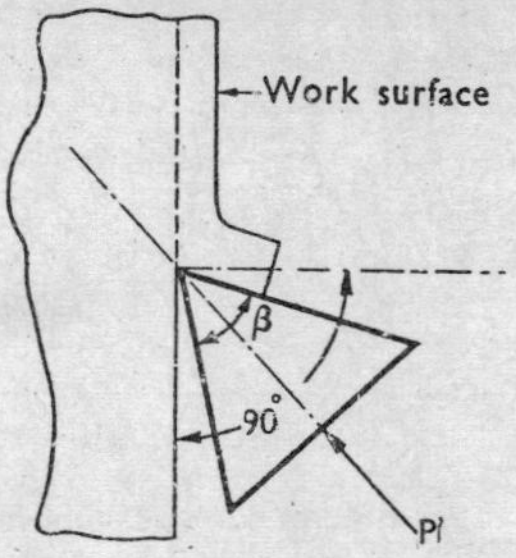

Fig. 8.4: Orientation of the wedge during the separation of chips.

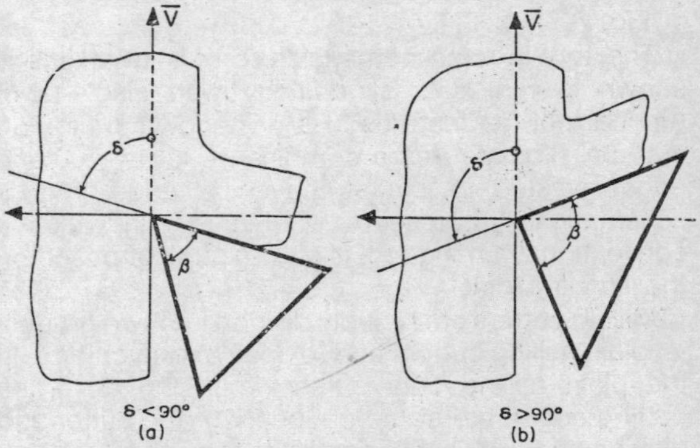

Fig. 8.5: Wedge angle and axis of symmetry.

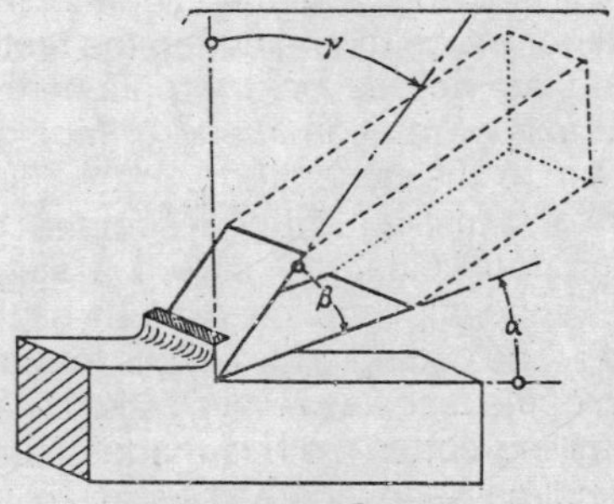

Fig. 8.6: Chipping action.

in hand or in space have only a few surfaces or edges to be defined, the first step is to designate and identify functional and non-functional surfaces and edges with no reference to their orientation or inclination. These are usually described as 'tool-in-hand' nomenclatures.

The geometry of a cutting tool consists of the following elements: face or rake surface, flank, cutting edges and the corner. Face is the surface of the cutting tool along which the chips flow out. Flank surfaces are those facing the workpiece. There are two flank surfaces, principal and auxiliary flank surface. Principal cutting edge performs the major portion of cutting and is formed by the intersecting line of the face with the principal flank surface. Auxiliary cutting edge (often called end cutting edge) is formed by the intersection of the rake surface with the auxiliary flank surface. Corner or cutting point is the meeting point of the principal cutting edge with

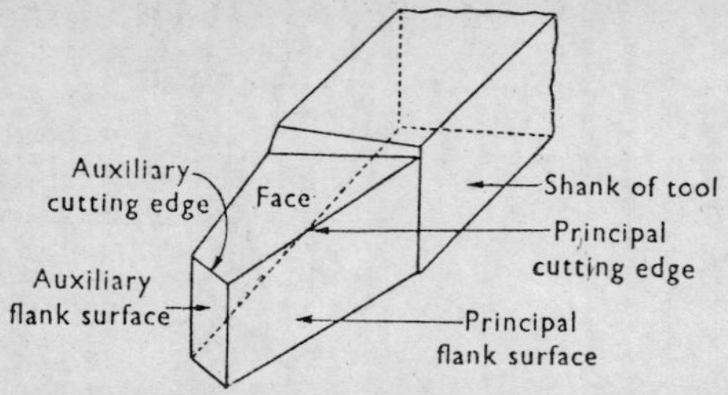

Fig. 8.7: Tool-in-hand nomenclature of a turning tool.

the auxiliary cutting edge. The tool-in-hand nomenclature of a single-point turning tool is shown in Fig. 8.7.

The tool-in-hand nomenclature of a twist drill is shown in Fig. 8.7. By analogy with single-point turning tool the features of a drill shown in Fig. 8.8 may be simulated as a combination of two cutting edges twisted to join at the chisel edge. This is schematically represented in Fig. 8.9 in which a conventional turning tool is shown superimposed on the lip of a drill.

Milling cutters are broadly divided into two groups: (*a*) plain milling cutters and (*b*) face milling cutters. In the plain milling cutters, shown in Fig. 8.10 the cutting edges are at the periphery of the cutter and may be arranged either parallel to the axis of rotation or helically. During cutting operation the axis of the cutter is parallel to the generated surface. In an end milling cutter or a face milling cutter, the teeth are on the one end as well as on the periphery. During cutting, the axis of rotation in end or face milling is perpendicular to the generated surface (Fig. 8.11). Each tooth of a milling cutter resembles a single point cutting tool. Fig. 8.12 shows a single-point cutting tool superimposed on a tooth of the plain milling cutter. The cutting edge or lip is formed by the intersection of the face and relieved land. The cutting edge of a milling cutter tooth is made actually as a narrow cylindrical land about 0.1 mm in width to ensure correct sharpening of the milling cutter. The tool-in-hand nomenclature of a milling cutter includes the external diameter of the milling cutter, the size and form of the chip space, the height and profile of teeth, the number and pitch of the teeth.

8.5. KINEMATIC ELEMENTS OF MACHINING OPERATIONS

In spite of dissimilarities of operations or product configurations, a basic similarity exists in all operations such as :

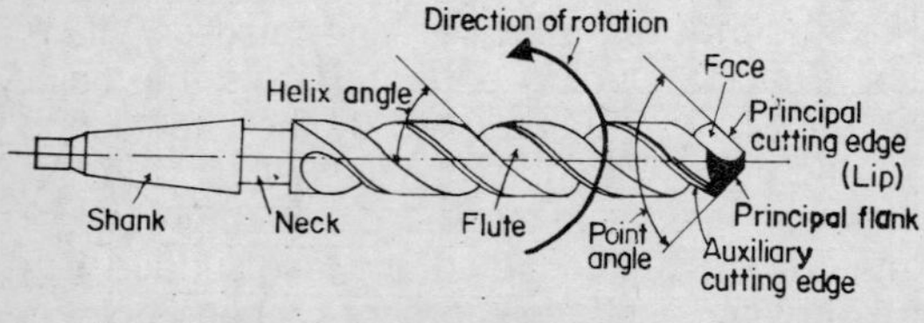

Fig. 8.8: 'Tool-in-hand' nomenclature of a twist drill.

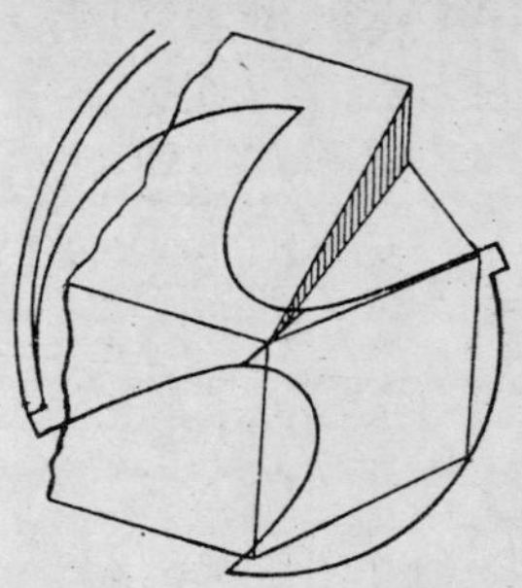

Fig. 8.9: Simulation of a twist drill with a truning tool

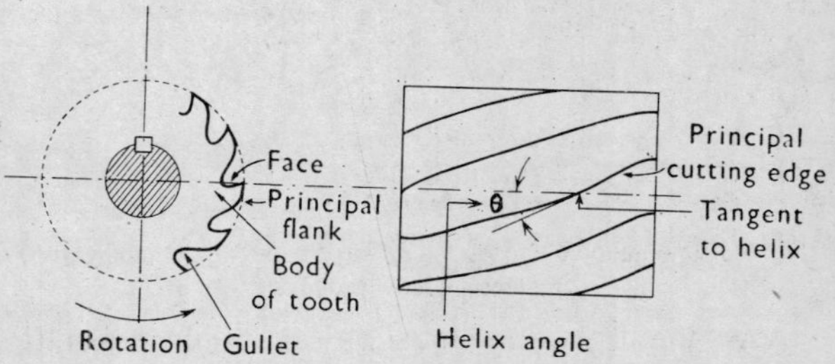

Fig. 8.10: Plain milling cutter nomenclature

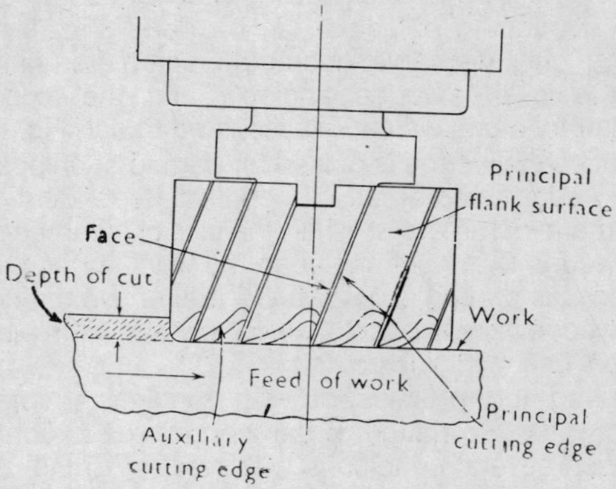

Fig. 8.11: Face milling cutter nomenclature

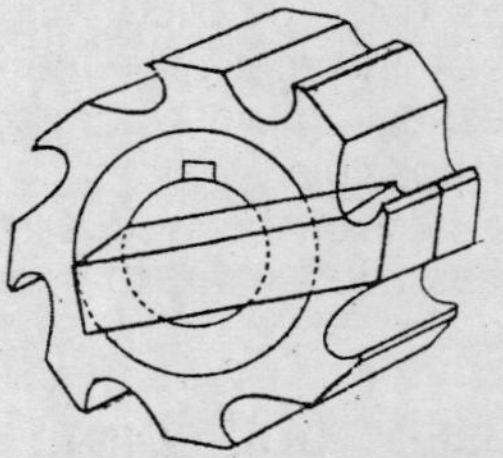

Fig. 8.12: Single-point turning tool superimposed on plain milling cutter

(a) An wedge-shaped stereometric feature performs metal cutting through kinematic interaction.

(b) Excess materials are removed in the form of chips.

(c) The basic elements of kinematic action are cutting speed (V), feed (s), and depth of cut (t).

These similarities of various machining operations lead to a comprehensive development of the science of metal cutting and related design of cutting tools. It is possible to develop an unified stereometric nomenclature of tool shapes which enable one to investigate the cutting actions of various machining operations.

Until now the stereometry of the basic wedge has been discussed, particularly with respect to the velocity vector. Before an analysis of the shapes of actual cutting tools are undertaken, it is intended to show that the basic wedge shape can be located in all types of cutting tools, if an appropriate cross-sectional view is obtained. Fig. 8.13 shows the action of a single-point right-hand turning tool, set at an angle ϕ with respect to feed direction with a depth of cut of t mm and a feed of s mm/revolution. The plan view does not indicate the chip-making action of the wedge. When a cross-sectional view at the cutting edge is taken (orthogonal or normal, defined later), the wedge is located. The true feed is a_1 mm.

Fig. 8.14 shows the action of a shaping tool. Similar views can also be obtained in the cases of drillng (Fig. 8.15) and milling (Fig. 8.16).

Turning Operation (Fig. 8.13):

$$\text{Cutting velocity} = \frac{\pi D\mathcal{N}}{1000} \text{ m/min}$$

Feed = s, mm/rev.

$$\text{Feed rate} = s_m = s\mathcal{N} \text{ mm/min}$$

$$\text{Time of cut} = \mathcal{T}_c = \frac{L}{s\mathcal{N}}$$

$$\textit{Depth of cut} = t = \frac{D_1 - D_2}{2} \text{ mm}$$

$$\text{True feed} = a_1 = s \,.\, \sin \phi$$

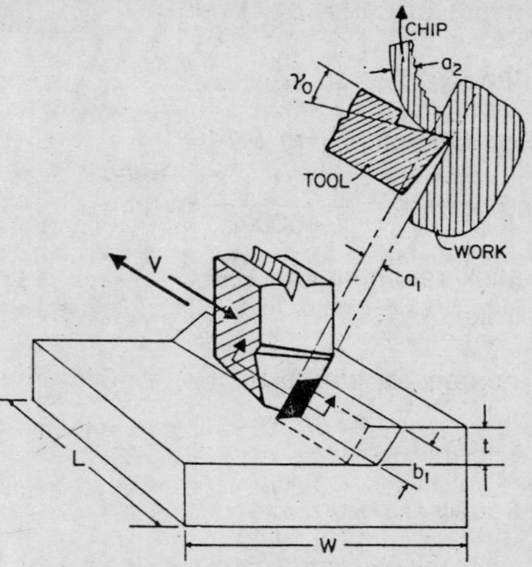

Fig. 8.14: Shaping kinematics

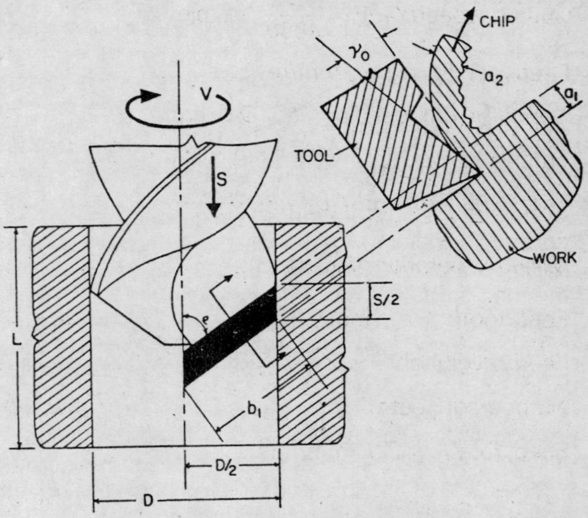

Fig. 8.15: Drilling kinematics

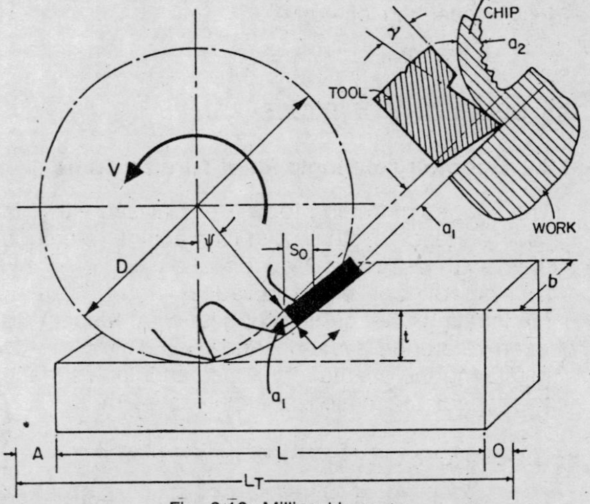

Fig. 8.16: Milling kinematics

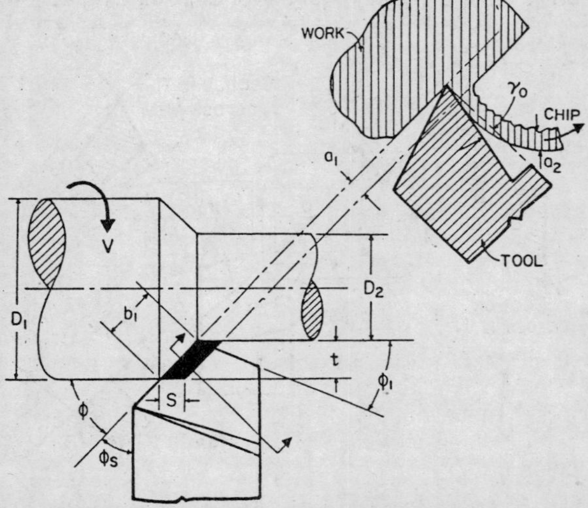

Fig. 8.13: Turning kinematics

True depth of cut $= b_1 = t/\sin\phi$

Chip thickness $= a_2$

Shaping Operation (Fig 8.14) :

Cutting speed $= \dfrac{N_s L(1+Q)}{1000.Q}$ m/min

Q = quick return ratio $= \dfrac{V_R}{V_f}$

L = stroke

N_s = number of strokes/min

$a_1 = s\cos\phi$

$b_1 = t/\cos\phi$

Drilling Operation (Fig. 8.15)

Cutting speed $= V = \dfrac{\pi DN}{1000}$ m/min

Feed per lip $= s/2$, mm/rev

Depth of cut per lip $= t = D/2$ mm

True feed per lip $= a_1 = (s/2)\sin\rho$

True width of cut $= b_1 = \dfrac{D}{2\sin\rho.}$

Milling Operation (Fig. 8.15):

Feed/tooth $= s_o$, mm

Feed/revolution $= s_o Z$

Number of teeth $= Z$

Feed rate $= s_o ZN$, mm/min

$a_1 = s_o\sin\psi$

A = approach of cutter

O = over-run of cutter

$L_T = L + A + O$

8.6. GEOMETRY OF TOOLS

8.6.1. Geometry of Single Point Turning Tools

The shapes of cutting tools are analysed either in (i) tool reference system or in (ii) machine reference system with assumed feed directions.

Analysis of tool shape includes

(a) locating the cutting edges with respect to chosen reference system ; and

(b) stating the orientation of face and flank surface with respect to chosen reference system.

A. *LOCATION OF CUTTING EDGES*

Consider the case of a straight turning tool shown in Fig 8.17. The principal cutting edge lies at angle ϕ with respect to X_m^+ axis in *assumed feed system,* while the auxiliary cutting edge lies at an angle ϕ_e with respect to X_m^- axis. Angle ϕ is called principal cutting edge angle and ϕ_e is called auxiliary cutting edge angle. In American system, the principal cutting edge is located with respect to Y_m^- axis and angle ϕ_s is called 'side cutting edge angle'.

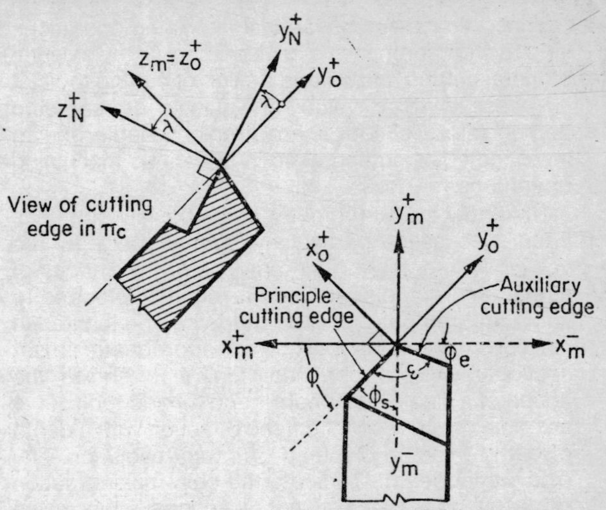

Fig. 8.17 : Locating the cutting edge with respect to assumed feed and tool reference systems.

From the geometry of Fig. 8.17,

$$\phi_s = 90° - \phi$$

$$\phi + \epsilon + \phi_e = 180°$$

where, ϵ = plan angle of the tool

The location of the principal cutting edge in the cutting plane (π_c) is defined by inclination angle λ which is the angle between Y_o axis and principal cutting edge.

B. *ORIENTATION OF FACE AND FLANK SURFACES*

In the continental system of nomenclature, the basic wedge shape required for chip-forming action

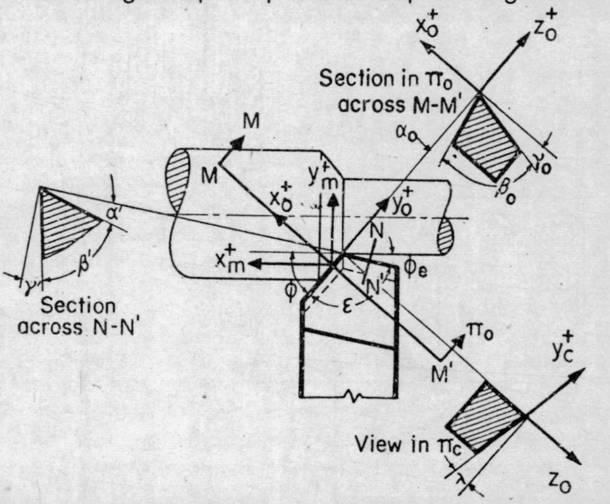

Fig. 8.18 : Orientation of face and flank surfaces with respect to tool reference system (orthogonal).

is defined in orthogonal plane (π_o) as shown in Fig. 8.18. The co-ordinates of reference are Y_o along the cutting edge and X_o perpendicular to cutting plane thus constituting what is known as *orthogonal tool reference system*. The entering angle of the cutting edge is still defined with respect to machine reference system and is called principal cutting edge angle, ϕ. Y_o-X_o system when rotated by ($90°-\phi$) counter-clockwise reduces to Y_m–X_m system in machine reference co-ordinates.

The rake angle, γ_o, known as *orthogonal rake* is defined in orthogonal plane π_o and is the angle formed by the face of the tool with X_o axis. The inclination angle, λ, is the angle made by the principal cutting edge in cutting plane π_c with Y_o axis. The clearance angle α_o, is the angular amount by which the tool wedge is relieved from the tangent at the cutting edge along Z_o axis. In orthogonal tool reference system

$$\gamma_o + \beta_o + \alpha_o = 90° \qquad \text{Eqn. 8.1}$$

where β_o = wedge angle, γ_o = orthogonal rake, α_o = clearance angle in π_o plane.

In the auxiliary orthogonal plane, i.e., plane perpendicular to the plane containing the auxiliary cutting edge, another clearance angle is defined, α_o' The angles, α_o' and α_o result in a nose clearance of α_1.

Fig. 8.19 shows the orientation of face and flank surface in machine reference system, i.e., with respect to X_m–Y_m co-ordinates. The inclination of face is defined by two rake angles ; back rake γ_y, and side rake γ_x. The principal flank orientation is described by front clearance, α_y and side clearance, α_x.

The back rake γ_y is defined in machine transverse plane (π_y) and is the angle made by the face of the tool with respect Y_m axis. Similarly, the side rake γ_x, is defined in machine longitudinal plane and in the direction of longitudinal feed (in π_x) and is the angle made by the face with respect to X_m axis.

Similarly, the side clearance, α_x, is the angle by which the wedge surface is relieved from the velocity axis (Z) in machine longitudinal plane (π_x). The back clearance, α_y, is the angle by which the same point at the wedge is relieved from velocity axis in machine transverse plane (π_y). The angle α_x and α_y together define the orthogonal clearance (α_o) by which the principal flank surface is relieved. Similarly, the auxiliary flank surface is also relieved in planes π_x and π_y. In the USA, the tool shape is described by this method, i.e., in machine reference system.

There is another way of describing the face orientation and clearances. The British system is to specify the maximum rake (γ_m) and the plane along which γ_m is contained. This plane is located with respect to machine transverse plane (π_y).

Table 8.1 shows the different ways by which the tool geomatry is specified as regards the location of cutting edges, orientations of face surface and clearances of the flank surfaces.

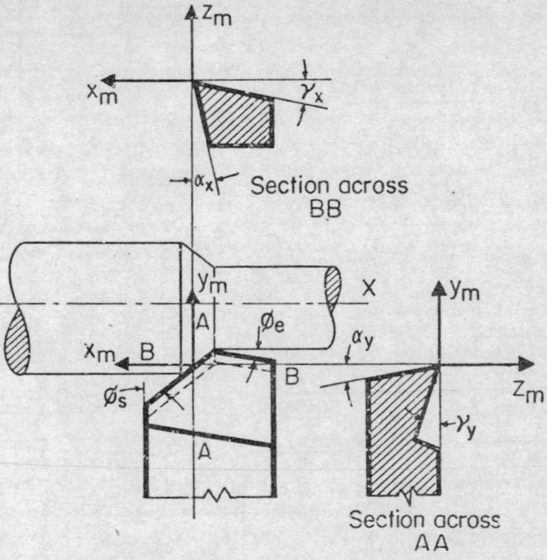

Fig. 8.19 : Orientation of face and flank surfaces with respect to machine reference system (American System of Tool Nomenclature).

TABLE 8.1

DIFFERENT NOMENCLATURE SYSTEMS FOR FACE AND FLANK ORIENTATIONS

Item	*Tool Reference System : Orthogonal*	*Tool Reference System : Normal*	*Machine Reference System : American*	*British System*
Location of cutting edges	ϕ, ϕ_e	ϕ, ϕ_e	ϕ_s, ϕ_e	ϕ, ϕ_e
Orientation of face	γ_o, λ	γ_n, λ	γ_x, γ_y	γ_{max}, ϕ_y
Orientation of principal flank	α_o	α_n	α_x, α_y	α_x, α_y
Orientation of auxiliary flank	α_o'	α_n'	–	–
Nose radius	r	r	r	r

The shape of a tool may be specified in a special sequence as given

1. American system

$$\gamma_y–\gamma_x–\alpha_y–\alpha_x–\phi_e–\phi_s–r \text{ (ASA)}$$

2. Orthogonal rake system

$$\lambda–\gamma_o–\alpha_o–\alpha_o'–\phi_e–\phi–r \text{ (ORS)}$$

3. Normal rake system

$$\lambda–\gamma_n–\alpha_n–\alpha_n'–\phi_e–\phi–r \text{ (NRS)}$$

4. Maximum rake system

$$\phi_y–\gamma_{max}–\alpha_y–\alpha_x–\phi_e–\phi–r \text{ (MRS)}$$

The most important tool designation is the stating of normal rake, γ_n, and normal clearance, α_n. The view along L-L of the cutting edge shows the inclination angle λ as seen in Fig. 8.20. If a cross-sectional view is taken at N-N, the angles γ_n and α_n are seen. Though λ is same both for tool orthogonal system and tool normal system, γ_n is different from orthogonal rake, γ_o.

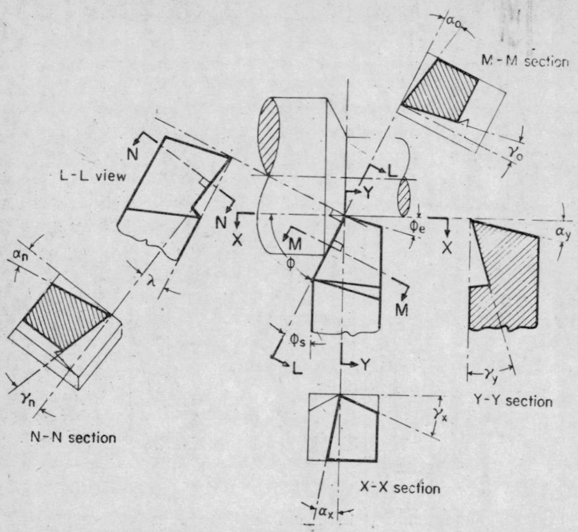

Fig. 8.20 : View along *LL* of the cutting edge

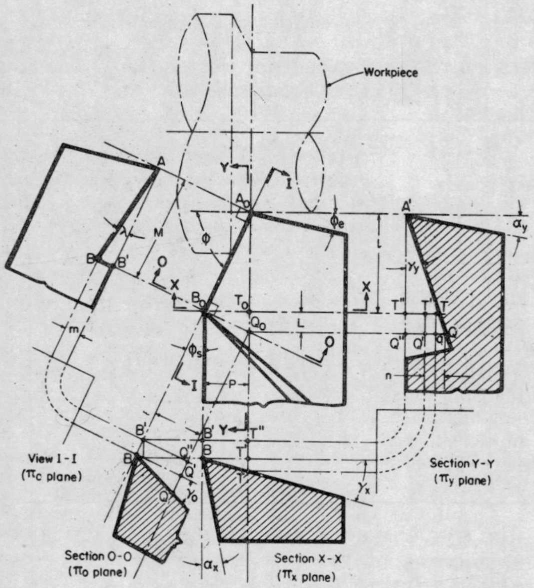

Fig. 8.21: Projections for rake angles

C. *INTERRELATION BETWEEN DIFFERENT SYSTEMS OF RAKE ANGLE NOMENCLATURE*

Interrelations can be established between ASA and ORS or between ORS and NRS systems and *vice versa*. Various methods are used for developing such interrelationships such as (*i*) method of projection, (*ii*) method of slopes, (*iii*) method of master line, (*iv*) circle diagram, (*v*) vector methods, etc.

Method of Projection for Rake Angles

Consider Fig. 8.21 in which the projections in sections over the planes π_o, π_c, π_x and π_y are shown. The cutting edge has gone down in π_c from B' to B. The points B′ and B′ are again located in orthogonal plane, π_o and machine longitudinal plane, π_x.

The projections are further carried on rotation to π_y plane. A_o, B_o, T_o, Q_o are points on the plan view, i.e., tool reference plane, while A, B, T and Q are the actual locations shown in the sectional views.

Now, with respect to Fig. 8.21,

$$\tan\lambda = \frac{BB'}{AB'} = \frac{TT'' - TT'}{A_oB_o}$$

$$= \frac{A_oT_o \tan\gamma_y - B_oT_o \tan\gamma_x}{A_oB_o}$$

But, $\dfrac{A_oT_o}{A_o B_o} = \sin\phi$ and $\dfrac{B_oT_o}{A_oB_o} = \cos\phi$

Hence

$$\tan\lambda = \tan\gamma_y \sin\phi - \tan\gamma_x \cos\phi \qquad \text{Eqn. 8.2}$$

Similarly

$$\tan\gamma_o = \frac{QQ'}{BQ'} = \frac{Qq + qQ'}{BQ}$$

$$= \left(\frac{T_oQ_o}{B_oQ_o}\right)\tan\gamma_y + \left(\frac{B_oT_o}{B_oQ_o}\right)\tan\gamma_x$$

$$= \tan\gamma_y \cos\phi + \tan\gamma_x \sin\phi. \qquad \text{Eqn. 8.3}$$

Equations 8.2 and 8.3 can be rewritten in the form

$$\begin{bmatrix}\tan\lambda \\ \tan\gamma_o\end{bmatrix} = \begin{bmatrix}\sin\phi & -\cos\phi \\ \cos\phi & \sin\phi\end{bmatrix}\begin{bmatrix}\tan\gamma_y \\ \tan\gamma_x\end{bmatrix} \qquad \text{Eqn. 8.4}$$

As the rotational matrix has unity determinant

$$\begin{bmatrix}\tan\gamma_y \\ \tan\gamma_x\end{bmatrix} = \begin{bmatrix}\sin\phi & \cos\phi \\ -\cos\phi & \sin\phi\end{bmatrix}\begin{bmatrix}\tan\lambda \\ \tan\lambda_o\end{bmatrix} \qquad \text{Eqn.8.5}$$

However, these relationships embodied in Eqn. 8.2 through Eqn. 8.5 relating γ_o and λ with γ_x and γ_y are dependent on the signs of the angles γ and λ.

Sign Conventions

Fig. 8.22 shows how positive and negative rake angles are denoted. The sign of rake angle can be determined from the condition

$$\delta_o = \alpha_o + \beta_o \lesseqgtr 90° \qquad \text{Eqn. 8.6}$$

The rake angle γ_o is *positive* when

$$\gamma_o = 90° - \delta_o = 90° - \alpha_o - \beta_o > 0,$$

i.e., when $\alpha_o + \beta_o < 90°$

The rake angle is *zero* when

$$\gamma_o = 90° - \delta_o = 90° - \alpha_o - \beta_o = 0,$$

i.e. when $\alpha_0+\beta_0=90°$

Thus the rake angle is *negative* when

$$\gamma_0 = 90°-\delta_0 = 90°-\alpha_0-\beta_0<0,$$

i.e., when $\alpha_0+\beta_0>90°$

In the left-hand cork-screw rule, δ_0, is always positive. In another way the sign of γ_0 can be determined. This is done with respect to the auxiliary co-ordinate Y_0 shown in Fig. 8.23.

If X_0 axis is rotated clockwise by an amount γ to reach extended rake surface, the angle is considered positive. When the X_0 axis is rotated counter clockwise, the rake angle is considered negative.

Similarly, the inclination angle, λ, is described in cutting plane π_c contained by Y_0-Z_0 coordinates. Again by using left-hand cork screw rule, λ is positive when Y_0 is rotated towards Z_0 to intercept the extended cutting edge. Here the cutting edge, like $(+\gamma_0)$, slopes down from the cutting point at the nose. Fig. 8.24 shows the sign convention for inclination angle which will be used althrough in this book.

Fig. 8.25 shows the sign convention used in the USSR and other countries. This convention was in vogue earlier and has been abandoned in preference to cork screw rule convention which brings in uniformaty as regards sign when a face or flank slopes down from the edge or nose.

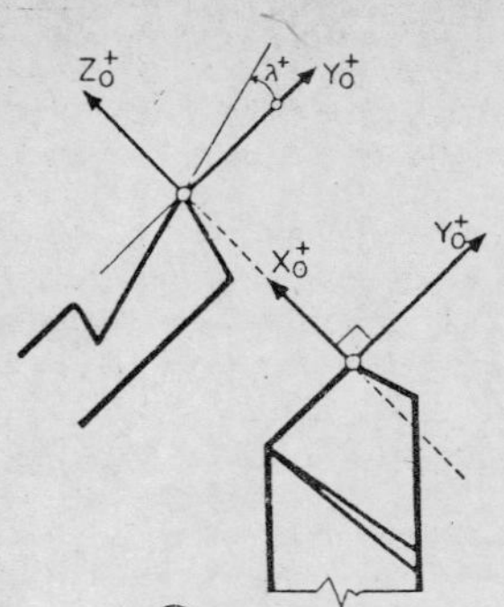

Fig. 8.23: Auxiliary coordinate, γ_0

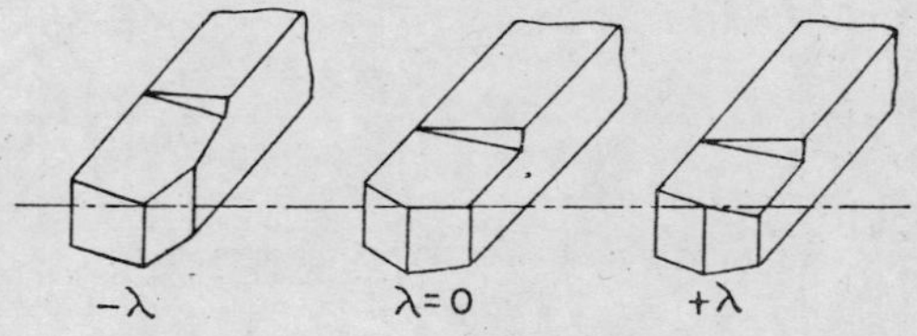

Fig. 8.24: Sign convention for inclination angle.

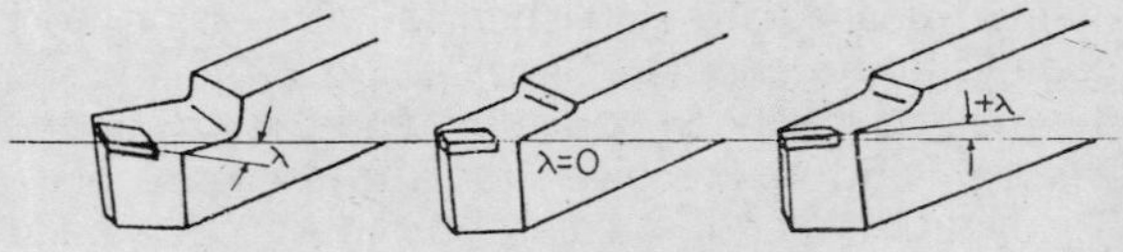

Fig. 8.25: Russian sign convention

Method of Slope for Rake Angles

This is also an elegant geometrical method for determining rake angles or clearance angles. To illustrate this method, the auxiliary rake γ_0', auxiliary inclination λ', auxiliary normal rake γ_n', etc., will be determined. Fig. 8.26 shows the several auxiliary views. The plane π_c' is along the auxiliary cutting edge while the plane π_0' is the orthogonal plane perpendicular to the auxiliary cutting plane. The view in π_c' shows the angle λ' while the section at NN in π_c' gives the sectional view in π_n' in which the auxiliary normal rake, γ_n', is determined.

To determine the interrelationships, consider the nose point A on the auxiliary cutting edge and a point M on the principal cutting edge. Proceeding from M by a distance *b* along the cutting edge (shown in Fig. 8.24), the edge slopes up by BB_1 at point A from where it slopes down along AB by the amount B_1B_2 to reach the point B. From B, it slopes up over a distance *a* to retrace to point M. From the geometry of Fig. 8.27

$$\overline{B_1B_2} = \overline{BB_2} + \overline{BB_1}$$

Thus,

$$m \tan \gamma_0' = a \tan \gamma_0 + b \tan \lambda \qquad \text{Eqn. 8.7}$$

In Fig. 8.27

$$a = m \cos (\phi + \phi_e)$$
$$b = m \sin (\phi + \phi_e)$$

Re-arranging and substituting in Eqn. 8.7

$$\tan \gamma_0' = \tan \gamma_0 \cos (\phi + \phi_e) + \tan \lambda \sin (\phi + \phi_e) \qquad \text{Eqn. 8.8}$$

Similarly it can be shown that

$$\tan \lambda' = \tan \gamma_0 \sin (\phi + \phi_e) - \tan \lambda \cos (\phi + \phi_e) \qquad \text{Eqn. 8.9}$$

Grouping

$$\begin{bmatrix} \tan \lambda' \\ \tan \gamma_0' \end{bmatrix} = \begin{bmatrix} \sin (\phi+\phi_e) & -\cos (\phi+\phi_e) \\ \cos (\phi+\phi_e) & \sin (\phi+\phi_o) \end{bmatrix} \begin{bmatrix} \tan \gamma_0 \\ \tan \lambda \end{bmatrix} \qquad \text{Eqn 8.10}$$

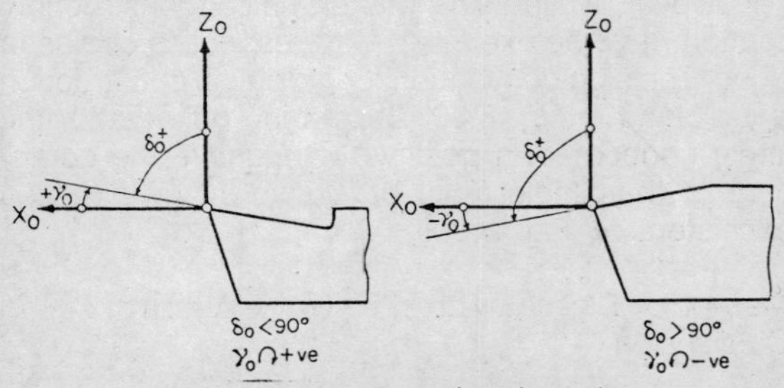

Fig. 8.22: Positive and negative rake angles

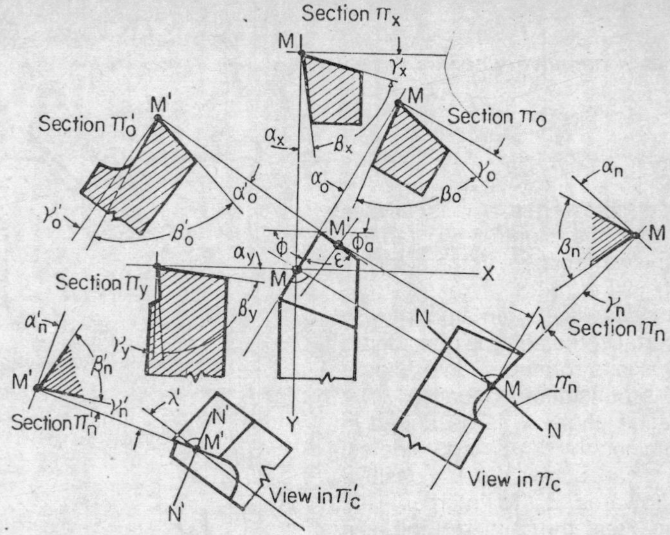

Fig. 8.26: Several auxiliary views of rake angles

METHOD OF MASTER LINE FOR RAKE ANGLES

Based on the foregoing analyses, a graphical method has been established for conversion from one system to another. Consider Fig. 8.28 in which the method has been applied to find γ_o, γ_{max} and λ, when γ_x and γ_y are given for a given value of ϕ.

Construct X and Y axes and cut off cot γ_x and cot γ_y to some chosen scale along X and Y such that,

$$OA = \cot \gamma_y$$
$$OC = \cot \gamma_x$$

Join AC and extend it to meet the trace of the cutting edge in plane view at D. Construct a perpendicular on OD at O to meet CD at B. Now,

$$OB = \cot \gamma_o$$
$$OD = \cot \lambda$$

The maximum rake, γ_m, is obtained when a perpendicular is dropped from O to E. Then

$$OE = \cot \gamma_m$$

The significance of this construction becomes apparent from Fig. 8.29. The line ABCDE is the master line on which all the projection points are lying, where

$$OC = T \cot \gamma_x$$

$$OA = T \cot \gamma_y$$

$$OD = T \cot \lambda$$

$$OB = T \cot \gamma_o$$

$$OE = T \cot \gamma_m$$

where T = depth of the cutting tool.

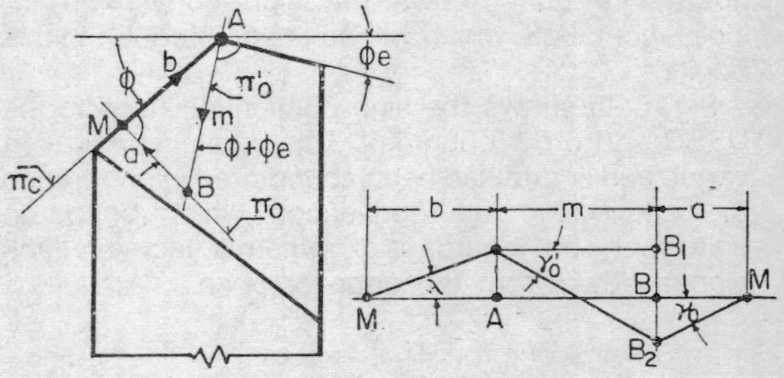

Fig. 8.27: The geometry.

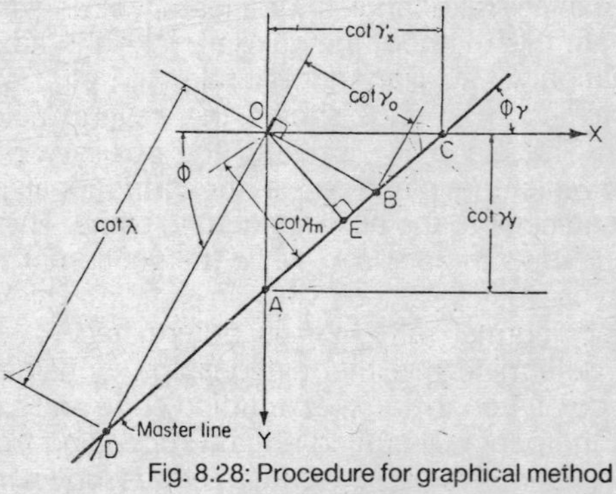

Fig. 8.28: Procedure for graphical method

However, while using this graphical construction, caution must be exercised with respect to choice of signs. The sign convention for graphical construction is indicated in Fig. 8.30. Depending on whether the angle concerned is positive or negative the cotangent of that angle must be drawn in the direction consistent to Fig. 8.30.

METHOD OF CIRCLE DIAGRAM FOR RAKE ANGLE CONVERSION

At a point M, the length MC is drawn in the trace of

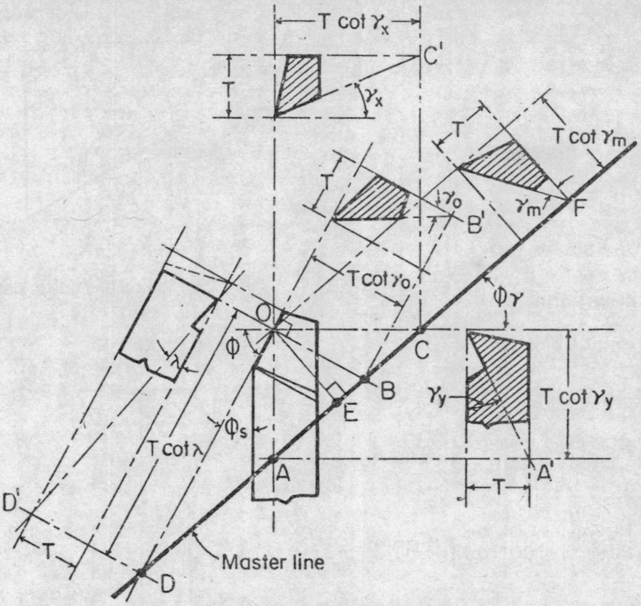

Fig. 8.29: The master line ABCDE

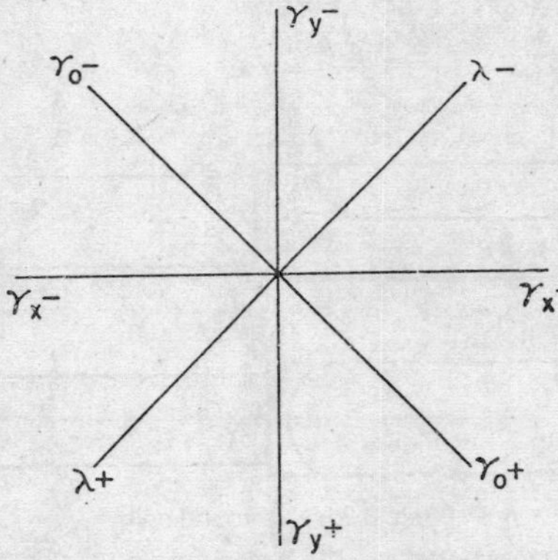

Fig. 8.30: Directions for cotangents

plane π_x equal to tan γ_x. Similarly, MF is drawn equaling to tan γ_y along the trace of plane π_y. These constructions are indicated in Fig. 8.31. Now, a circle is drawn with O as centre and CF as diameter. If the traces of the planes π'_c, π_0, $\pi_0{}'$, π_c are drawn, then the intersection of these traces with the circle determine the points D, A, G and B. Measure MD MA, MG and MB to indicate the following

MD = tan λ' $\qquad$ MG = tan γ_o'
MA = tan γ_o $\qquad$ MB = tan λ

Proof

In Fig. 8.32, constructions have been made for finding the equation for tan γ_x. In this figure

MA = BE = tan γ_o $\qquad$ MB = AE= tan λ

Now,

$$\tan\gamma_x = MC = MA' - CA'$$
$$= \tan\gamma_o \sin\phi - \tan\lambda\cos\phi$$

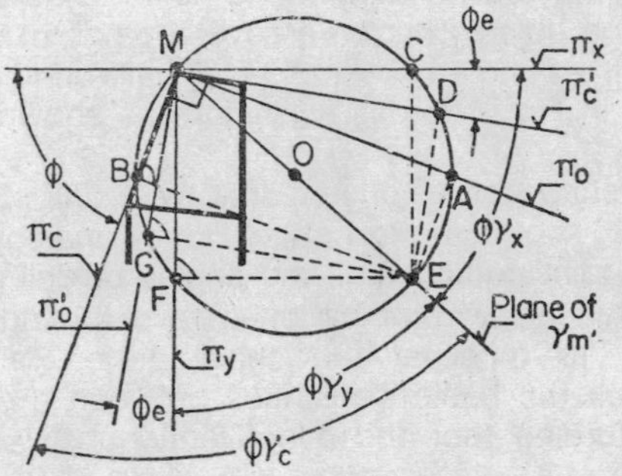

Fig. 8.31: Method of circle diagram

This expression is identical with the analytical expression derived in Eqn. 8.5.

D. INTERRELATIONSHIPS BETWEEN CLEARANCE ANGLES OF TURNING TOOLS

The clearance angles, α_x and α_y, in machine reference system can be converted to orthogonal clearance, α_0 and auxiliary clearances $\alpha_0{}'$ and normal clearance, α_n, by several techniques, and the interrelationships can be established. Fig 8.20 given earlier, depicts the orientations of flank surfaces in various reference frames whereupon α_x,α_y, α_0, α', and α_n are established.

Fig 8.33 shows the method of slope applied for determining the interrelationship of clearance angles defined in the orthogonal and machine reference systems. Consider a point M from where moving over to point A by traversing the distance *b* means a drop of AA_1 which is regained by the path AB−BM in steps of rise B_2B_1 and B_1B. From Fig. 8.33(b)

$$AA_1 = B_1B_2 + B_1B$$

$$b \cot \alpha_y = m \cot \alpha_O + a \tan \lambda$$

From Fig. 8.33(a)

$$a = b \sin \phi \text{ and } m = b \cos \phi$$

Hence,

$$\cot \alpha_y = \cos \phi \cot \alpha_O + \sin \phi \tan \lambda$$

Similarly it can be shown that

$$\cot \alpha_x = \sin \phi \cot \alpha_O - \cot \phi \tan \lambda$$

Combining,

$$\begin{bmatrix} \cot \alpha_x \\ \cot \alpha_y \end{bmatrix} = \begin{bmatrix} \sin \phi & -\cos \phi \\ \cos \phi & \sin \phi \end{bmatrix} \begin{bmatrix} \cot \alpha_0 \\ \tan \lambda \end{bmatrix} \quad \text{Eqn. 8.11}$$

The inverse relationship is

$$\begin{bmatrix} \cot \alpha_O \\ \tan \lambda \end{bmatrix} = \begin{bmatrix} \sin \phi & \cos \phi \\ -\cos\phi & \sin\phi \end{bmatrix} \begin{bmatrix} \cot \alpha_x \\ \cot \alpha_y \end{bmatrix} \quad \text{Eqn. 8.12}$$

8.6.2. Geometry of Twist Drills

Drill is a tool for originating holes in solid or for enlarging holes. There are several types of drills: (i) flat drills as shown in Fig. 8.34, (ii) half round drills as shown in Fig. 8.35, (iii) twist drills as shown in Fig. 8.36.

The shape of a conventional twist drill shown in Fig. 8.36 is of complex nature. But by analogy with a single point cutting tool, the angles of drill may be simulated as a combination of two turning edges twisted to meet at the chisel edge. Table 8.2 indicates the basic similarities between a conventional turning tool and a conventional twist drill.

TABLE 8.2

Conventional Turning Tool	*Twist Drill*
Principal cutting edge angle (ϕ)	½ (Point angle), ρ
Inclination angle (λ)	$\lambda = -(\sin \phi_N) \left(\frac{d_w}{d}\right)^*$
Back rake angle (γ_x) at outer edge	Helix angle (θ)

* *Explained later.*

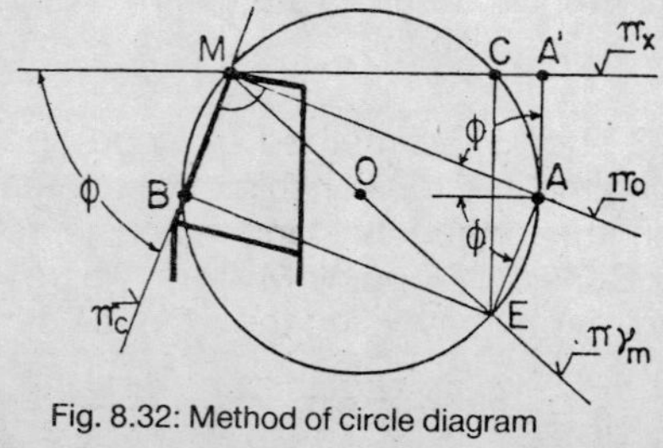

Fig. 8.32: Method of circle diagram

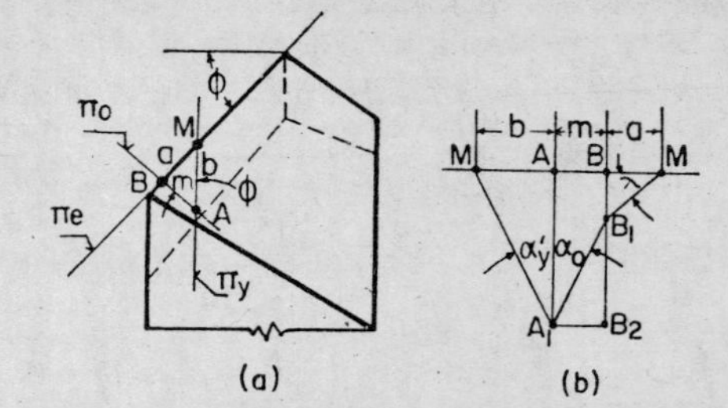

Fig. 8.33: Method of slope to determine clearance angles

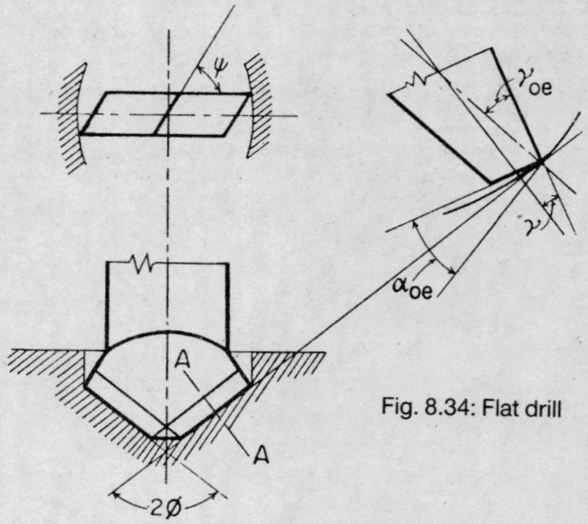

Fig. 8.34: Flat drill

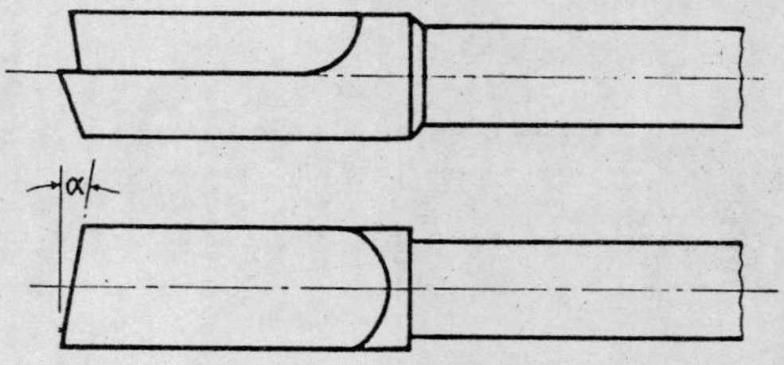

Fig. 8.35: Half-round drill

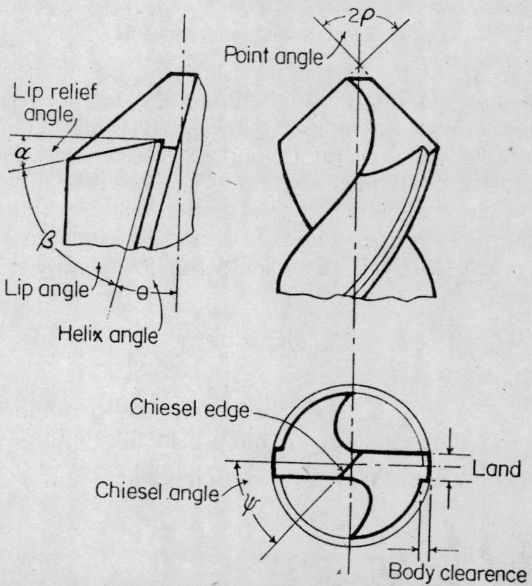

Fig. 8.36: Conventional twist drill

The features of the drill are indicated in Fig. 8.37. Two lips are involved in chip-forming action and the cutting edges are offset from this axis of symmetry by $(d_w/2)$. This offsetting generates a deviation of the velocity vector at the point A, by an angle φ_A (Fig. 8.38). This angle of deviation φ changes all over the cutting edge as indicated in Fig. 8.39. This will result in variation of inclination angle all along the cutting edge.

Fig. 8.40 shows the variation in rake angle (γ_x) at various diametral locations (d). At the outer edge, the rake angle is the helix angle (θ). But at any other location

$$\tan \gamma_{xi} = \left(\frac{d}{D_A}\right) \tan \theta$$

where D_A = nominal diameter of drill, mm.

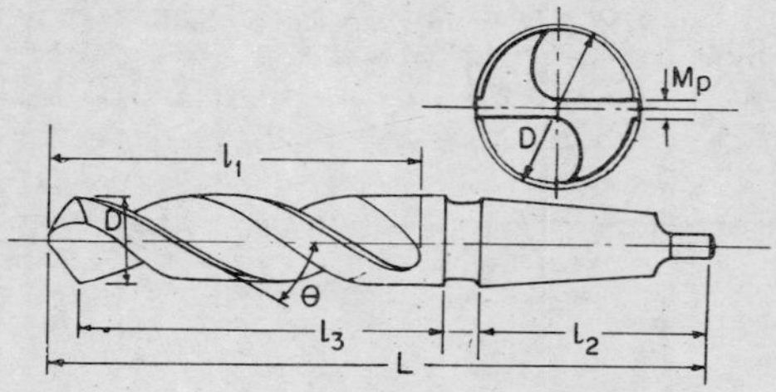

Fig. 8.37: Twist drill features

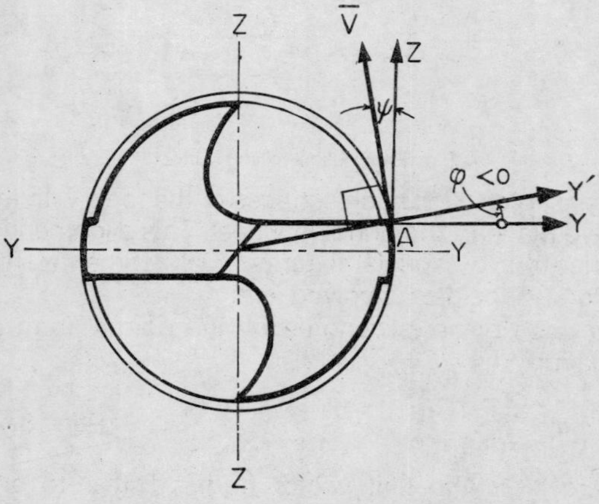

Fig. 8.38: Off-setting of cutting edges

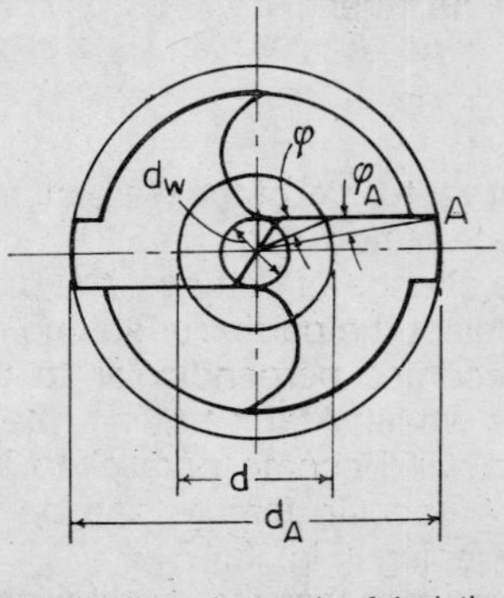

Fig. 8.39: Changing angle of deviation, ψ

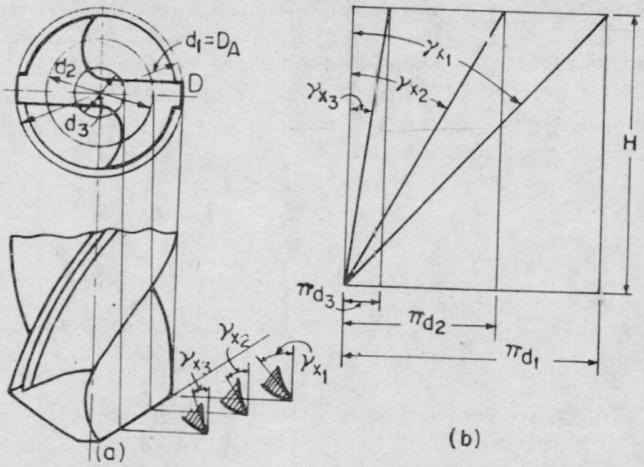

Fig. 8.40: Variation of side rake along cutting edge

Using the same nomenclatures of a single point turning tool, twist drill features can be simulated by taking appropriate sectional views. Fig. 8.41 shows the manner in which sections are to be taken in simulation to a turning tool.

The graphical technique using the master line concept can be extended to the geometrical analysis of twist drills. If λ is small, the inclination angle can be ignored and simplified forms for rake angle conversion and clearance conversion can be developed. Fig. 8.42 shows the method of determining γ_o and α_o from given values of γ_x, α_x and ϕ.

8.6.3. Geometry of Plain Milling Cutter

In the basic plain milling cutters, the cutting edges are at the periphery of the cutter and may be arranged either parallel to the axis of rotation or helically. During cutting, the axis of the cutter is parallel to the generated surface.

Plain milling is characterized as the type of milling in which there is a contact between the cylindrical surface of a plain milling cutter and the machined surface.

Fig. 8.43 shows the cross-section of a plain milling cutter with no helix angle in which the teeth are wedges characterized by the rake and clearance angles. The face, a, and flank, b, constitute the wedge of any tooth of the cutter. The rake angle is specified in a plane perpendicular to the cutter axis. This is called the radial rake angle and can be compared to the rake, γ_x, of the single point turning tool as

$$\tan \gamma_o = \sin \phi \tan \gamma_x + \cos \phi \tan \gamma_y$$

From Fig. 8.44, $\phi = 90°$

and hence $\gamma_o = \gamma_x$ and $\alpha_o = \alpha_x$

Many milling cutters employ helical teeth where the leading cutting edge forms a helix on the cylindrical surface of the cutter as shown in Fig. 8.45.

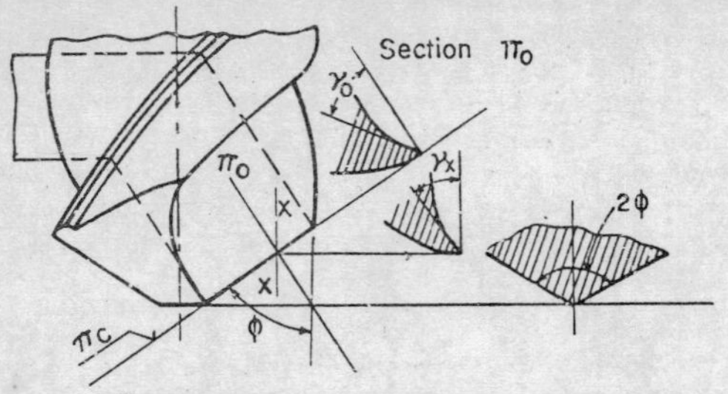

Fig. 8.41: Drill edge simulated to a turning tool

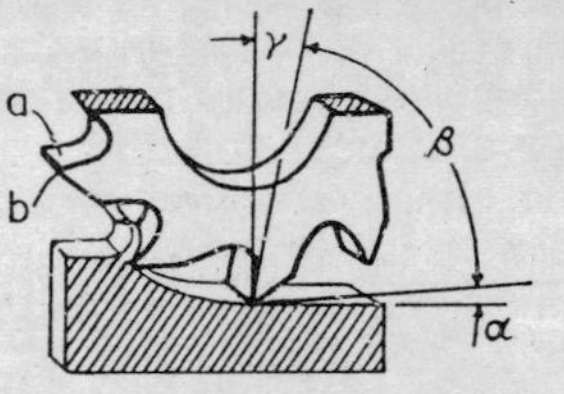

Fig. 8.43: Cross-section of a plain milling cutter

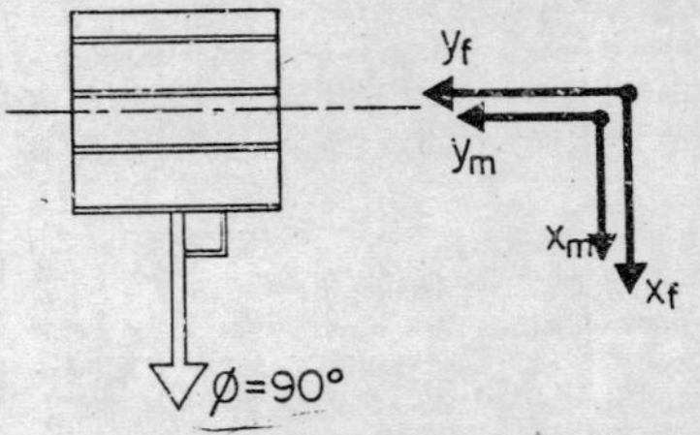

Fig. 8.44: From a direction $\phi = 90°$

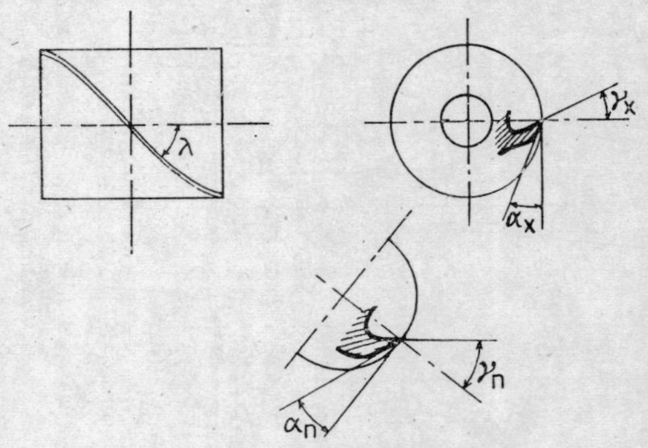

Fig. 8.45: Helical milling cutter

The helix angle of the cutting edge increases the number of teeth engagement during cutting action which in turn brings about the uniformity of chip load on the teeth. The helix angle can be simulated to 'inclination angle' of single-point turning tool. Hence, the normal rake, γ_n, measured in the plane π_n, at right angles to the cutting edge will be given by

$$\tan \gamma_n = \tan \gamma_x \cos \lambda \quad \text{as } \gamma_o = \gamma_x \quad \text{and}$$

$$\tan \alpha_n = \frac{\tan \alpha_x}{\cos \lambda}$$

8.6.4 Grinding Tool Angles

A. *GENERAL PRINCIPLES*

Most of the cutting tools are ground at the face and the flank surface in order to impart the required stereometric features at the initial stage and also for resharpening the features for restoring their machining capability. The elements to be ground may be flat, conical, spiral or helical in nature. It is necessary to orient the element to be ground with respect to the grinding wheel in order that the required element can be correctly ground.

Face or flank surfaces of single point turning tools or other similar tools are usually flat. These flat surfaces may occupy many possible positions. It is necessary to align such surfaces of the tools with the active plane of the grinding wheel. This alignment is carried out by special fixtures or attachments. The general principles involved are:

(i) Each plane is denoted by the normal form of equation

$$\bar{r}.\bar{N} = p \qquad \text{Eqn. 8.13}$$

where $\bar{N}$ is the unit vector perpendicular to and directed from the origin to the plane and p is the perpendicular distance from the origin. If the origin 0 be on the plane, then

$$\bar{r}.\bar{N} = 0 \qquad \text{Eqn. 8.14}$$

This normal vector, $\bar{N}$, of the element to be ground has to be determined (Fig. 8.46).

(ii) With the change of settings, the direction of $\bar{N}$ are altered. Select the particular setting in which the vector, $\bar{N}$, becomes perpendicular to the grinding surface of the wheel. At this setting, the flat surface to be ground will become parallel to the grinding plane and the specific feature can be reproduced through sharpening (Fig. 8.47).

In view of the fact that the rotation of an element

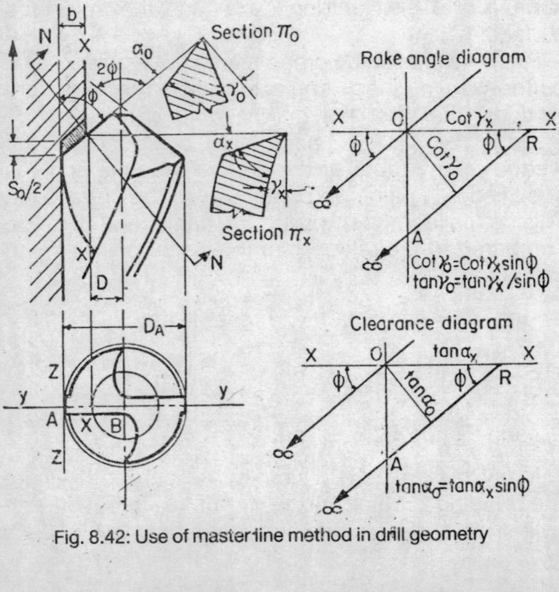

Fig. 8.42: Use of masterline method in drill geometry

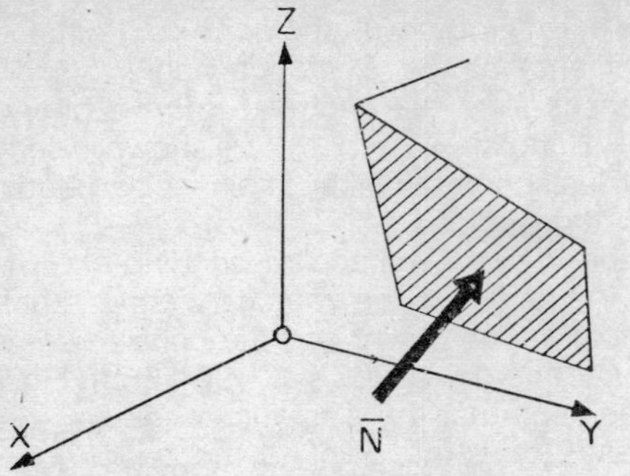

Fig. 8.46: Normal vector to the element to be ground

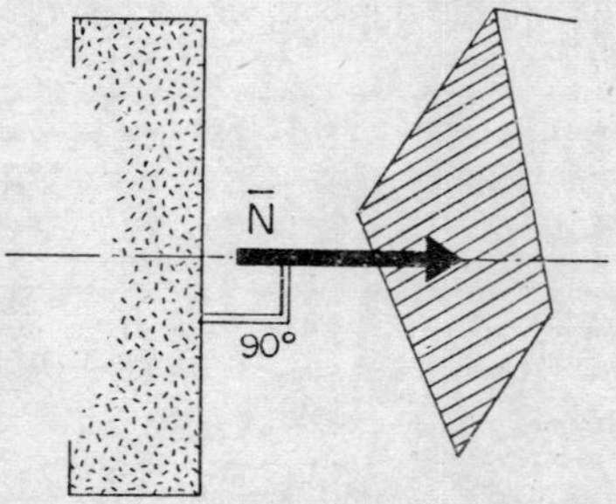

Fig. 8.47: Surface to be ground, parallel to the grinding plane

about a chosen axis can be represented by the sum of two motions, *viz.* rotation about a parallel axis through the same angle along with a translation. The latter movement is provided by bringing the element closer or away from the grinding face of a wheel and is usually obtained by moving the work table. These parallel axis rotations can be obtained either (i) in an attachment with three swivel setting shown in Fig. 8.48. or (ii) in a swivelling work table of a grinder shown in Fig. 8.49.

B. *GRINDING OF THE RAKE FACE OF A SINGLE POINT TURNING TOOL*

The tool is rotated in horizontal plane I–I by an angle ϕ_γ, the plane giving γ_m inclination is brought to coincide with O–Z axis. Thereupon, it is tilted so that AR becomes parallel to the horizontal plane, Fig. 8.50. The amount of tilt is γ_m. The relevant setting parameters are given by

$$\tan \gamma_m = \sqrt{(\tan^2\gamma_o + \tan^2\lambda)}$$

$$\phi_\gamma = \phi - k_\gamma$$

$$\tan k_\gamma = \frac{\tan \lambda}{\tan \gamma_o}$$

C. *METHOD OF TWO SETTINGS TO GRIND PRINCIPAL FLANK IN A SWIVELLING WORK TABLE*

In this case, the tool is set as shown in Fig. 8.51 in a grinding machine similar to one shown in Fig. 8.49. Initially set as shown, the first swivel is made through an angle ϕ_α about axis I-I followed by an angle of tilt of α_m about axis II-II. These angles are determined by the master line construction or from the circle diagram. These angles can also be determined from

$$\cot \alpha_m = \sqrt{(\cot^2\alpha_o + \tan^2\lambda)} = \sqrt{(\cot^2\alpha_x + \cot^2\alpha_y)}$$

$$\phi_\alpha = \phi - k_\alpha$$

$$\tan k_\alpha = \tan \alpha_o \, . \tan \lambda$$

8.6.5. Grinding The Lip Clearances of Twist Drills

Twist drills are sharpened by grinding the clearance surfaces of the point lips. Various methods are employed either using special attachments or with conventional universal tool grinders. These sharpening teachniques are

(i) Washburne's method of conical grinding,
(ii) Weisker's method of conical grinding,
(iii) Blowa's method of conical grinding,
(iv) Flat plane grinding,
(v) Oliver's method of grinding
(vi) Suchow's method of planetary sharpening,
(vii) Elliptical grinding.

A. *CONICAL SHARPENING METHODS*

In these methods each lip clearance surface of the twist drill is part of a conical surface. The three different methods listed above for conical grinding

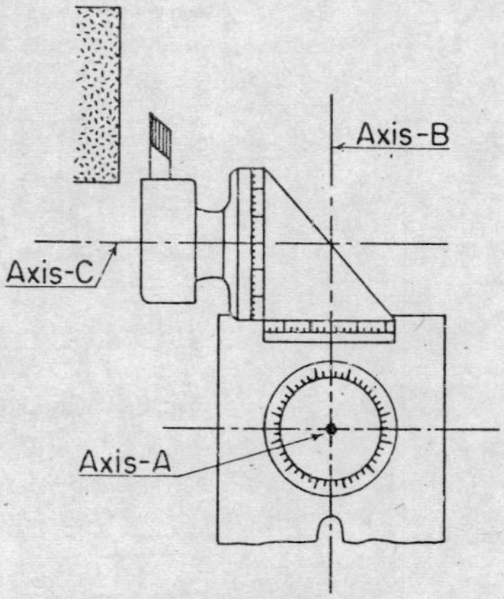

Fig. 8.48: Three sweivel setting

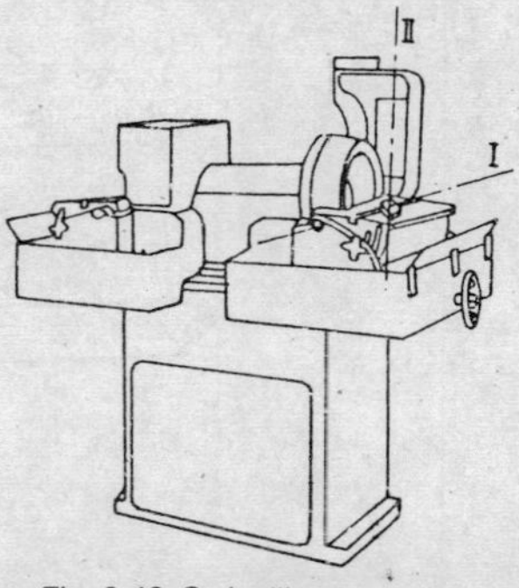

Fig. 8.49: Swivelling work table

differ only in the location of the imaginary cone in reference to the drill axis. The position of the imaginary cone in reference to the drill is given by the following particulars :

(a) the offset, k, of the cone axis from the drill axis in order to form clearance angle,
(b) the angle, σ, between the axis of the cone and the drill axis,
(c) the distance, h, from the vertex of the cone to the drill axis,
(d) the semi-apex angle, δ, of the imaginary cone.

In the Washburne method of conical grinding (Fig. 8.52), the cone vertex is above the point of the drill (apex up) and its axis makes an acute angle, δ (usually 20° or 45°) with the drill axis. From Fig. 8.52, it is seen that the clearance angle, α_n, gradually decreases from the centre to the periphery of the drill. The angles, α_{nI} and α_{nII}, are shown in sections I-I and II-II respectively taken in planes perpendicular to the cutting edge.

In the Weisker method of grinding shown in Fig. 8.53, the vertex of the imaginary cone is below the drill point (apex down) and the cone axis is at right angles to the drill axis, i.e. $\delta = 90°$. From Fig. 8.53, it can be seen that the clearance angle, α_n, gradually increases from the point to the periphery.

The clearance surface of each lip is ground in separate settings by indexing the drill through 180°. The side face of the grinding wheel coincides with an

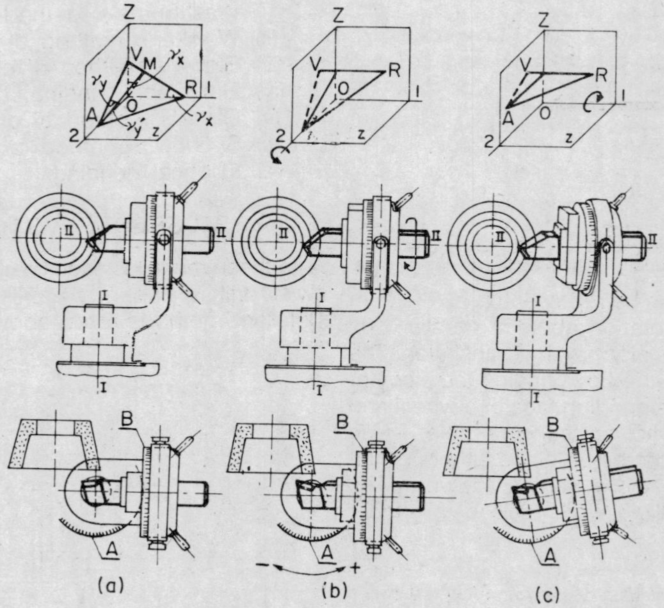

Fig. 8.50: Grinding the rake face of a single point turning tool

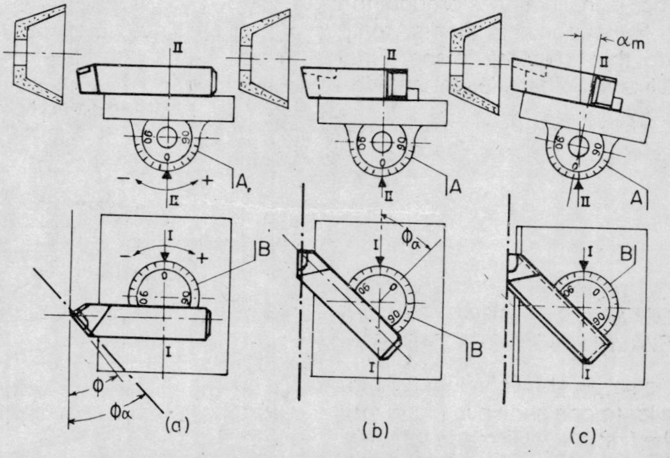

Fig. 8.51: Grinding the principal flank by the method of two settings

element of the imaginary cone where grinding is carried out. The drill is mounted in a holder in a definite position which is, in turn, mounted on a spindle whose axis coincides with that of the imaginary cone. With an oscillating motion of the drill holder about the spindle axis, the clearance surface moves in reference to this side of the wheel along a conical surface. The dimension of the imaginary cone can be adjusted suitably. The tool holder set-up for first two methods is shown in Fig 8.54.

B. *CHOOSHING SETTING PARAMETERS FOR CONICAL GRINDING*

The clearance angle along the principal flank of a drill depends not only on the helix angle, θ, or radial location but also on the setting parameters for lip grinding. The clearance angle should be such that the actual angle is never zero. At the critical location α should be greater than zero.

In the Washburne method of grinding, consider the

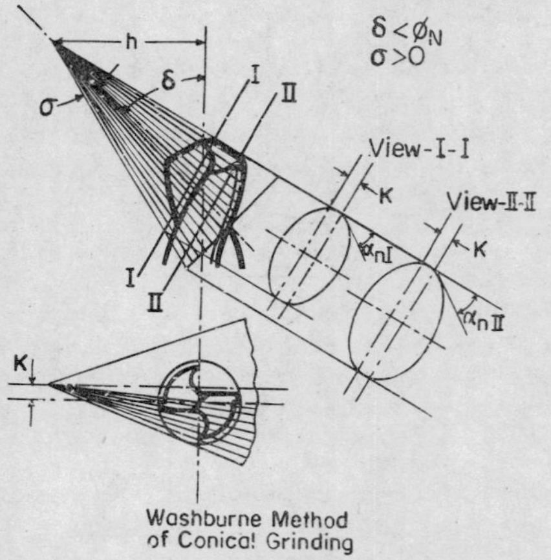

Fig. 8.52: Washburne method of conical grinding

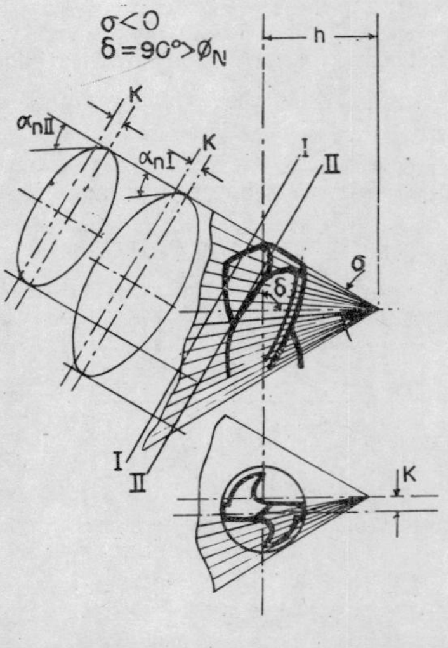

Fig. 8.53: Weisker method of conical grinding

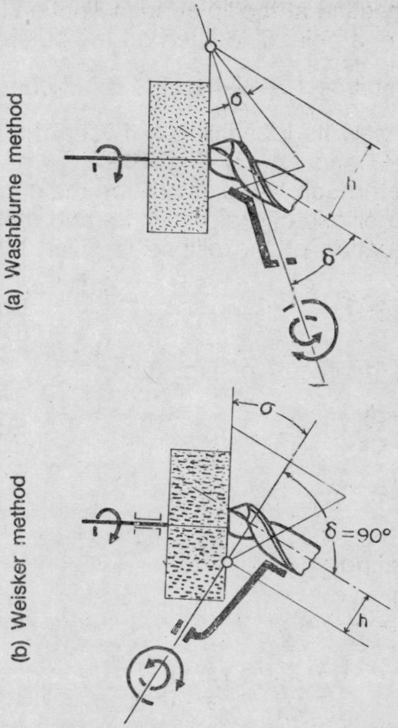

Fig. 8.54: Tool holder set-up for Washburne and Weisker methods

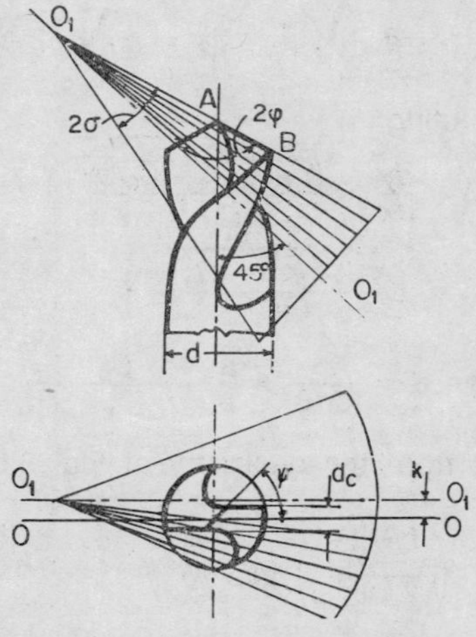

Fig. 8.55 : Parameters for conical grinding in Washburne method

cutting edge, AB, shown in Fig. 8.55 lying along the generetrix of a cone whose axis is O_1-O_1 and is inclined by 45° with respect to drill axis. The cone angle is 2σ, while the point angle is 2ϕ. In the plan view, it is seen that the cone axis, O_1-O_1 is offset from the drill transverse axis O–O by an amount k. Any point on the cutting edge is at a distance $\left(\frac{d_c}{2}\right)$ from the axis O–O and is also at a distance C_o from O_1-O_1 in the plan view, when

$$k = C_o + \frac{d_c}{2}$$

It is observed that the two generatrices of the

cones are meeting at the lip at an inclination to each other. Hence, a point R located on the cutting edge lies at a distance $\left(l + \frac{d_i}{2}\right)$ on the grind cone, while in the plan view its location is between the projections of O–O and O_1–O_1 (Fig. 8.56).

A section through Y-Y axis at R on the grind cone indicates an ellipse of major axis $2a$ and minor axis $2b$. The equation of the ellipse is given by

$$\frac{x^2}{a^2} + \frac{y^2}{b^2} = 1 \qquad \text{Eqn. 8.15}$$

From the geometry of Fig. 8.56,

$$\tan \beta = \frac{dy}{dx} \qquad \text{Eqn. 8.16}$$

$$\text{As } y^2 = \left(1 - \frac{x^2}{a^2}\right) b^2$$

Differentiating

$$-y\frac{dy}{dx} = \frac{b^2}{a^2} . x$$

Hence, $$-\frac{dy}{dx} = \frac{b^2}{a^2} . \frac{x}{y}$$

Again $y = C_0$

$$\text{and } -x = a\sqrt{1 - \frac{y^2}{b^2}} = a\sqrt{1 - \frac{C_0^2}{b^2}}$$

Substituting

$$\tan \beta = \frac{dy}{dx} = \frac{b^2 a\sqrt{1-(C_0/b)^2}}{a^2 C_0}$$

$$= \pm\frac{b}{a}\frac{\sqrt{(b^2 - C_0^2)}}{C_0}$$

$$\text{and } \tan \alpha = \frac{1}{\tan \beta} = \frac{a}{b}\frac{C_0}{\sqrt{(b^2 - C_0^2)}}$$

Again, from the geometry of Fig. 8.56

$$a = \left(\frac{l + \frac{d_i}{2}}{2 \sin \rho}\right) \tan 2\sigma$$

and $b = a(1 - \tan^2 \sigma)$

Substituting and re-arranging

$$\tan\alpha = \frac{C_0\ [\sqrt{(1 - \tan^2\sigma)}]^{-1}}{\sqrt{\left[\left\{\frac{l+(d_i/2)}{2 \sin \phi}\right\}^2 / \tan^2 2\sigma\ (1 - \tan^2\sigma) - C_0^2\right]}}$$

Eqn. 8.17

where 2σ = grind cone angle,
2ϕ = point angle of drill,
$C_0 = k - \frac{1}{2} d_c$.

It is seen that the clearance angle, α, at any location $r_i = \frac{1}{2} d_i$ varies. The grind cone must be so designed that, the condition $6 \geq \alpha > 0$ is satisfied.

It is seen that if the following parameters are chosen for the Washburne technique :

$\phi_c = 60°$, $\sigma = 45°$, $h = 1.9.D$,
$d_c = 0.2.D$, $k = \frac{1}{15}$. D, $\sigma = 15°$

the various values of lip clearance, when considered at various radii of location, r_i, gradually decreases from the centre to the periphery of the drill and varies from 5.5° to 4.5°.

It can be shown that for Weisker' method of grinding, the lip clearance angle gradually increases from the point to the periphery.

The results are graphically indicated in Fig. 8.57. It is better if the relief angles are greater at the centre than at the periphery of the drill. This requirement has made Washburne's method of lip grinding more acceptable compared to Weisker's method as it grinds a drill point with clearance angle decreasing from the centre to the periphery of the drill.

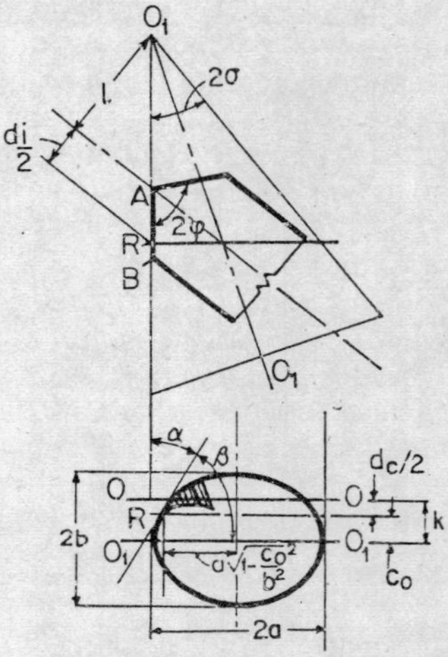

Fig. 8.56: Location of a point R on the cutting edge

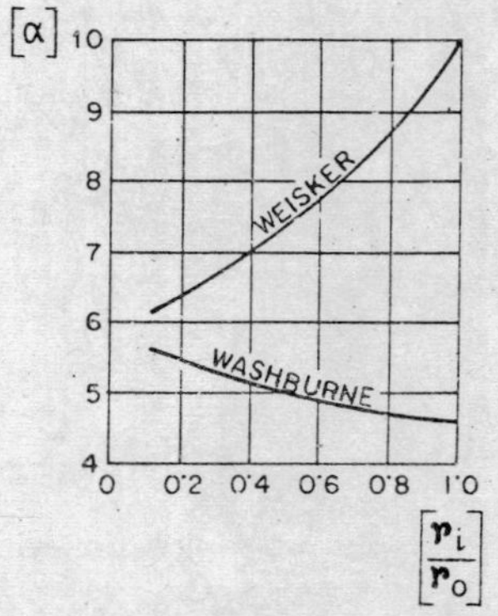

Fig. 8.57: Comparison of Washburne and Weisker methods

8.7. TOOL MATERIALS

8.7.1. Historic Development

Since the stone using days of pre-palaeolithic era, man has used cutting tools as his most intimate companion in his continuous quest for progress and civilization. The general sequence of human cultures to which these elements or implements pertain is in many cases well established and use of such cutting tools are often used to signify the stages of cultural development and progress of human civilization as indicated in Table 8.3

TABLE 8.3

TOOLS AS INDEX OF PROGRESS OF CIVILIZATION*

Rough estimates of time (in years)	*Cultural stages*	*Types of tools*
1,000,000	Pre-Palaeolithic	Crude usage of stone
800,000	Lower-Palaeolithic	Stone implements chopping tools
100,000	Middle-Palaeolithic	Transitional blending of ancient tool types
40,000	Upper-Palaeolithic	Blades, files, fine flakes
BC 8,000	Mesolithic	
BC 5,500	Neolithic	Agricultural implements and animal domestication
BC 3,000	Bronze Age	Use of bronze elements
BC 1,400	Iron Age	Use of metal and beginning of modern cutting tools

*From *Man and the Vertebrates* by *A. S. Romer*

The extent to which cutting tools have increased our standards of living is seen from Table 8.4. Greater output per man-hour means lower cost per product as well as higher wages. Cutting tools are the key implements by which natural resources can be converted for useful purposes. With better cutting tools used efficiently, the productivity per man-hour can be considerably increased.

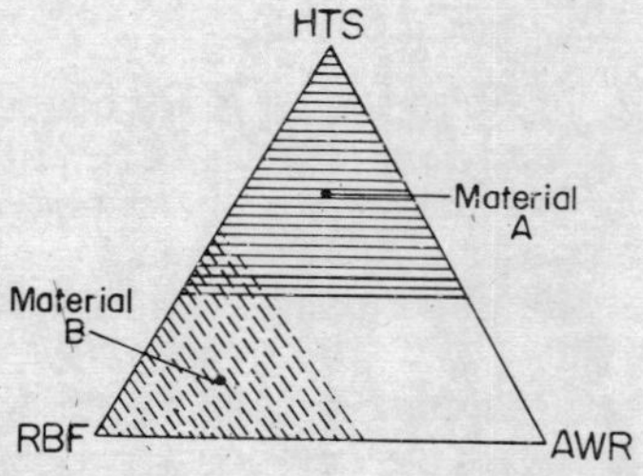

Fig. 8.58: HTS, AWR and RBF in a ternary plot

TABLE 8.4

CORRELATION BETWEEN TOTAL INVESTMENT AND OUTPUT PER MAN-HOUR

Year	*Energy supplied by*			*Total tool invest-ment (billion dollars)*	*Output per man-hour (dollars)*
	men %	*animal %*	*power, %*		
1850	23	53	26	–	0.27
1900	15	33	52	7	0.56
1920	8	11	81	15.5	0.70
1940	4	3	93	20	0.97
1950	4	2	94	33	1.21

(Based on the USA data: 'Civilization through Cutting Tools')

Developments of metal cutting techniques and machine tool industry are dependent mainly on the development of high quality tool materials. History of machine tool development shows that the improvement in machining and design of machine tools have been achieved chiefly through the discovery of new tool materials which can meet the requirements of cutting process efficiently.

8.7.2. Requirement for Cutting Tool Materials

During the process of chip formation, the cutting tools are subjected to very high pressures and temperatures (600 to 800° C or higher). Further 'wear' will take place at chip tool and work tool interfaces. Due to wear the shape of cutting elements changes and the cutting tool is said to have failed after a certain period of time requiring replacement or regrinding.

To prevent failure the cutting tool material must possess certain essential requirements :

(a) The tool material should be strong and hard enough to withstand the high pressures without having brittle fracture (RBF),

(b) The tool material should be able to retain its strength and hardness at high operating temperature (HTS),

(c) The tool material must have sufficient wear resistance, i.e., resistance against adhesion or diffusion at high temperatures (AWR).

These qualities, RBF (resistance to brittle fracture), HTS (high temperature stability) and AWR (abrasive wear resistance) do not always have concurrent high attributes in any given tool material. Generally, if a material is made more refractory to gain HTS, it becomes more brittle with low RBF. If it becomes more abrasion resistant (high AWR), it also becomes brittle (low RBF). Fig. 8.58 shows the relative HTS, AWR and RBF for two arbitrary materials in a ternary plot. Such a plot may be used to show the comparison between various tool materials. Fig. 8.59 shows schematically the usefulness of this ternary plot to evaluate the merits of HSS, WC and TiC tool materials. However, the exact location of the boundary lines for each material will depend on many other parameters which include : (a) relative hardness of tool and work materials, (b) chemical compatibility of tool-work pair, (c) interface temperature, (d) type of operation, etc.

Developmental activities in the area of cutting tool materials are guided by knowledge of the extreme conditions of stress and temperature produced at the tool work interfaces. Tool wear occurs by one or more complex mechanisms which include abrasive wear, chipping at the cutting edge, thermal cracking, adhesion and seizure at interfaces, plastic deformation, oxidation, diffusion, etc. Since most of these processes are greatly accelerated by increased temperatures, the more obvious requirements for new tool materials are improvements in physical, mechanical and chemical properties at elevated temperatures.

As the rate of metal removal has increased over the years, there has been need to develop tool materials which have more HTS. A look into the Periodic Table locates a few materials as shown in Fig. 8.60 which can render tool materials more refractory rendering HTS along with AWR or AWR with HTS. This accounts for progression from HSS to tungsten carbide, to titanium carbide and to ceramic tool materials.

The major classes of cutting tool materials may be listed in order of HTS, i.e., hardness at high temperature as shown in Fig. 8.61. As this list is ascended, it is seen that toughness decreases, i.e., the strain at fracture goes down. All of these materials except for diamond possess a two-phase structure -a softer continuous phase separating very hard particles. In case of HSS, the spacing of hard particles (Cr or Mo or W or composite) is so large that the Young's modulus of elasticity corresponds to that of the substitutional solid solution of iron. In the group of cemented carbides, the hard particles are quite close and the Young's modulus approaches that of the particles (W or Ti) instead of the cobalt binder phase. In the case of ceramics no binder material is used and the space between the hard particles is essentially that corresponding to crystal boundaries.

Natural diamond is a single phase tool material and is extremely brittle due to absence of crack

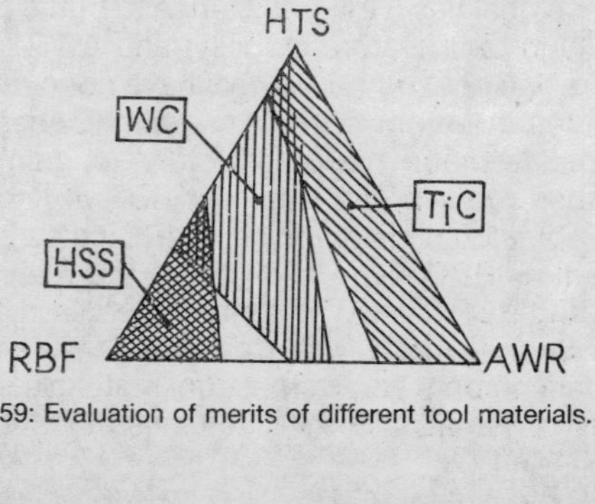

Fig. 8.59: Evaluation of merits of different tool materials.

Ti	V	Cr
Zr	Nb	Mo
Hf	Ta	W

Fig. 8.60: Tool alloying metals from periodic table

arresting boundaries which are present in polycrystalline diamond. The crystal structure of the cubic boron nitride (CBN) is same that of the diamond with covalent bonding of the atoms. It has excellent HTS and high stability against chemical reaction and oxidation at elevated temperature.

8.7.3. An Overview of Existing Tool Materials

A. CARBON TOOL STEEL

Up to 1870, tools in use were all of this type. The chief characteristic of carbon tool steels are low hot hardness and poor hardenability. A typical composition of carbon tool steel would be: carbon 0·8~1·3%, silicon 0.1~0·4%, manganese 0.1~0·4%. The higher the carbon content, the greater will be the hardness and the wear resistance.

B. MEDIUM ALLOY STEELS

These steels differ from the plain carbon steels by the presence of alloy elements added to improve hardenability (Table 8.5) and are used for drills, taps and reamers.

TABLE 8.5

TYPICAL MEDIUM ALLOY TOOL STEELS, %

C	S	Mn	Cr	Mo	W	Fe
1.2	0.3	0.6	0.5	–	–	Balance
1.2	0.3	0.7	0.5	0.5	–	Balance
1.2	0.3	0.3	0.7	0.3	1.5	Balance
1.3	0.3	0.3	0.7	–	4.0	Balance

C. HIGH-SPEED STEELS (H S S)

Around 1870, Robert Mushet introduced a steel having the composition: 2% C, 1·6% Mn, 5·5% W and 0·4% Cr. This steel was air-hardening and could retain its hardness at higher cutting temperatures. Mushet steel was later modified by replacing Mn with Cr. About 1900, F. W. Taylor and M. White introduced a heat-treatment procedure for a modified Mushet steel having the composition 1·9% C, 0·3% Mn, 8% W and 3·8% Cr. This steel had high hot hardness and could operate at higher cutting speeds (about 20 m/minute) and became known as high-speed steel (HSS). The period of 1900-1906 witnessed the rapid development of HSS tools with increase in tungsten (W) and reduction of carbon. The decrease in carbon was necessary to make the steel forgeable while allowing for hot hardness. By 1903, carbon decreased to 0·7% with tungsten increasing to 14%. In 1904, Mathews observed that addition of vanadium (V) provides AWR. By 1910, the well-known T-1 type HSS tools emerged having the composition of 18% W, 4% Cr and 1% V. The cutting speed capability for machining steel increased to about 33 m/minute (100 f.p.m.). It was also found in 1912 that addition of cobalt (Co) increases the high temperature stability of the carbide matrix which led to the development of T-4 type HSS tools having the composition of 18% W, 4% Cr, 1% V and 5% Co. Increased content of vanadium and cobalt led to so-called super-HSS in 1923 (T-6 type) having the composition of 4% Cr, 20% W, 2% V and 12% Co. However difficulty of forging of this type led to other modifications by compromising W content with increase in C. This led to T-15 HSS tools having the composition 1·5% C, 4% Cr, 5% V, 12% W and 5% Co.

By 1923, it was known that molybdenum can be substituted for tungsten in HSS. Since the atomic weights of molybdenum and tungsten bear a ratio 1:2, only half the weight of molybdenum be used in the substitution process. A typical M-2 steel has the following composition: 0·8% C, 4% Cr, 2% V, 6% W, and 5% Mo. The capability of HSS tools rose to about 100 m/minute for machining steel by 1950. Table 8.6 shows the composition and types of several HSS tools.

TABLE 8.6

COMPOSITIONS AND TYPES OF POPULAR HIGH SPEED STEELS

Type	C	W	Mo	Cr	V	Co	R_c
M-2	0.80	6	5	4	2	–	64.7
M-1	0.80	1.5	8	4	1	–	
M-10	0.85	–	8	4	2	–	
T-1	0.70	18	–	4	1	–	
T-4	0.75	18	–	4	1	5	
T-6	0.80	20	–	4	2	12	
T-15	1.50	12	–	4	5	5	
M-3(I)	1.00	6	5	4	2.5	–	
M-3(II)	1.15	6	5	4	3	–	
M-7	1.00	1.5	8	4	2	–	
M-4	1.30	6	5	4	4	–	
M-15	1.55	12	–	5	5	5	
M-15	1.55	6	3	5	5	5	
M-42	1.08	1.5	9.5	4	1.1	8	62.4

D. *CAST ALLOY TOOLS*

Around 1915, non-ferrous high temperature alloys containing significant amount of cobalt, chromium and tungsten were developed by Elwood Haynes.

A number of such non-ferrous alloys have been used as cutting tools known as 'stellites'. These materials cannot be heat-treated and are used as cast from a temperature of about 1300°C. A typical composition for the materials is as follows :

Co-40 to 50%, Cr-27 to 32%, W-14 to 29% and C-2 to 4%.

E. *CEMENTED CARBIDE TOOLS*

Around 1928, tungsten carbide was developed in Germany. Originally, these comprised of finely ground tungsten carbide (WC) particles sintered together with a cobalt binder. These materials in early stages were brittle, difficult-to-grind and had tendencies to caused adhesion craters when machining steels. Such tools were mainly used to machine cast iron and non-ferrous materials. Around 1938, it was found that addition of TiC and TaC reduces these adhesion tendencies while machining steel, Table 8.7.

TiC and TaC have been found to be more stable than WC and have greater resistance to decomposition in presence of FCC γ-iron. TaC provides greater crater resistance with less loss of impact strength than TiC since it gives rise to less grain growth during sintering.

Today, there are three very general groups of metal cutting cemented carbides in use :

(a) WC + Co (for use in machining cast irons and non-ferrous metals)

(b) WC + Co +(WC–TiC-TaC–NiC) (for steels)

(c) TiC + Ni + Mo (for high temperature high strength metals)

F. *GENERAL CHARACTERISTICS OF CARBIDE TOOLS*

In comparison with other tool materials (except oxide tools) these are as follows :

1. high hardness over a wide range of temperature (up to 900°C),
2. very stiff (Young's modulus is about three times that for steel),

TABLE 8.7

TYPICAL COMPOSITIONS OF CARBIDE TOOLS

Type	WC	Co	TaC	TiC	*Use*
883	94.0	6.0	–	–	Cast iron
350	70.7	4.5	12.2	12.6	Steels
330	46.4	–	9.5	30.1	Steels { Ni-12
370	72.0	8.5	11.5	8.0	Mo-2

3. no plastic flow to stress as high as 352 kg/mm^2,
4. low thermal expansion (compared to steel),
5. high thermal conductivity,
6. low specific heat,
7. weaker in tension than in compression.

G. *CLASSIFICATION OF CEMENTED CARBIDES*

There are four different systems of classification for cemented carbides that are in use today:
(i) ISO system, (ii) US (C-system),
(iii) UK (BHMA) system, (iv) USSR system.

There are three categories in ISO classification system, P for steel-cutting grades, K for cast iron-cutting grades and M for ductile irons, hard steels, HSTR alloys, etc, (Table 8.8).

TABLE 8.8

ISO CLASSIFICATION SYSTEM

ISO System	Usage	Grades
P	Ferrous materials	P-01, P-10, P-20, P-30, P-40, P-50
M	Intermediates	M-10, M-20, M-30 M-40
K	Non-ferrous materials, Cast Iron	K-01, K-10, K-20 K-30, K-40

The C-system in the USA is used to compare the tool materials produced by various manufacturers. The equivalence is shown in Table 8.9.

TABLE 8.9

CLASSIFICATION AND TYPES OF CEMENTED CARBIDE TOOLS (AMERICAN)

Grade	*Chip Removal Application*	*Class*	*Carboloy*	*Kennametal*
Cast iron grade	Roughing	C-1	44A	K-1
	General finishing	C-2 C-3	883 860 905	K-6 K-8
	Precision boring	C-4	999	K-11
Steel grade	Extreme roughing	–	–	
	Roughing	C-5	370	K-M, K-25,
	General Finishing	C-6 C-7	78 350	K-2-1 K-3-H
	Precision boring	C-8	330	K-4-H, K-5-H K-7-H, K-165

The BHMA designation consists of three numbers arranged in sequence $X_1X_2X_3$ each ranging in a scale between 1 and 9, i.e. from low to high. X_1 stands for hardness and reflects wear resistance. X_2 gives the transverse rupture strength (1 standing for 1,50,000 p.s.i. and 9 for 5,00,000 p.s.i.) and denotes the degree of shock resistance. The third one, X_3, denotes low or high content of TiC or TaC or both reflecting crater resistance.

The Russian system is based on composition of carbide, (Table 8.10).

TABLE 8.10

RUSSIAN SYSTEM

Equivalence of	*Type*	*Designations*	*Explanation*
WC	BK	BK-2, BK-4, BK-3-M, BK-6, BK-6-M, BK-8, BK-8-B	Figure after K denotes % of Co
WC-TiC	T-K-	T-5-K-10, T-14-K-8, T-15-K-6, T-30-K-4	Figure after T is the % of TiC and the figure after K is the % of Co
WC-TiC-TaC	TT-K-	TT-7-K-12, TT-7-K-15	TT-7=TiC-4% + TaC-3%.

H. *OXIDE TOOL MATERIALS (CERAMICS)*

Today's commercially available cutting tools are basically aluminium oxide (alumina, or Al_2O_3); consequently, they are referred to as oxide cutting tools. Oxides are resistant to heat and do not soften at elevated temperatures; oxide cutting tools reflect the high hardness and good wear resistance of the prime constituent, Al_2O_3. Since its first introduction in the USA in 1955, the properties have been improved a great deal, (Table 8.11).

TABLE 8.11

COMPARISON OF PROPERTIES OF OXIDE TOOLS OF 1956 and 1964

Property	*1956 Tools*	*1964 Tools*
Rockwell-A	86-90	93-94
Transverse rupture strength, kg/mm^2	28-42	63-78
Impact fracture strength, 10^4 kg-mm	6-8	11-13
Compressive strength, kg/mm^2	212-247	300-335
Young's modulus, kg/mm$^2\times10^3$	30-32	40-42

I. *GENERAL REMARKS*

In general, materials having a high Young's modulus are more refractory, i.e., they have more hot hardness characteristic. Fig. 8.62 gives the hardness values for several classes of existing tool materials as measured over a wide range of temperatures. The hardness and strength of HSS fall off rapidly above 500°C. It is interesting to note that the cast cobalt alloys while softer than HSS at room temperature are harder at temperature above 500°C. The wide spectrum of carbide characteristic is due to the variation in cobalt content which may vary between 3% and 12%. When the cobalt content is increased, the carbide becomes more ductile and shock resistant but less refractory.

Fig. 8.63 indicates recovery hardness for these tool materials. The values of hardness measured at room temperature often are not regained following exposure to elevated temperatures. The ceramics, carbides, cast cobalt alloys recover their hardness completely upon cooling. But HC steel and HSS soften irreversibly above a given temperature (about 500°C for HSS). This is one reason why HSS tools are limited to operations giving interface temperature below 500°C.

There are other properties which are of importance in tool material selection and utilization. These include coefficient of thermal expansion, thermal conductivity and volume specific heat. The coefficient of thermal expansion determines the influence of thermal stresses and thermal shock on a material. Carbides have a coefficient of expansion about one-half that of HSS and this value increases with cobalt content.

Again, the carbides are very brittle and are much more sensitive to thermal shock compared to HSS. This is the reason that when milling with carbides cutting fluids are rarely used while with HSS tools, a copious supply of cutting fluid is made.

Another aspect of tool material selection is the feature of *grindability.* As the wear resistance is improved upon, the grindability generally decreases. Very often one variety of HSS tools is chosen over another though the latter gives better tool life, due to differential cost in grinding finished tools from the two steels.

The problems of low fracture toughness and low thermal shock resistance still persist in carbides and ceramics. The brittle fracture and damage of the tools with composite materials like carbides are classified into chipping, fracture, breakage, flaking and cracking from the view-point of morphology as indicated in Fig. 8.64. It is also classified to impulsive fracture, mechanical fatigue fracture, thermal fatigue fracture and adhesion fracture from the view-point of fracture mechanics. The impulsive fracture is caused when the applied stress exceeds the limiting fracture stress. The mechanical fatigue fracture is caused by the propagation of microcracks inherent in the tool material or created by repeated loading. The thermal fatigue fracture takes place when the thermal stress due to repeated heat generation during cutting exceeds the limit of fracture stress. All these factors affecting carbide and ceramic tool performance need to be tackled, if their performance level is to improve through appropriate modifications.

8.7.4. Development of New Tool Materials

One of the most important elements in the design of cutting tools is the material of construction of tools and their judicious selection. The properties those are needed for a tool material include not only the capacity to retain form stability at high temperature during high cutting speeds but also some other properties including the cost and ease of fabrication for complex tool shapes, high resistance to brittle fracture, resistance to diffusion and resistance to thermal and mechnical shock.

During the machining of high strength temperature resistant (HSTR) alloys commonly known as aerospace materials, it is essential to improve mechanical and thermal strength of the materials. The development potentialities of these materials are schematically shown in Fig. 8.65.

Though high speed steel remain at the bottom for high speed machining, yet it is a necessary tool material for designing complex form-tools, drills, reamers, etc. Hence, improvement in high speed steel is continuously sought for. Improvements in HSS have already been reported where hard phase contents are greatly increased and new manufacturing technique adopted. At Illionois Research Institute, addition up to 15% V, 5% Ti to M-1, HSS steel has been done and the tools have been prepared by directly casting into copper chill moulds. These tools are reported to have a very large carbide volume compared to regular HSS tools. Highly modified HSS tools and cobalt base alloys are also being prepared from pre-alloyed atomized powders which are canned and best worked into dense bar stock. Such material properties are midway between the properties of commercial cast cobalt base alloys and cemented WC-Co alloys as regards carbide volume, hardness and abrasion resistance.

An important development in HSS involves the atomization of a pre-alloyed HSS followed by consolidation into shanks by powder metallurgy (PM) technique. This gives smaller carbide size and more uniform carbide distribution. The smaller carbide sizes give better grindability.

An improved procedure consists of a cold impact fracture process as the final communication step before sintering. This 'cold stream impact communication process' is cheaper and complex shapes can be produced more easily.

Often, a new strategy is worked upon high speed

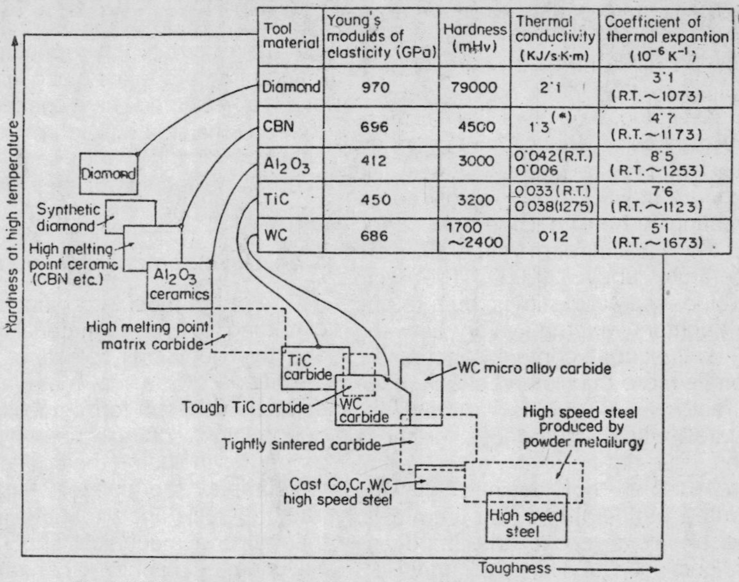

Tool material	Young's modulus of elasticity (GPa)	Hardness (mHv)	Thermal conductivity (KJ/s·K·m)	Coefficient of thermal expantion ($10^{-6}K^{-1}$)
Diamond	970	79000	2·1	3·1 (R.T.~1073)
CBN	696	4500	1·3(*)	4·7 (R.T.~1173)
Al_2O_3	412	3000	0·042(R.T.) 0·006	8·5 (R.T.~1253)
TiC	450	3200	0·033(R.T.) 0·038(1275)	7·6 (R.T.~1123)
WC	686	1700 ~2400	0·12	5·1 (R.T.~1673)

Fig. 8.61: Comparative study of hardness at high temperature and toughness of different tool materials.

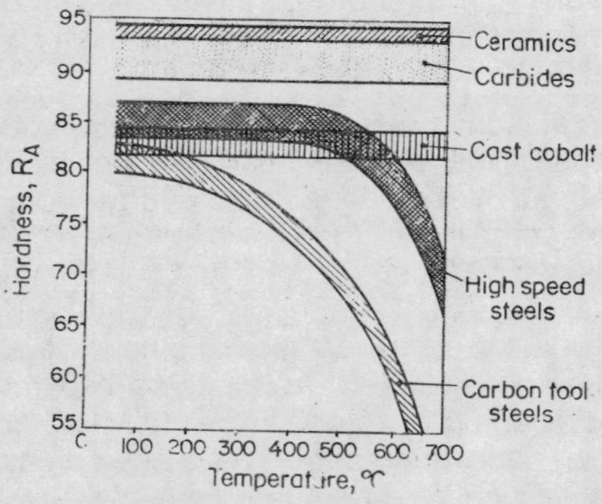

Fig. 8.62: Hardness values of several classes of tool materials.

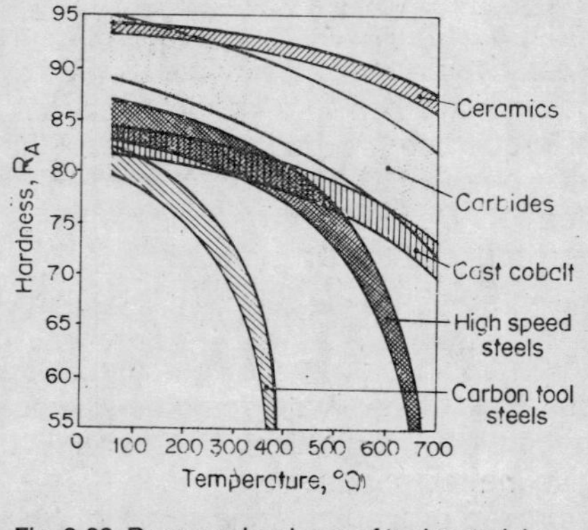

Fig. 8.63: Recovery hardness of tool materials

steels for increased wear resistance. A hard surface layer which may contain up to 15 to 20% V or Ti may be produced on tough substrata. Often electrophoretic deposition of WC-Co on to molybdenum

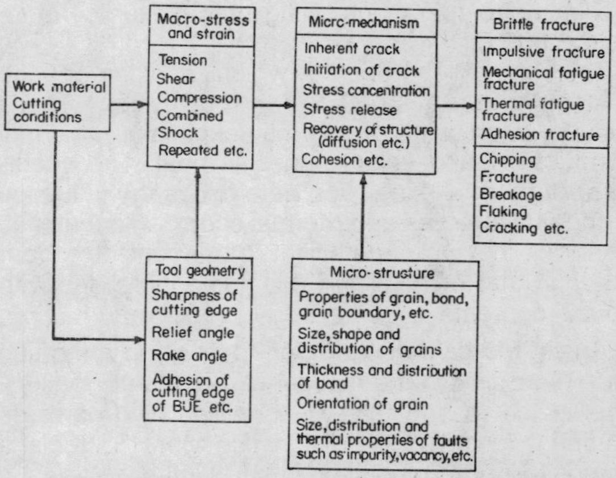

Fig. 8.64: A morphology of the brittle fracture and damage of the tool

alloy or tungsten alloy (W-Ni-Co) substrata is made whose carbide rich layer is subsequently sintered under high pressure to result good binding and full density. Extremely hard but very thin surface layers have also been produced by vapour deposition of carbides.

All such new developments with HSS already indicate that there are still quite a number of possibilities for further improvement of HSS.

Experiments are being conducted at many laboratories to improve present grades of carbides by having a wide of variety of refractory alloy binders as matrix candidates. Much of this work is being centered on high-melting matrix alloys of columbium-cobalt and columbium-nickel systems and the carbides being explored include WC, TaC, TiC and solid solution alloy of these.

Ford Motor Company has developed and is still

improving titanium carbide cutting tools. With one grade of TiC tools four-times tool life has been realized compared to several commercial grades of carbides. The reason for having TiC tools is their low density and more oxidation resistance. Best results have been found with TiC when it is bonded with nickel. However, these materials are very brittle. Humenik has shown the effect of wetting in the liquid-phase sintering of TiC.

While liquid cobalt was found to have a wetting angle of zero against WC indicating perfect wetting, the wetting angle of nickel against TiC is about 30 degree. With 10% molybdenum powder mixed with nickel before sintering TiC the wetting angle has been found to have improved with consequent uniform dispersion of bonding material. The resulting tool material is relatively tougher.

The oxide cutting tools are good as regards their form stability at high temperature. Tools made from pure Al_2O_3 are more refractory than TiC but highly brittle. Mixtures of hot pressed Al_2O_3 and TiC give cermets which are useful in high speed machining of hard cast irons and hard steels under conditions of moderate mechanical shock.

Carborrundum Co. in association with the Pennsylvania State University developed and tested new refractory tool materials. This work area involves the hot pressing of various nitrides, borides and carbides. Some of these new material systems include ($TaN + ZrB_2$) or ($TaN + TiB_2$) or ($TaN + TiC$) or ($TaN + ZrB_2 + TaC$). One of the most successful combinations has been (60% TaN + 40% ZrB_2).

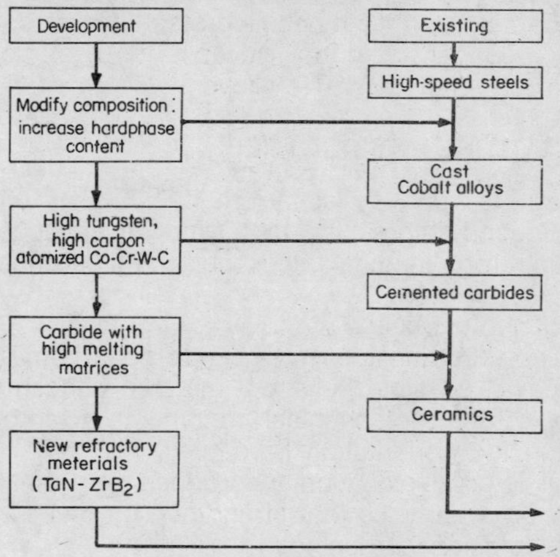

Fig. 8.65: Development potentialities for several existing tool materials

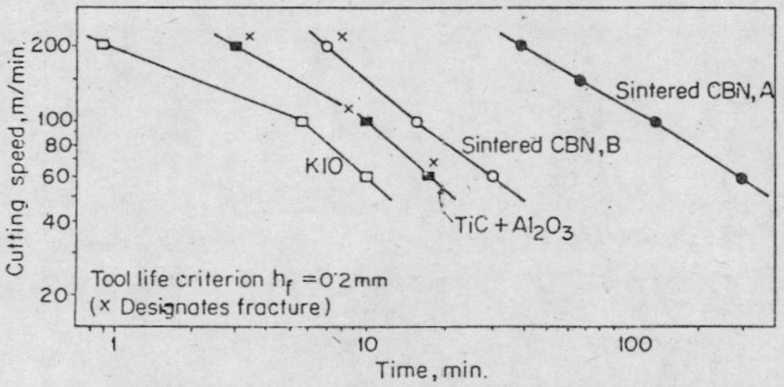

Fig. 8.66: Cutting speed — tool life characteristics of few latest tool materials.

A. *COATED CARBIDES*

The most important tool development in the decade of sixties is the introduction of coated carbides when a thin film of TiC was deposited on the steel cutting grade of carbide by vapour phase deposition. A coating of the order of 0.005 mm reduced the tendency of a cemented carbide tool to crater. It has been observed that the coating material diffuses into the tool surface as wear proceeds. The usual coatings are from TiC, TiN and Al_2O_3. While TiC and Al_2O_3 appear to provide the most chemically stable screening layer between chip and tool, TiN appears to offer the lowest tool friction. This is due to the surface strengthening action of nitrogen in the presence of austenitic surface of the chip. Other new candidates for coating are hafnium carbide and hafnium nitride.

Coatings may be applied by sputtering or other physical deposition techniques. Better bonding is realized by chemical vapour deposition where vapours such as $TiCl_4$ and methane are reached on the hot tool surface at 650°C or higher to produce the coating.

Composite or multiple-layer coatings are also used. The layer next to the carbide substrate is TiC which provides a stable diffusion barrier to prevent carbon leaving the substrate or iron from entering the cobalt phase. The outer layer is TiN to provide low interface friction. The transitional layer is of titanium carbo-nitride.

Coated tools are inherently more brittle than uncoated tools and are used when crater formation is to be countered.

B. *CUBIC BORON NITRIDE* (CBN)

Around 1973, a coating of polycrystalline diamond or cubic boron nitride (CBN) was introduced on the surface of a sintered tungsten carbide substrate having a cobalt binder. The CBN coated tools are used primarily to machine stainless steel, HSS and HSTR alloys. An example shows, (Fig 8.66) that the tool life of CBN tool is excellent for machining quenched alloy steel for dies.

C. *NIOBIUM BASED* 'UCON'

Union Carbide developed UCON based on columbium (niobium) consisting of 50% columbium, 30% titanium and 20% tungsten. These are cast, rolled and slit into blanks. Nitrogen is finally diffused into the surface at a temperature of 1650°C to produce a hard surface layer. The blanks are not reground or brazed and are used as clamped inserts.

These are used for machining steel at high speeds and feeds and are not suitable for cast iron or stainless steel or HSTR alloys.

8.8. FUTURE POSSIBILITIES

From the above discussions it is evident that the continuous interest is there for developing new cutting tool materials to meet the challenges of aero-space age.

To quote Shaw :

"The following ideas appear to have some relevance to possible new tool material concepts :

(i) Thermo-mechanical treatments to produce texturals or orientation hardening,

(ii) Internal oxidation to produce very small (10 nm) hard particles that strengthen by precipitation or dispersion hardening,

(iii) Multiphase materials with each phase interconnected produced by spinodal decomposition with a phase-wave length of 10 nm or less.

(iv) Improved fracture toughness of ceramics by precipitation of a crack arresting phase,

(v) Directional solidification of cast tool materials and casting techniques that eliminate the embrittling action of micro-shrinkage cracks and centres of tensile stress due to the dimensional change associated with the last material to solidify between dendrites,

(vi) Application of SPLAT cooling techniques to produce materials of very fine grain size approaching amorphous structure,

(vii) Continued and extended application of the composite concept."

It is seen that there are many exciting new possibilities for further improvement of cutting tool material capability.

CHAPTER 9

Machining of Surfaces

9.0 INTRODUCTION

Surfaces can be considered to belong to generic classes as shown in Fig. 9.1

9.1. PRODUCT PROCESS LINKAGE

Every manufactured product has got to have its specifications fulfilled through the operation and control of production system parameters such as machines, toolings, fixtures, measuring systems, etc. Plant layout, organisational structure, management objectives, order schedules also influence operations involved in the technological transformation of a given production system. The effectiveness of any production system is measured by *productivity* ($\hat{P}$), which is defined as

$$\hat{P} = \frac{R\{v_i\}}{R\{u_i\}}$$

where operator R is the rupee equivalent for inputs and outputs.

All the characteristics of the product are interdependent on each other as well as other exogenous factors involved in the manufacturing process. It is difficult to isolate them and characterize them as product-process parameters. However, it is useful to have a basic appreciation about their characteristics and methods by which they are delineated.

9.2. PRODUCT CLASSIFICATION THROUGH 'GROUPING'

While developing 'group technology' (GT), Mitrofanov, Opitz, Koloc and many others have investigated size, shape and technologies in order that a rationale be developed for selection of machines, tools, fixtures, etc. This systematic search for product groups has led to classification systems which will be useful in understanding machining technologies conformable to size and shape of products.

Noting that size, shape and machining technologies are the three guiding parameters to be considered in order to group parts having similarity, several classification systems for products have been developed.

Fig. 9.2 shows how Opitz has developed the broad field of application of the part classification system as a systematic procedure. It indicates the preconditions, supplementary means and aims of product classification system commonly known as group technology. The use of this technique in manufacturing system for developing adapted machine tools, higher degree of automation or design rationalisation is a separate topic not being covered in this treatise. The basic theme is to show that certain technological processes are interrelated with sizes and shapes of products. This theme can be examined from the Opitz classification chart of Fig. 9.3 which aims at a product coding system with a decimal base.

With respect to Fig. 9.3 it is seen that the form elements are arranged according to the sequence of operations. The shape of the part is characterized by the five digits of the code, while dimensions, raw materials, required accuracy are indicated in the additional code. In the first digit, this class of parts—rotational or non-rotational—is distinguished. Specific parts are found in position 5 and 9. Each of these 10 groups can be further subdivided systematically. The second digit shows the outer contour and form elements for rotational parts or the principal shape for variations of non-rotational parts. The third digit characterizes the internal shape and its elements, the principal bores, etc. The fourth digit describes the machined surfaces while the fifth describes drill holes, gear generation and chipless forming.

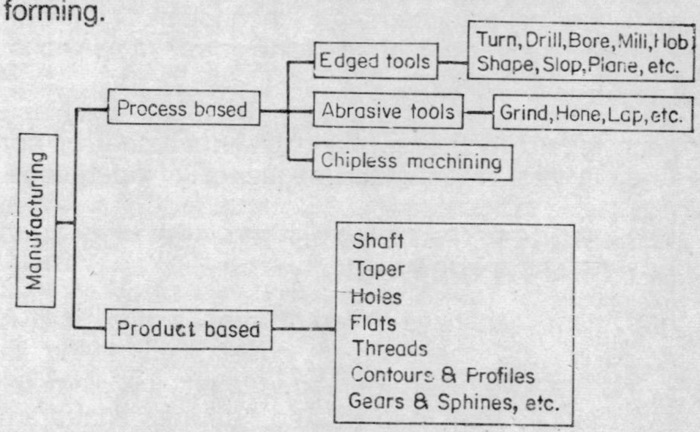

Fig. 9.1: Generic classification of surfaces.

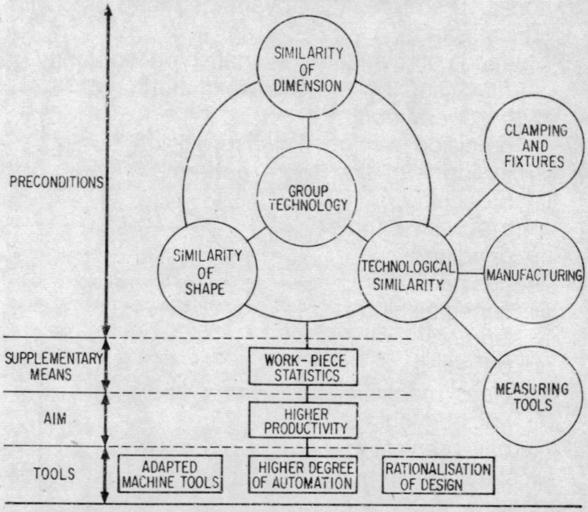

Fig. 9.2: Pre-conditions and aims of Group Technology (GT).

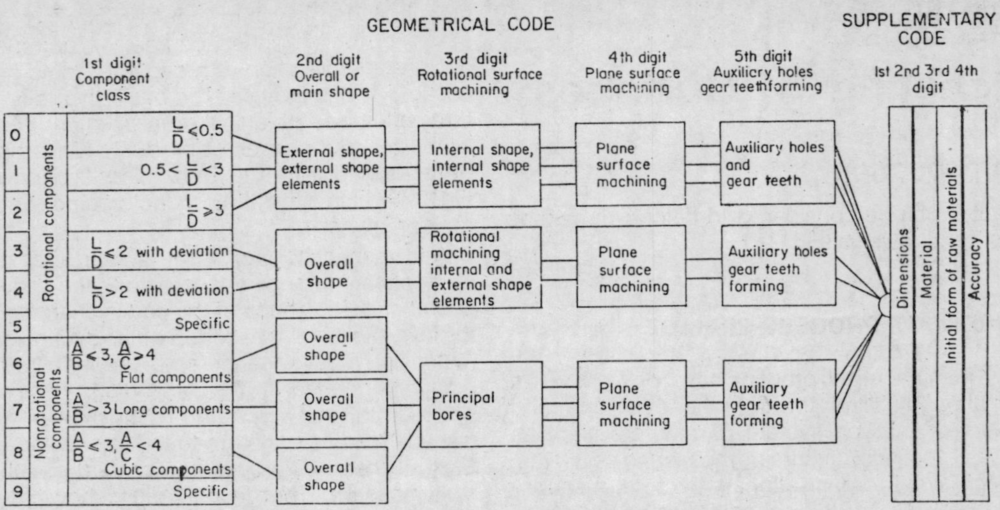

Fig. 9.3: Opitz classification chart – geometrical coding system.

In the process of finding 'groups' or 'part families', the following are the main steps:

(i) the components must be coded and sorted in increasing order of code numbers;

(ii) a selection according to technological criterion follows, the main factors being form, dimensions, material, accuracy;

(iii) the pre-sorted groups have to be examined with regard to their working plans whether or not they fit into the intended sequence of operations.

9.3. PROCESS-PRODUCT SYNTHESIS FOR SHAFT-TYPE WORK

(a) Shafts can have different shapes, Fig. 9.4. A cylindrical shaft has an equal diameter on its entire length. On a stepped shaft, the diameter changes from step to step. Stepped shafts are more expensive to manufacture than cylindrical ones. Therefore, it is advantageous to use cylindrical shafts.

(b) Surfaces of revolution on shaft-type work may be finish-machined by the following methods :

With edged tools:

(i) precision turning (finish precision turning) ;
(ii) milling with fine tooth cutters ;
(iii) broaching.

With abrasive tools :

(i) grinding ;
(ii) honing ;
(iii) superfinishing ;
(iv) lapping ;
(v) polishing.

Chipless machining with metal tools :

(i) roll burnishing ;
(ii) ball burnishing.

Few specific examples of shaft-type work are described.

9.3.1. Turning Stepped Shafts

Stepped shafts may be turned on multiple-tool lathes by the following methods (Fig. 9.5):

(i) longitudinal feed, i.e. by parallel (simultaneous) machining of the steps (Fig. 9.5 a) ;

(ii) by an angular approach cut with subsequent longitudinal feed (Fig. 9.5 b);

(iii) turning with cross-feed.

In all case, all the tools must finish their cuts simultaneously.

When a stepped shaft is machine by the first method, the tools begin their cut one after the other (Fig. 9.5 a).

9.3.2. Burnishing Process

The burnishing process consists of pressing hardened steel rolls (one or three arranged opposite each other) into the surface of the workpiece and imparting a feed motion to the rolls. A case of ball burnishing is shown in Fig. 9.6.

Considerable radial forces appear in roll burnishing, the surface layers of the work are compressed and residual compressive stresses are developed. In some cases, the fatigue strength is increased by this process. The wear-resistant qualities are also increased by compressing the surface layer.

9.3.3. Honing External Surfaces of Revolution

Honing consists in machining surfaces with abrasive sticks *1*, *2*, *5* and *6* (Fig. 9.7) held in the holders *3* and *4* which, in turn, are mounted in the contracting yokes. The abrasive sticks envelope the workpiece, which is rotated. This method has found limited application and is used for finishing automobile crankshaft journals.

9.3.4. Super Finishing External Surface of Revolution.

Superfinishing consists in removing the surface irregularities of a rotating workpiece by very fine-grain (grain size 400 to 600) abrasive sticks which continuously travel slowly along the length of the work. At the same time, the sticks oscillate rapidly in short strokes (2 to 5 mm) (Fig. 9.8).

In this process, the work rotational speed is low (2 to 20 m per min), the longitudinal feed ranges from 0.1 to 0.15 mm per workpiece revolution, the abrasive sticks oscillate with a frequency from 500 to 1,800 strokes per minute, and the springs hold the sticks against the work with a force from 2 to 10 kg. A special lubricant, having a definite viscosity, is used to obtain a high quality of surface finish (usually a mixture of kerosene and oil). The force exerted by the spring is changed during the course of operation. It is increased as the area of contact between the abrasive stick and work increases. The latter is due to the gradual removal of the ridges on the machined surfaces.

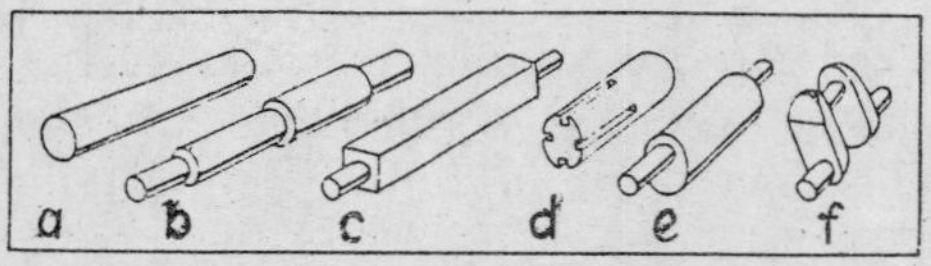

Fig. 9.4: Various shapes of shafts. (a) Cylindrical shaft, (b) stepped shaft, (c) shaft with square cross-section, (d) multiple splined shaft, (e) eccentric shaft, (f) crank-shaft.

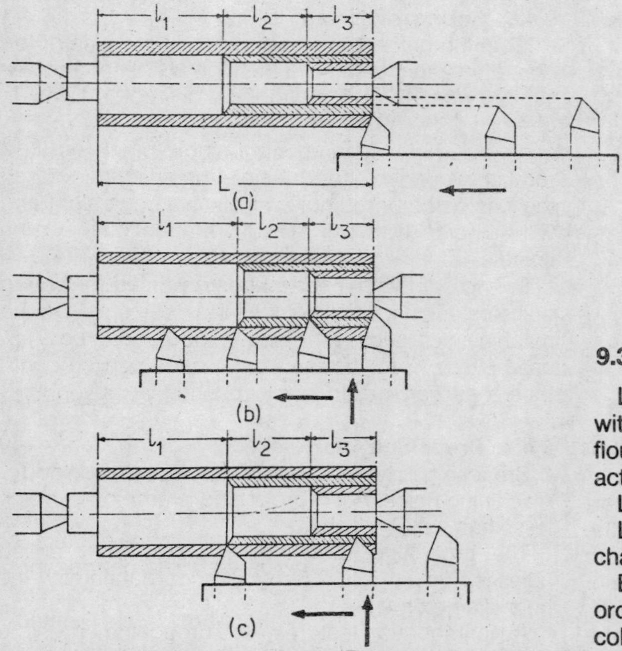

Fig. 9.5: Turning of stepped shaft on multiple-tool lathes.

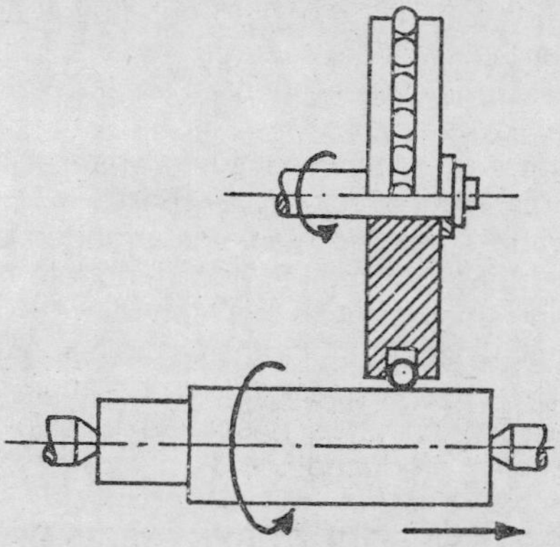
Fig. 9.6: Schematic diagram of the ball burnishing process.

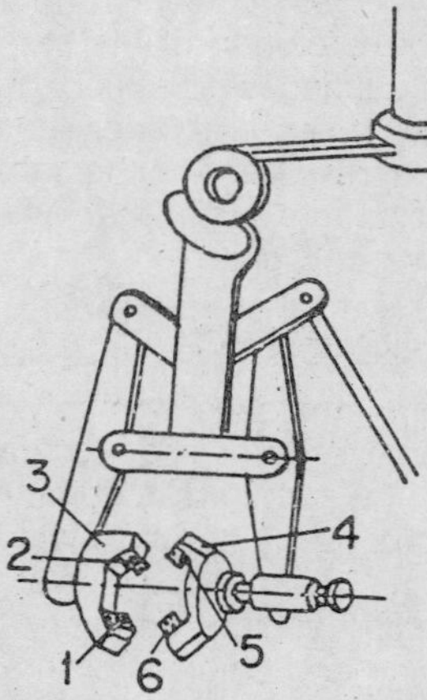

Fig. 9.7: Hone for finishing shaft journals.

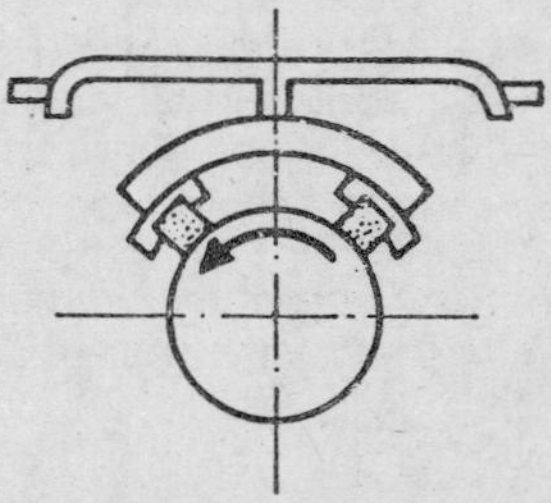
Fig. 9.8: Super-finishing to remove surface irregularities

9.3.5. Lapping of External Surfaces of Revolution

Lapping of external surfaces of revolution is done with a lap which is a tool covered by an abrasive flour, mixed to a paste with some carrier. Chemically active fluids are widely used for this purpose.

Laps are usually made of cast iron or bronze.

Lapping may be done by two methods: (a) with a charged lap and (b) with an uncharged lap.

External surfaces of revolution may be lapped on ordinary lathes with hand laps in the form of a split collar (for small-lot production conditions) or on special lapping machines. This machine has two cast-iron discs (laps) mounted on vertical spindles. The lower lap is rigidly secured to its spindle while the upper lap is swivel-mounted so as to align with the lower lap surfaces.

The upper and lower spindles rotate at different speeds.

A movable spider or retainer is arranged between the laps. It has openings to accommodate the work (usually of small size) to be lapped.

A complex path of the abrasive grains on the work surface (Fig. 9.9) is observed in operation of these machines. This provides, to some extent, for the fine surface finish produced.

9.4 PROCESS-PRODUCT SYNTHESIS OF HOLE MACHINING

It is more difficult to machine a hole than an external surface of revolution of the same size. For this reason, the tolerances specified by the standards for accurate holes are wider than those for external surfaces of revolution of the same diameter and grade of accuracy. Hole machining can be grouped as shown in Fig. 9.10.

9.4.1. Drilling to Obtain a Hole

Drilling is usually employed primarily to obtain a hole in solid material. After drilling, accurate holes (in accordance with the required accuracy and other factors) are machined by the following methods: boring with a single-point tool or enlarging with a core drill, reaming,broaching, grinding, honing, superfinishing, lapping and chipless sizing. Machining is carried out by one or several methods.

Holes are most frequently drilled with twist drills.

Holes may be drilled either with a rotating tool (drilling machines) or with a rotating workpiece (for example, deep-hole drilling machines of the lathe type).

Very rigid and powerful machines are required to drill large holes. Springing of the work is possible in these cases. For this reason, holes are drilled in solid material only with drills up to 25 or 30 mm in diameter. Holes of larger diameter are enlarged by one or two drills of the corresponding size. The diameter of the first drill is selected about one-half of that of the drill used for enlarging the hole.

Various drilling and core-drill etc. are shown in Fig.9.11.

9.4.2. Reaming of Holes

The principal aims in reaming a hole are to impart the correct form to the hole and to ensure the specified accuracy and surface finish but not to change the position of the hole axis. If a reamer floats freely in a previously drilled or bored hole, i.e., the reamer is guided by the hole surface, it will be easier to meet the above-mentioned requirements than if the reamer is rigidly mounted. This explains

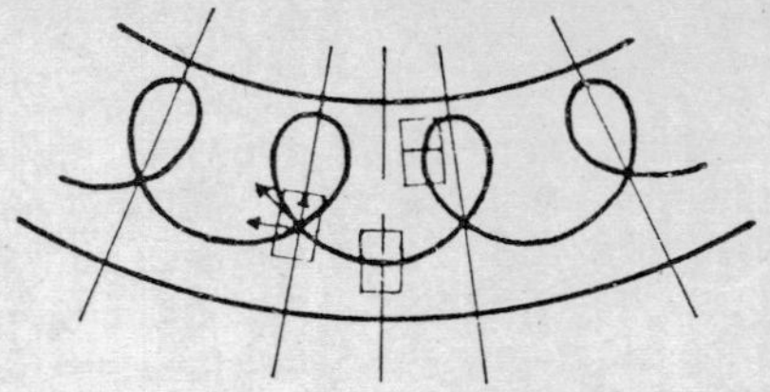

Fig. 9.9: Complex path in a lapping process.

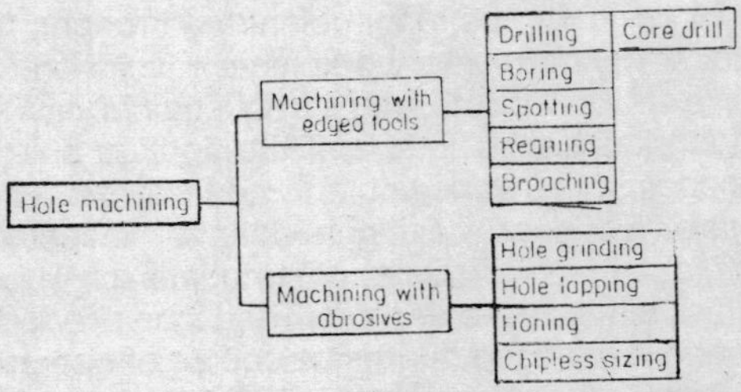

Fig. 9.10: Classification of hole machining

the wide use of floating finishing reamers or reamers with floating blades.

Reamers are usually made of carbon tool steel, and high speed steel. Frequently, the cutting edges are tipped with cemented carbide. Depending on the operation of the tool, it is distinguished as hand and machine reamers, (Fig. 9.12). The long cutting edges provide the hand reamers with a good guidance. Machine reamers have shorter cutting edges. A cylindrical or tapered shank is used for chucking in the machine. Bigger holes will be reamed with shell reamers.

9.4.3. Boring of Holes

In boring holes, the straightness of the axis of the bored hole and its relative position will differ according to whether the workpiece or tool rotates in operation, Fig. 9.13 (a).

When a hole is bored by a single-point tool in a rotating workpiece (for example, on an engine lathe), the axis of the bored hole will always be straight and will coincide with the axis of rotation of the lathe spindle.

Boring bars with inserted bits are used for boring of holes after predrilling. On the horizontal boring machine, self-supporting and guided boring bars are used Fig. 9.13 (b). These are hardened and ground, thus they can run in guide bushes at any point.

9.4.4. Broaching of Holes

Broaching is a very widely employed method for machining holes. Push-broaching is employed much less than pull-broaching.

The broaching allowance for previously drilled holes is from 0.15 to 0.6 mm in accordance with the hole diameter.

Advantageous features of broaching are the reasons for its extensive use, especially in mass production.

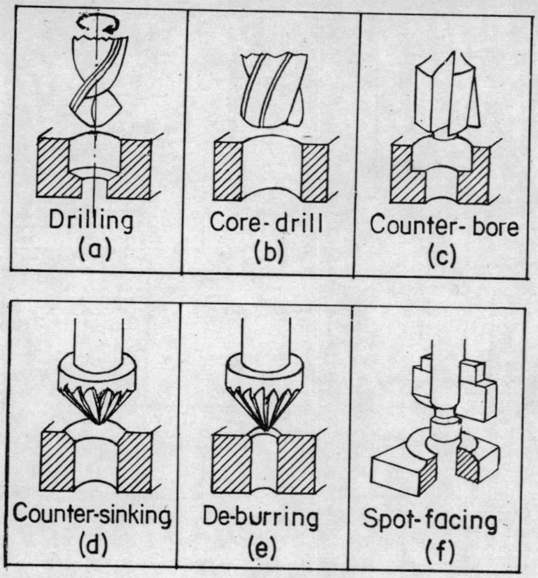

Fig. 9.11. Different types of reamers:

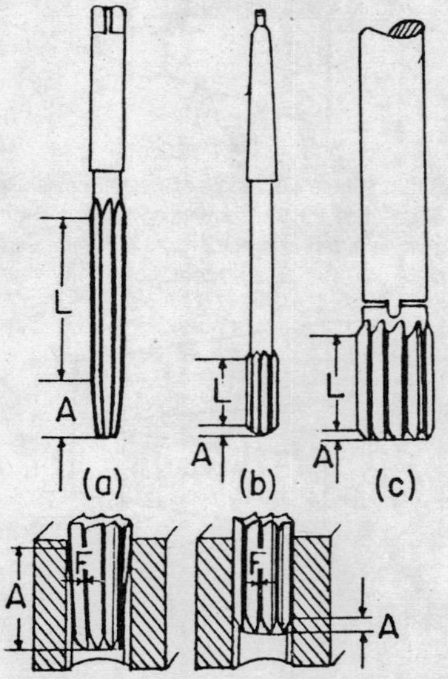

Fig. 9.12: Reamers. (a) Hand reamer with long cutting edges and a long taper lead (A); (b) Machine reamer with short cutting edges and a short taper lead (A); (c) Shell reamer.

Push-broaching is usually performed on hand or power presses. Its purpose is to size a previously machined hole.

9.4.5. Machining Holes with Abrasives

(i) Grinding of holes

Holes are ground in production stage to a limited extent only. This is so because it is a low productive and expensive method even if a high accuracy and surface finish may be obtained. It is therefore employed only: (1) for machining very hard surfaces, for instance hardened parts, or surfaces of non-

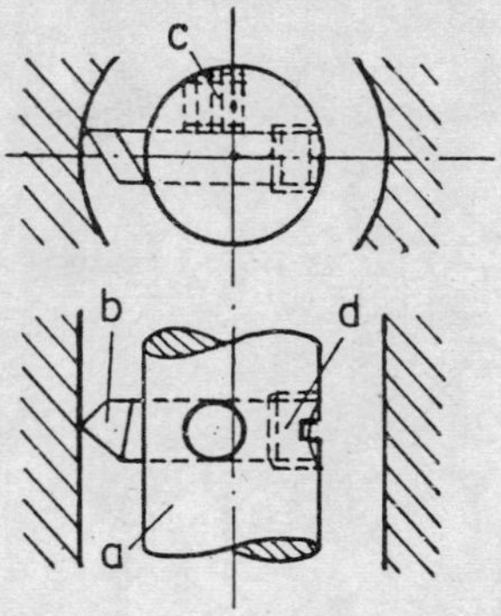

Fig. 9.13 (a) Boring with boring bars.

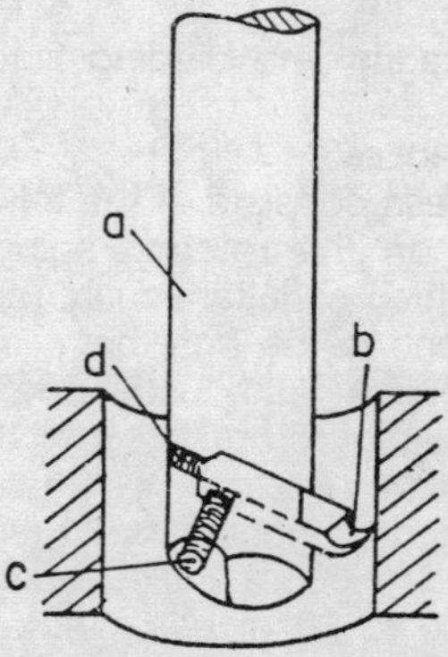

Fig. 9.13 (b) Boring of a predrilled hole, (a) Boring tool, (b) boring head, wobble box for clamping (c) and setting (d) of the boring tool.

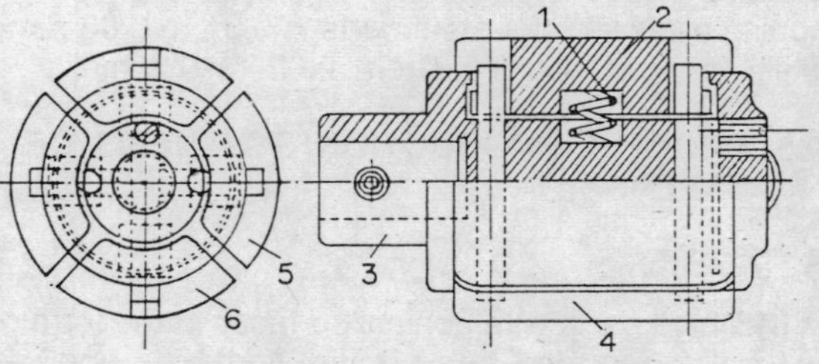

Fig. 9.14 (a) A lap for lapping holes.

uniform hardness, (2) for machining precise holes with precisely located axes in parts difficult to machine by other methods (thin-walled parts, blind holes, etc.), and also interrupted holes (with keyways or splines, etc.).

(ii) Lapping of holes

Lapping of holes consists in removing irregularities on the surfaces of holes by friction of a lap charged with fine abrasive material. Laps are usually made of cast iron, Fig 9.14 (a), or copper and have the form of sectors *2, 4, 5* and *6*, mounted on the arbour *3* and expanded by springs *1* or by cones. The lap is inserted into the hole where it is rotated and reciprocated.

The surfaces of laps are charged with an abrasive, usually applied as a paste or embedded in the

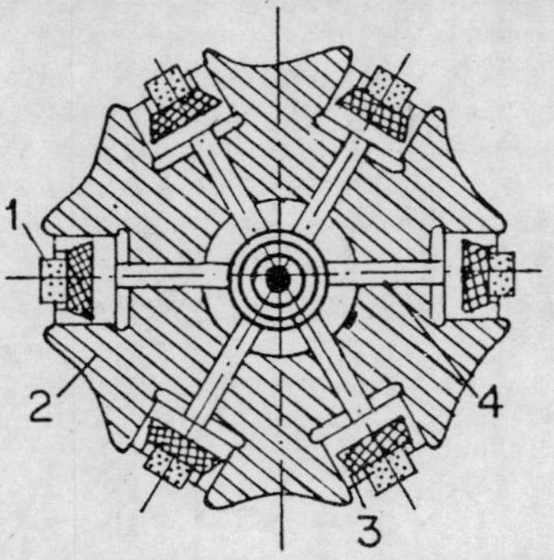

Fig. 9.14 (b) A hone for honing of holes

surface of the lap. In the latter case, oil or kerosene is used as the lubricating and binding fluid.

(iii) Honing of holes

The honing process consists of the following: the hone is mounted on the rotating spindle of the honing machine which simultaneously reciprocates and rotates the hone in the hole being finished. A lubricating and cooling fluid, usually kerosene or soap-and-water solutions, are used in the honing process Fig 9.14 (b).

The spindle speed in honing is larger than the number of strokes. The peripheral speed of the hone for steel is from 40 to 60 m per minute, the axial speed is from 10 to 12 rn per minute. For cast iron, the peripheral speed is 60 to 75 m per minute and the axial speed is from 15 to 20 m per minute. The honing allowance for steel parts is 0.01 to 0.06 mm, while for cast iron it is from 0.02 to 0.2 mm.

9.5. PROCESS-PRODUCTION SYNTHESIS OF TAPERS

A. *TAPERS*

Machine parts with conical surfaces in the form of extenal tapers and taper holes, find wide applications.

Tapers are turned bodies with symmetrical reduction. In the workshop also the conical frustrum is designated as "taper". Workpieces with tapers and taper holes (Fig 9.15) are used for all purposes, e.g. for fastening and sealing

B. *DESIGNATION OF TAPERS*

The most customary tapers are standardized. The designations on the taper have also been shown in Figs 9.16-9.19.

9.5.1. Taper Turning

Taper surfaces are turned in a lather by one of the following methods :

(a) off-setting the tail-stock
(b) swivelling of compound rest

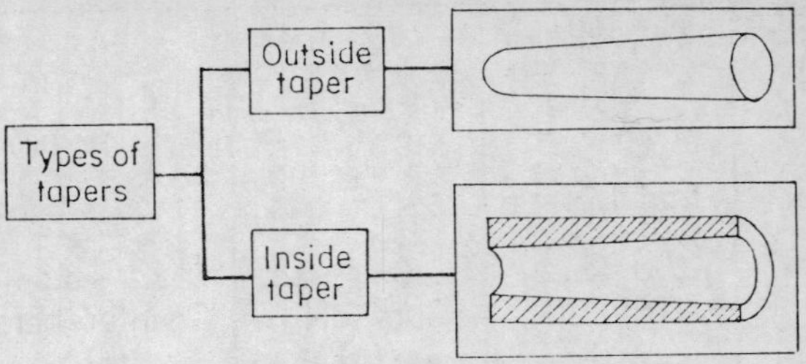

Fig. 9.15 Types of taper surfaces.

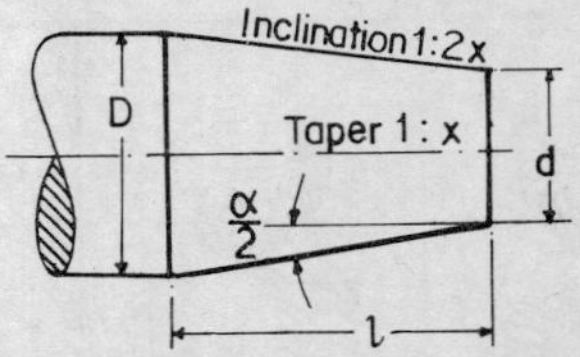

Fig. 9.16: Designation on the taper. D=big diameter, l=length of taper, 1:x=taper, d=small diameter, 1:2x=inclination, α/2=angle of inclination (setting angle on machine tools).

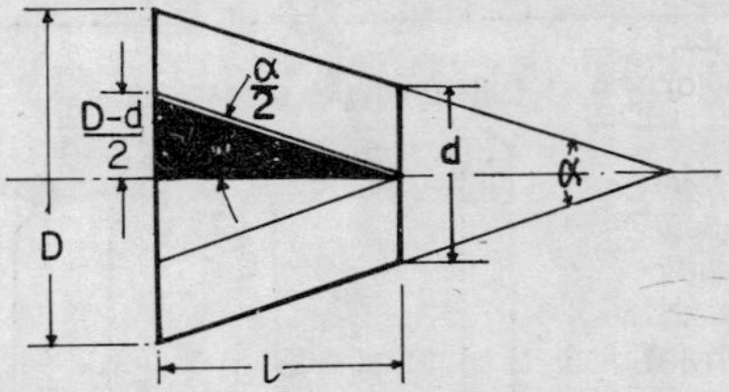

Fig. 9.17: Setting angle α/2 (angle of inclination) is used for setting of the compound slide for the turning operation. The tangent (tan) of the setting angle is tan α/2=(D/2–d/2): l.α is the vertex angle of the taper.

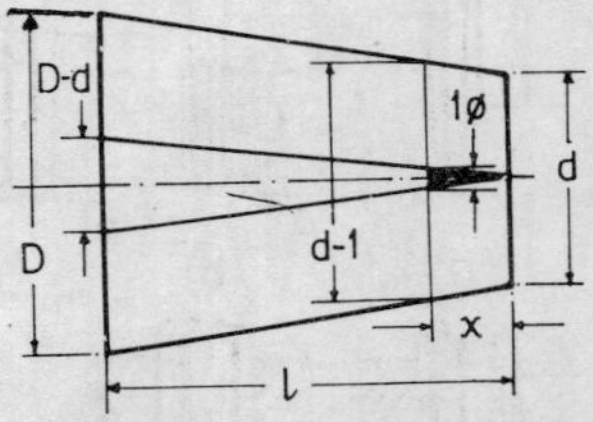

Fig. 9.18: Taper, (D–d) : l abbreviated to 1:x. Taper 1:x means – at a length of x mm, the diameter of the taper varies by 1 mm.

(c) using a taper turning attachment, or
(d) Using two feeds

(i) Taper-Turing by Off-Setting The Tail Stock

The body of the tailstock is off-set crosswise by an amount equal to *h* in reference to the base (Fig 9.20). Consequently, the axis of the work mounted between centres makes a certain angle α with the direction of longitudinal feed of the carriage. It follows from the diagram in Fig. 9.20 that

$$\frac{h}{L\cos\alpha} = \frac{D-d}{2l} = \frac{k}{2}$$

where $k = \dfrac{D-d}{l}$ = the *taper* of the surface to be

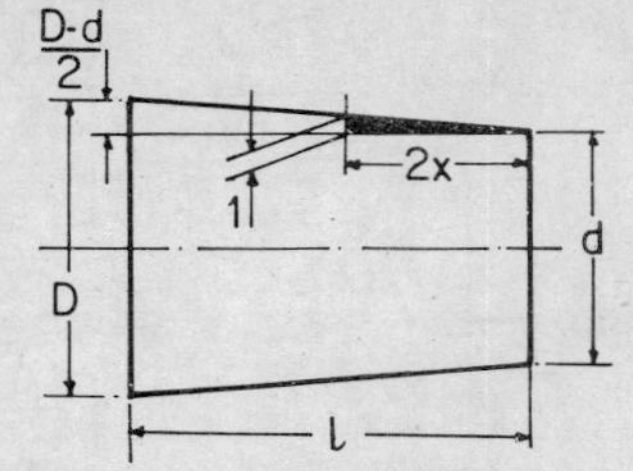

Fig. 9.19: Inclination (D/2–d/2): l abbreviated to 1:2x. Inclination 1:2x means – At a length of 2x mm the radius of the taper varies by 1 mm.

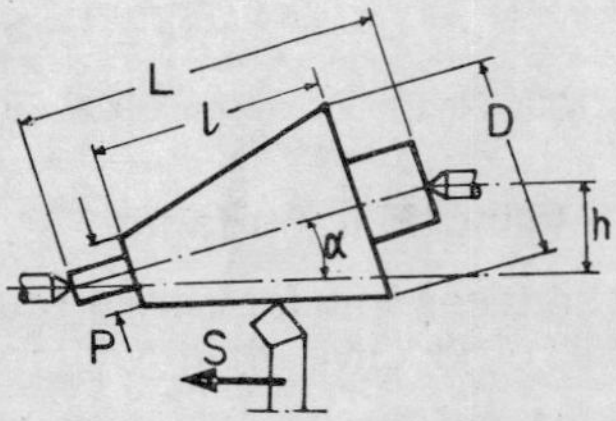

Fig. 9.20: Turning a taper by off-setting the tailstock

turned. Therefore, the off-set of the tailstock to obtain a taper of $k = 2 \tan \alpha$ is

$$h = k\frac{L \cos \alpha}{2}$$

This method can be applied for turning surfaces with a relatively small taper (up to about 1:4) and when the accuracy requirements are not particularly high since the misaligned position of the centres in the centre holes leads to rapid wear of the latter and loss of locating accuracy.

(ii) Swivelling of Compound Rest

This procedure involves swivelling the compound rest through an angle α in reference to the line of centers of the lathe and feeding the tool either by hand or power (in heavy-duty lathes). This method is suitable for turning accurate internal and external tapers of a length limited by the travel of the compound rest, Fig 9.21 (a),(b).

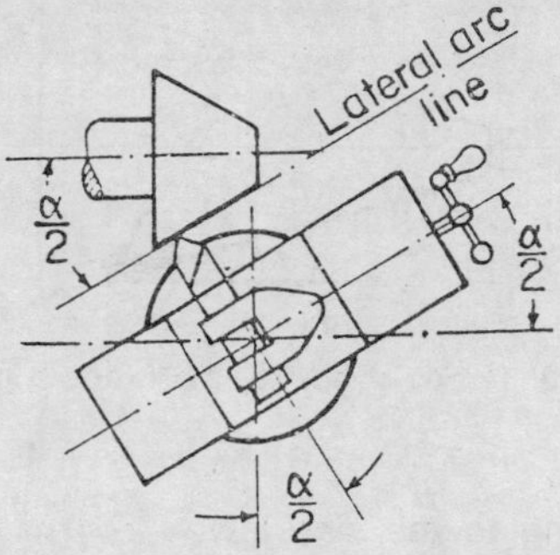

Fig. 9.21 (a) Taper turning (external) with the compound slide.

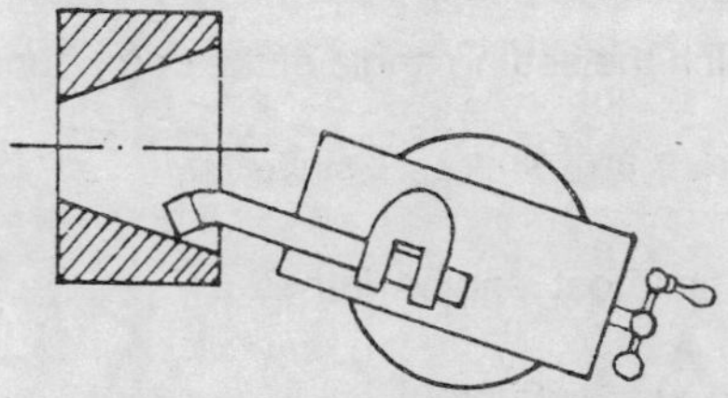

Fig. 9.21 (b) Turning of an inside taper.

(iii) By Taper-Turning Attachment

This method requires that a special device called a *taper-turning attachment* be installed on the engine lathe.

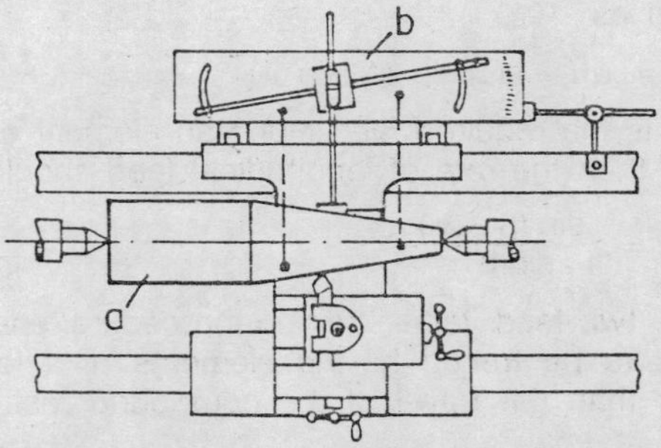

Fig. 9.22: Taper turning with the taper turning attachment. (a) Workpiece, (b) taper guide bar.

The simplest type of taper attachment is showen in Fig 9.22. It consists of a guide bar 2 secured on brackets 1 which are bolted to the rear side of the bed. The guide bar can be set to the required taper angle α in reference to the line centres of the lathe. Guide block 3, sliding along the guide bar, is linked to the cross slide. The latter is disconnecterd from saddle by removing the cross feed screw. If the carriage travels with longitudinal feed along the bed, a cross movement, due to the travel of block 3 along 2, is transmitted to the tool in addition to its longitudinal movement. The resulting motion of the tool will be at an angle α to the axis of the work, this angle being equal to that at which the guide bar is set. More universal taper attachments are available which permit the tool to pass over automatically from turning a cylindrical part of the workpiece to a tapered section.

Taper turning attachments enable accurate tapers with an angle up to 15° or 18° to be turned. The length of the taper must not exceed that of the guide bar.

(iv) Two-Feed Method

Taper turning by means of two feeds is limited to lathes having power feed to the compound rest, and the possibility of engaging this feed simultaneously with longitudinal feed of the carriage. Mainly these are heavy-duty lathes. The resultant feed of the tool (Fig. 9.23) in reference to the line of centres will be the geometrical sum, in the given case, of the longitudinal feed s_{lg} of the carriage and the feed s_c of the compound rest which is set at the angle β to the line of centres. Thus

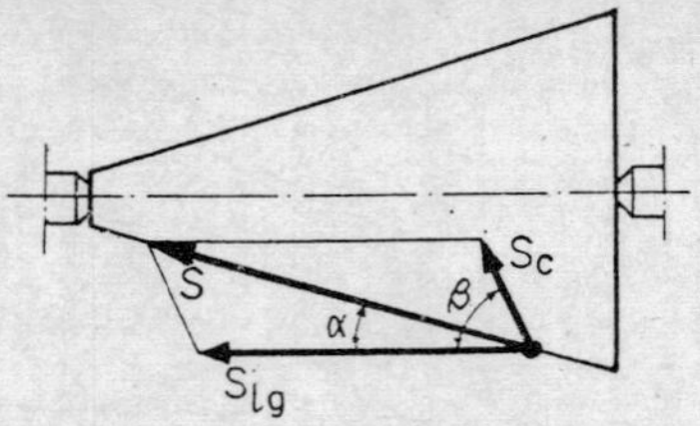

Fig. 9.23: Turning a taper by the two-feed method.

$\bar{s} = \bar{s}_{lg} + \bar{s}_c$

According to the sine law

$$\frac{s_{lg}}{\sin(\beta - \alpha)} = \frac{s}{\sin\beta} = \frac{s_c}{\sin\alpha}$$

from which the setting angle of the compound rest is

$$\beta = \alpha + \text{arc}\,\sin\left(\frac{s_{lg}}{s_c}\sin\alpha\right)$$

Since in most engine lathes

$$\frac{s_{lg}}{s_c} = \Delta = \text{constant}$$

the setting angle of the compound rest can be written as

$$\beta = \alpha + \text{arc}\,\sin(\Delta\sin\alpha)$$

If s is the required feed along an element of the taper, then the rate of longitudinal feed should be

$$s_{lg} = s\frac{\sin(\beta - \alpha)}{\sin\beta}$$

The two-feed taper turning procedure enables tapers to be turned having elements of a length longer than the travel of the compound rest.

9.6. PROCESS-PRODUCT SYNTHESIS FOR FLAT SURFACES

Flat surfaces are usually machined by the methods shown in Fig 9.24.

9.6.1. Planing of flats

The advantages of planing, in comparison with milling, are: cheaper cutting tools may be used, planers are less expensive and more versatile than milling machines, and planing is more flexible, i.e. a planer may be set up for a new operation with a minimum loss in time.

The following disadvantages are inherent to planing and limit its application:

(i) difficulty in machining to a preset size on a planer;

(ii) large time losses on idle return strokes;

(iii) limited cutting speeds due to large inertia forces arising in planer table reversal and impact loads on the cutting tool;

(iv) the necessity of employing more highly skilled operators, etc.

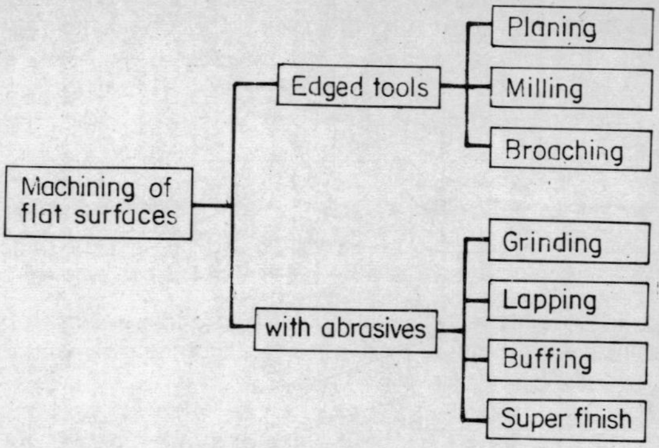

Fig. 9.24: Machining of flat surfaces-Classification.

9.6.2. Face Milling Flat Surfaces

Milling flat surfaces with face milling cutters or end mills is more productive than with plain milling cutters.

The advantages of face milling cutters are:

(i) extremely rigid cutter mounting;

(ii) simultaneous operation of several teeth or blades (providing for smoother operation);

(iii) possibility of machining a workpiece on several sides simultaneously. Fig. 9.25 illustrates schematically how all the surfaces of lathe bed ways are milled simultaneously by various cutters and gang milling sets, as well as the spindles 1, 2, 3 and 4 on which the cutters are mounted;

(iv) possibility of using large diameter cutters with a large number of teeth (or blades).

9.6.3. Plain Milling

The *advantages* of milling in comparison with planing are especially evident when several surfaces may be machined simultaneously with a set of cutters (gang milling). It is not always possible to plane a part with several single-point tools on a planer because of the difficulty in watching the operation of the tools.

In many cases, however, milling must be performed, not with face milling cutters, but with plain and side milling cutter, end mills, etc.

As mentioned before, flat surfaces may be milled with plain cutters either against the feed (conventional milling) or with the feed (climb milling).

In the first case, the chip thickness increases in the course of chip formation (up cut) while in the second case it decreases (down cut).

Higher production capacity may be attained in climb cutting and a finer surface finish than in conventional milling. The reason for this is that the cutter teeth, in conventional milling, rub over the machined surface for a time before beginning to cut, with the result that the teeth dull very rapidly and ridges appear on the machined surface.

Nevertheless, conventional milling is still more widely, used than climb milling since the latter requires a special backlash eliminating device in the table feed mechanism to ensure smooth operation and especially rigid milling machines are required.

Milling machines, specially designed for climb milling, are also available.

9.6.4. Gang Milling of Flats

In many cases a workpiece may be milled on several sides simultaneously (Fig. 9.25).

9.6.5. Multiple-station Milling of Flats

Small and medium-sized parts are often milled, one piece at a time or in groups, by the multiple-station method employing an index base 3 (Fig. 9.26) on which two identical fixtures 1 and 2 are mounted, one at each end. While one workpiece (or a group) is being milled in fixture 1 the next blanks are loaded into fixture 2. When the workpieces in fixture 1 have been milled, the base is indexed about a vertical axias to start milling the workpieces in fixture 2. At the same time, fixture 1 in unloaded of finished workpieces and reloaded with new blanks.

In this manner, handling time for loading and unloading the work is covered by the milling time.

9.6.6. Parallel Simultaneous Milling of Flats

Simultaneous (Parallel) milling of several workpieces 1, 2 and 3 (Fig. 9.27) may be performed with a face milling cutter 4. Fort this purpose, it is necessary that the diameter of the cutter exceeds the total width of all the surfaces of the workpieces set up on the table and that the milling machine is amply powered for the operation.

9.6.7. Consecutive Milling

Consecutive milling is the machining of a row of workpieces, set up one behind the other in the direction of table feed (Fig 9.28a). Parallel consecutive milling (Fig 9.28b) is also used.

9.6.8. Broaching Flats

Flat surfaces are broached, for the most part, in mass production. This is due to the high cost of the broaches and the necessity for having the corresponding broaching machines available. Flat surfaces, as well as other external surfaces, are broached on vertical broaching machines. Often flat surfaces are broached without any previous machining operations.

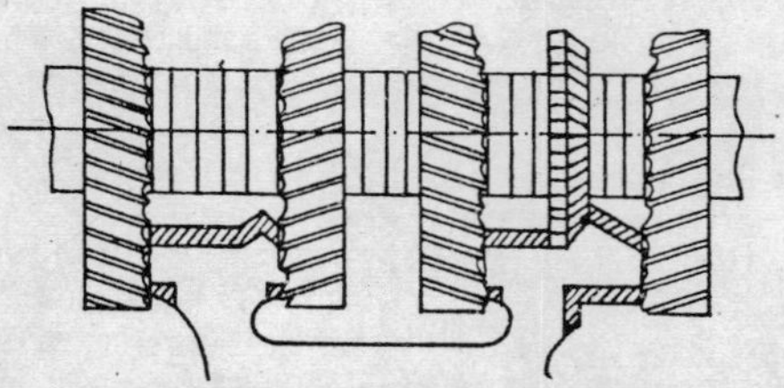

Fig. 9.25. Gang milling of lathe bed.

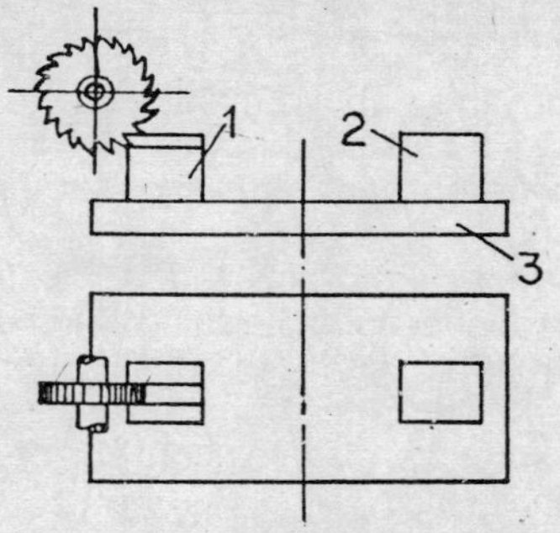

Fig. 9.26. Multiple station milling of small and medium sized parts.

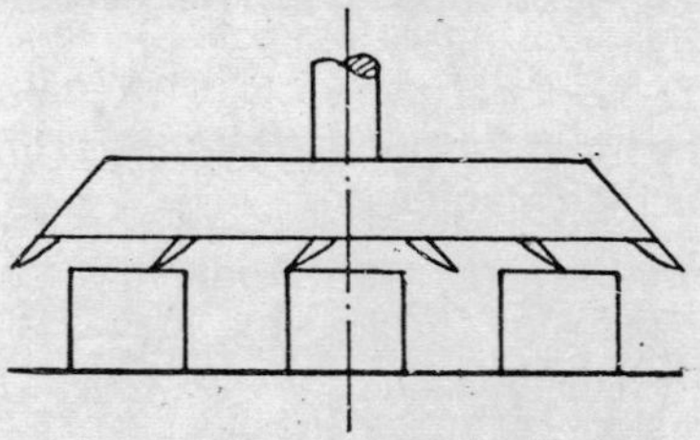

Fig. 9.27. Simultaneous milling of flats.

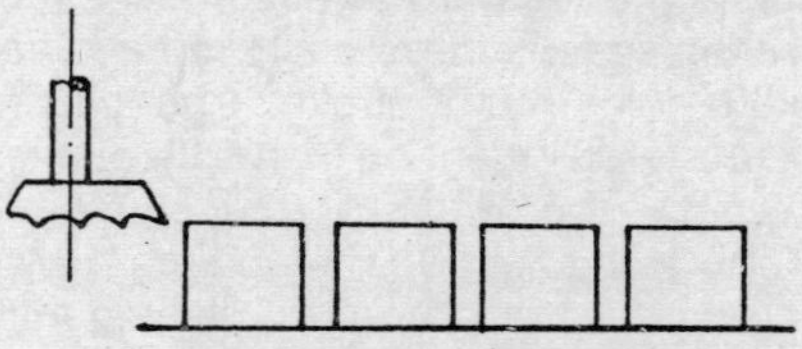

Fig. 9.28.(a) Consecutive milling.

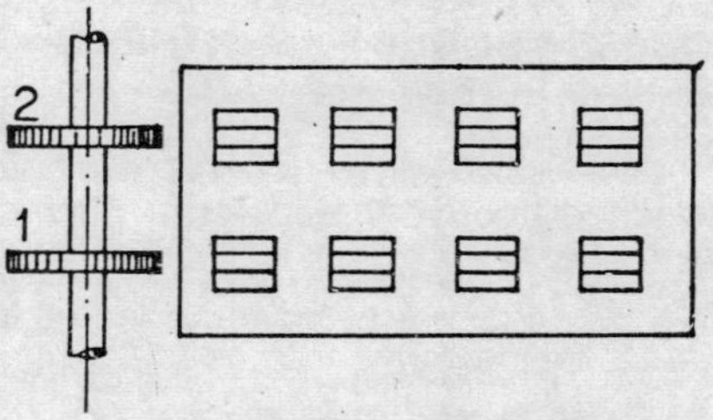

Fig. 9.28.(b) Parallel consecutive milling.

9.6.9. Grinding Flat Surfaces

Flat surfaces of work are usually ground in the following cases: (i) when milling is difficult due to the high hardness of the material, hard skin on the surfaces, or due to difficulties in clamping a thin workpiece for milling, etc.; (ii) when it is difficult to obtain the specified accuracy and surface finish with edged tools.

The most productive method of surface grinding is with the face of the wheel.

Grinding flat surfaces in one pass (depth method) is applied usually for rough machining of cast or forged blanks.

9.6.10. Finishing Flat Surface with Abrasives

Besides the above-mentioned methods for machining flat surfaces, they may be finished by one of the following methods: superfinishing, lapping, and buffing.

Flat surfaces are processed by these methods in the same manner as for finishing extrenal surfaces of revolution. Flat surfaces are superfinished on machines which differ from those used for superfinishing surfaces of revolution but operate on the same principles.

Flat surfaces of small parts are lapped on the machines used for lapping surfaces of revolution on small parts.

9.6.11. The Most Productive Methods for Flat Surface Machining

Flat surfaces are most frequently roughed by milling or planing. These same methods are often used for finishing.

In the majority of cases, the planing of flat surfaces is less productive than milling since in the latter case several cutting edges are in action simultaneously, higher cutting speeds may be applied and no time is lost on the idle return stroke. The planing time depends to a large extent on the width of the surface to be planed. The width of the surface have a considerably less effect on milling time. However, in machining narrow long surfaces, planing may prove more productive than milling.

9.7. PROCESS-PRODUCT SYNTHESIS FOR FORMING SCREW THREADS

In cutting threads, it will be necessary to obtain the correct pitch diameter, thread form, pitch, and a smooth surface.

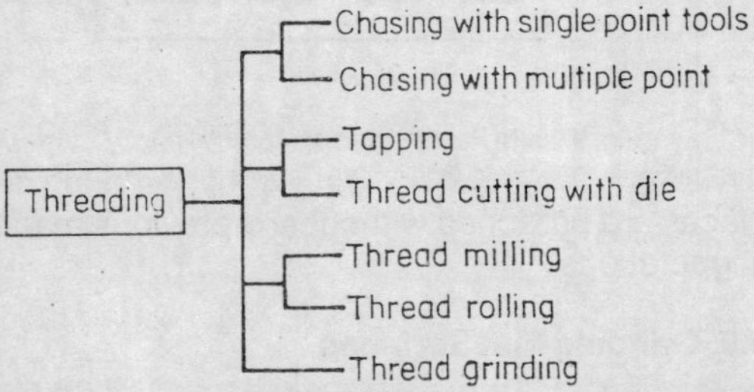

Fig. 9.29. Classification of thread machining.

External threads may be produced on machine tools by the methods shown in Fig. 9.29.

Threads can be produced by various methods, e.g. with taps, dies and die stocks by hand or on machines, with threading tools on the turning lathe, by milling, grinding and rolling, Fig. 9.30. Frequently, threads can also be manufactured by pressing or casting (e.g. die casting).

The selection of the manufacturing method depends on the number of workpieces and on the desired accuracy and surface quality.

Thread chasers (Fig. 9.31) have the advantage that several teeth are cutting simultaneously; hereby the time for thread cutting is reduced.

9.7.1. Principles of Thread Cutting

The thread-cutting gear train of an engine lathe is shown schematically in Fig. 9.32 where the diamond-shaped symbols denote the setting-up units, namely the change-gear quadrant, feed gearbox and coarse pitch unit with gearing ratios of i_q, i_f and i_{cp}, respectively. In the general case, the feed gear train can also incorporate transmissions with a constant total gearing ratio of p.

The setting-up units of the thread-cutting gear train must be calculated and set up so that the longitudinal travel of the carriage per spindle revolution exactly coincides with the pitch T of the thread to be cut. The equation of kinematic balance, linking together the motions of the final links of the kinematic chain, can be written as follows:

$$1_{\text{spin rev}}\; p\; i_{cp}\; i_q\; i_f\; t_{ls} = T$$

where t_{ls} is the pitch of lead screw and from which one obtains the formula for determining the required ration for the setting-up. Thus

$$i_{cp}\; i_q\; i_f = \frac{T}{p\; t_{ls}}$$

The pitch T of the thread to be cut and the pitch t_{ls}, of the lead screw should be specified in the same units of measurement.

Threads not provided for by the feed mechanism of the lathe and threads with nonstandard pitches can be cut with a special setup of the change-gear quadrant, either with or without the use of the feed gearbox mechanism.

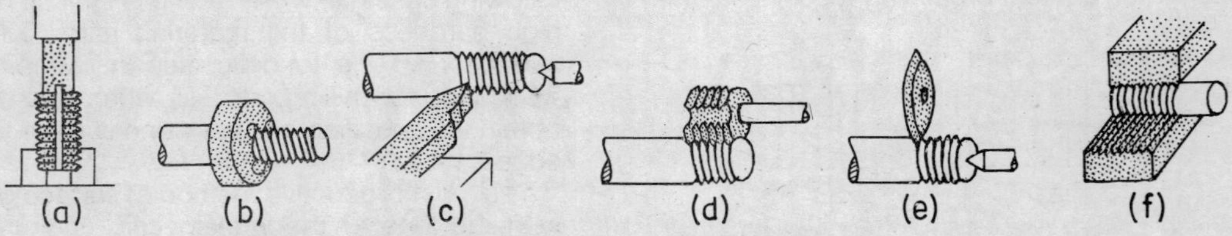

Fig. 9.30: Various methods for the manufacture of threads (examples).
(a) Thread cutting with tab, (b) thread cutting with die,
(c) thread cutting with threading tool,
(d) thread milling, (e) thread grinding, (f) thread rolling.

It is to be noted that in cutting multiple-start threads, the gear train is set up to the lead of the thread which is equal to the product of the pitch and the number of starts.

9.7.2. Change-gear Technique

Change gear calculations are made with following designations (Fig. 9.33).

T = lead (lead of the screw to be cut),
t_s = lead of the lead screw.
Z_1 = number of teeth of the driving gear,
Z_2 = number of teeth of the driven gear,
Z = compound or idle gear, has no influence on the gearing; any number of teeth.

The gear Z_1 is driven over the reverse plate and has the same numbers of revolutions as the working spindle.

The gearing ratio, calculated from the pertinent formula, must be set down as the ratios of the number of teeth on the change gears of the quadrant

$$i_q = \frac{A}{B} \times \frac{C}{D}$$

where A, B, C and D = numbers of teeth on the change gears.

Special tables are available to facilitate the selection of change gears to obtain a definite given gearing ratio.

The following inequalities are used to check whether the selected change gears can be installed on the quadrant and will properly mesh:

$(A+B) > C$ + (15 to 20)
$(C+D) > B$ + (15 to 20)

These formulae ensure that gears B and C do not interfere with shafts of other gears in the train. Depending upon the construction of the quadrant, it is also necessary that the sums of the numbers of teeth $(A+B)$ and $(C+D)$ exceed a definite value as otherwise the pairs of gears cannot be meshed together.

In cutting threads with pitch T_{tab}, provided for by the lathe feed mechanism, the change-gear quadrant, coarse thread unit and the feed gearbox have definite gearing ratios i_q^0, i_{cp}^0 and i_f^0. Thus,

$$i_{cp}^0 \, i_q^0 \, i_f^0 = \frac{T_{tab}}{p \, t_{ls}} \quad \text{Eqn. 9.1}$$

If it is necessary to cut a thread with a pitch T_{th} that differs from T_{tab} then, without changing the value of i_f^0 and i_{cp}^0, the gearing ratio i_q of the change-gear quadrant must be altered. Thus

$$i_{cp}^0 \, i_q \, i_f^0 = \frac{T_{th}}{p \, t_{ls}} \quad \text{Eqn. 9.2}$$

After dividing the members of equation (9.2) by those of equation (9.1) the setting-up formula for cutting thread with the pitch T_{th} is obtained.

$$i_q = i_q^0 \frac{T_{th}}{T_{tab}}$$

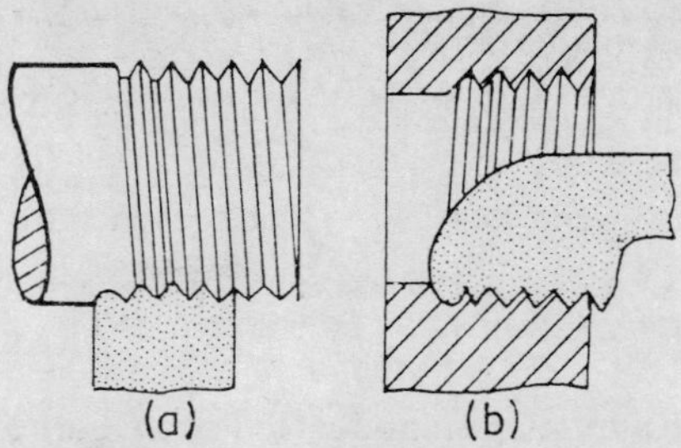

Fig. 9.31. Thread chaser for simultaneous cutting of several threads.

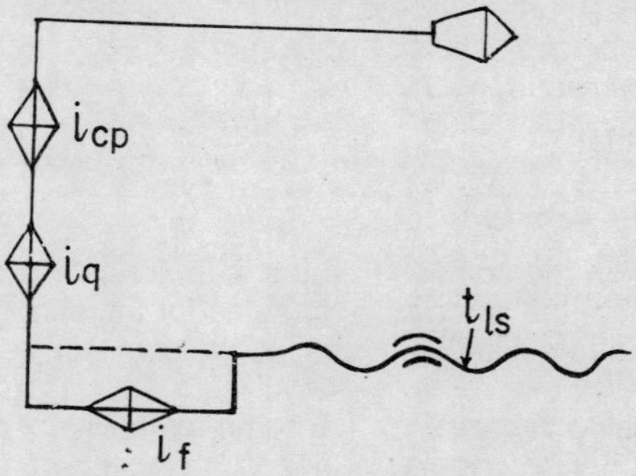

Fig. 9.32. A scheme of feed gear train in a lathe.

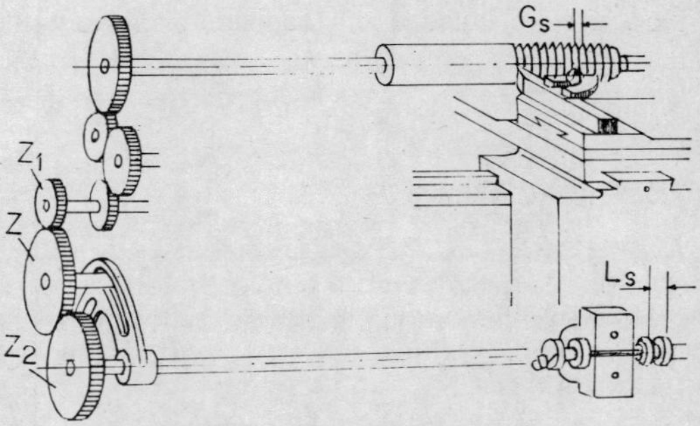

Fig. 9.33: Pick-off change gears.

9.7.3. Kinematic Balance for Thread Cutting

English thread is specified not by the pitch T, but by the number of threads per inch n. Thus

$$T = \frac{25.4}{n} \text{ mm}$$

Module worm thread is specified by the module m in mm and its pitch

$$T = \pi m \text{ mm}$$

Diametral pitch worm thread is specified by the diametral pitch DP and its pitch is

$$T = \frac{25.4\pi}{DP} \text{ mm}$$

In the case of some specific engine lathe, the various equations for setting up the change-gear quadrant, making use of the feed gearbox, can be written as follows:

Metric threads

$$i_q = \frac{42}{50} \times \frac{T_{th}}{T_{tab}}$$

English threads

$$i_q = \frac{42}{50} \times \frac{n_{tab}}{n_{th}}$$

Module worm threads

$$i_q = \frac{64}{97} \times \frac{m_{th}}{m_{tab}}$$

Diametral pitch worm threads

$$i_q = \frac{64}{97} \times \frac{DP_{tab}}{DP_{th}}$$

9.7.4. Kinematic Balance: No Feed Gearbox

If the feed gearbox is excluded from the feed gear train, the equation of kinematic balance is changed to

$$p\, i_q\, i_{cp}\, t_{ls} = T$$

so that

$$i_q = \frac{T}{p\, t_{ls}\, i_{cp}}$$

If on the engine lathe the lead screw has a metric thread, the setting-up formulae for the change gears in cutting various types of threads will be:

Metric threads

$$i_q = \frac{T_{th}}{p\, t_{ls}\, i_{cp}}$$

English threads

$$i_q = \frac{25.4}{p\, t_{ls}\, n_{th}\, i_{cp}}$$

Module worm threads

$$i_q = \frac{\pi\, m_{th}}{p\, t_{ls}\, i_{cp}}$$

Diametral pitch worm threads

$$i_q = \frac{25.4\, \pi}{p\, t_{ls}\, DP_{th}\, i_{cp}}$$

9.7.5. Thread Milling

Milling external threads is classified into two methods: (1) milling with a single-thread cutter, and (2) milling with a multiple-thread cutter.

In the first case, the thread is cut on the slowly rotating work *1* by means of the single-thread cutter *2* inclined, in relation to the work axis, at the helix angle. This cutter, having a form corresponding to the thread form, also rotates and, in addition, is fed along the work axis a distance equal to the thread pitch per revolution of the work (Fig. 9.34).

This method, in which the number of full revolutions of the work equals the number of thread pitches, is applied for rough cutting large threads on long workpieces (lead screws, etc.).

The second method of *milling thread with multiple-thread cutters* is used for cutting short threads and especially for hard materials which are machined with difficulty by other methods.

9.7.6. Thread Grinding

Unhardened, but in most cases hardened parts, can be provided with threads by thread grinding (Fig. 9.35). Thread grinding is selected, when high accuracy and high surface quality are required, e.g. with taps, thread plug guages, ring screw gauges, measuring screws, etc.

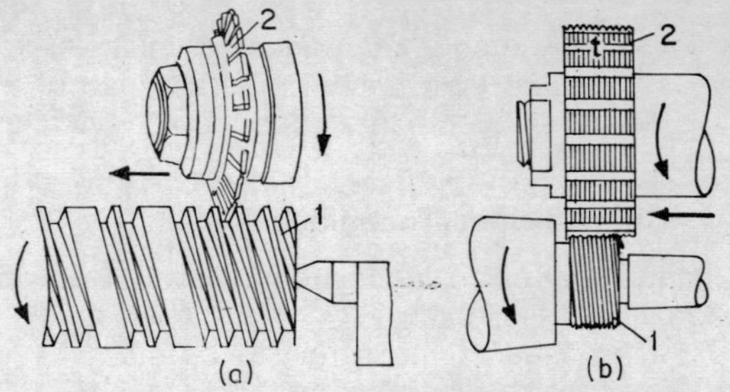

Fig. 9.34. Thread milling setups. (a) Single thread cutter, (b) Multiple thread cutter.

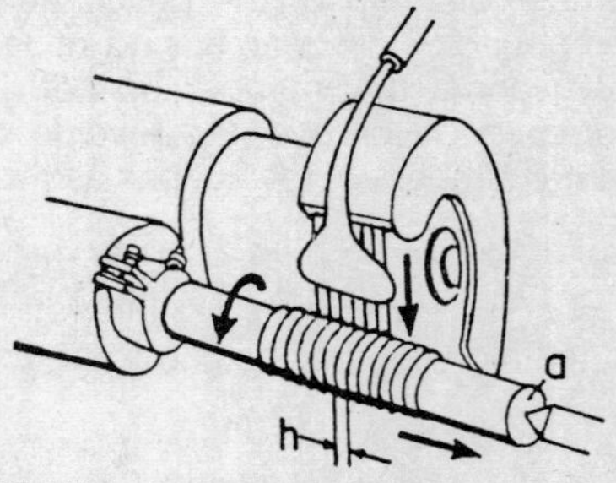

Fig. 9.35 : Thread grinding with multiple profile wheel. (a) Workpiece, (h) lead.

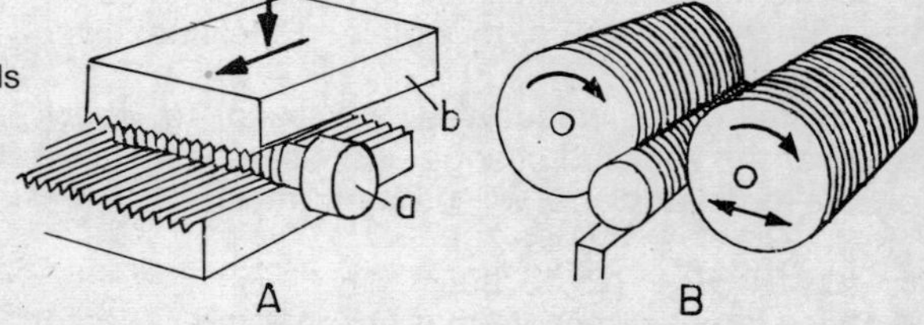

Fig. 9.36 (A) Thread rolling with flat rolling dies. a) Workpiece, b) flat rolling dies. The dies are fixed in the thread rolling machine. The upper die is rolled over the workpiece with high pressure. Thereby, the threads are pressed in.

Fig. 9.36 (B) Thread rolling with rollers. (a) Workpiece, (b) rollers, (c) guide-rail. The workpiece rests between a stationary roller and a roller shiftable with pressure. Short and long threads can be rolled.

Single and multiple profile grinding wheels are used for grinding. The profile of the wheels will be reground with a dressing diamond which is guided in an attachment on the thread grinding machine. The cutting motion is given to the grinding wheel by the thread grinding machine. The workpiece rotates with low rpm and executes the feed movement which corresponds with the lead. Threads with small lead are ground out of the solid material. Bigger threads are rough turned or rough milled before grinding.

9.7.7. Thread Rolling

The thread is rolled by grooved flat dies or rollers, Fig. 9.36. This is a chipless working method which is applied in mass production.

The thread surface is die-burnished. On rough ground parts the grade 'fine' can be achieved. Since

Fig. 9.37: Fibre structure. a) Cut thread, b) Rolled thread.

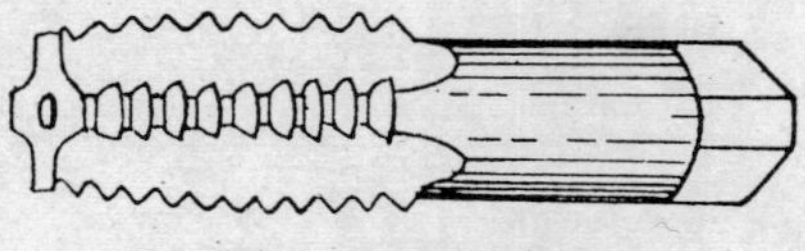

Fig. 9.38: Machine tap.

the fibre structure is not interrupted, the threaded parts have good qualities regarding strength, Fig. 9.37.

9.7.8. Cutting Internal Threads

Internal threads may be produced by cutting with single-point tools, tapping with solid or collapsible taps and milling with multiple-thread cutters.

Internal as well as external threads are not chased by tools in large-scale production of thread.

Threaded holes of small and medium sizes are usually cut with *taps*. Machine taps are classified as (i) straight soild taps (Fig. 9.38), (ii) adjustable inserted-blade taps, and (iii) tapper or nut taps.

Taps with inserted adjustable blades are used for cutting thread of comparatively large diameters (up to 300 mm).

9.8. PROCESS-PRODUCT SYNTHESIS FOR FORMS AND PROFILES

Forms of surfaces can be grouped (Fig.9.39) in two fundamental classes, one which includes cylinders, flats and tapers while the other which can cover all other of-forms of surfaces.

9.8.1. Forming with Form Tools

External forms of revolution are usually machined by turning or grinding.

Short surfaces (up to 50 or 70 mm in length) are usually turned with cross-feed (0.02 to 0.05 mm per spindle revolution) by a form tool (Fig. 9.40).

9.8.2. Template Profiling

Templates are used for contouring the end faces of workpieces.

In the first case (Fig. 9.41) the carriage is power-fed along the axis of rotation of the workpiece while, in the second case (Fig. 9.42), the cross slide is power-fed square to this axis.

In both cases, the path followed by the cutting edge of the tool is determined by two movements: the feed motion of the carriage (or cross slide) along or square with the axis of rotation of the work (main feed) and a motion, square with the first motion, of the cross or top slide controlled by a template and directed square with or along the axis of rotation of the work (supplementary feed).

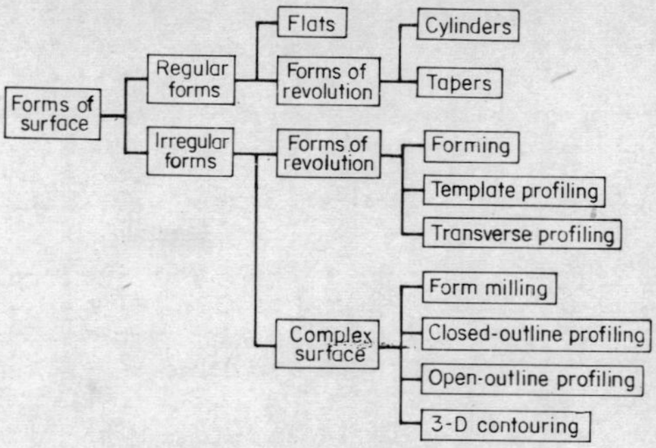

Fig. 9.39: Forming of surfaces-classification.

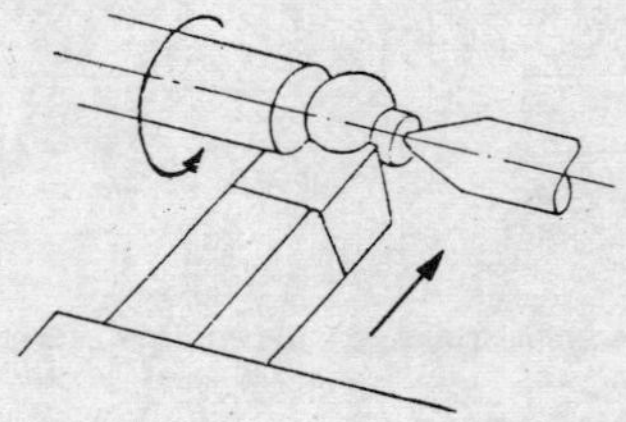

Fig. 9.40: Turning with a form tool.

The template 1 (Fig. 9.41) is secured on the plate 2. The action of the weight 5 holds a following roller 4 in constant with the profile of the template. Roller *4* rotates on a pin secured in a bracket which, in turn, is fastened to the cross slide (from which the cross-feed screw has been removed or disengaged).

In Fig. 9.42 the template 1 is fastened on plate 2 which is located by the bed ways and is clamped in the required position. Under the action of spring 4, the following roller 3 constantly contacts the template 1. Roller 3 rotates on a pin press-fitted in the holder 5.

9.8.3. Closed-Outline Profiling

Simple closed-outline profiles, generated by movement of their elements in a single plane are usually component parts of cams. They are machined by turning, milling, and grinding.

Such surfaces are turned on lathes using longitudinal feed and a rotating master (Fig. 9.43).

Special-purpose multiple-tool machines are available for simultaneously turning of several cams, such as camshafts of internal combustion engines. A feature of these machines is that, in addition to the rotating masters 1 (Fig. 9.43) which impart the required path to the cutting edges of the tools, they have a second set of rotating cams 2. These cams rock the tools 3 so as to provide a constant cutting angle. If the tools had a constant angular position, the cutting angle would vary in the same degree as the curvature of the profile varies at various points.

9.8.4. Profiling Milling of Closed-outline

Profiling milling of closed-outline surfaces, generated by the motion of an element in a single plane, is

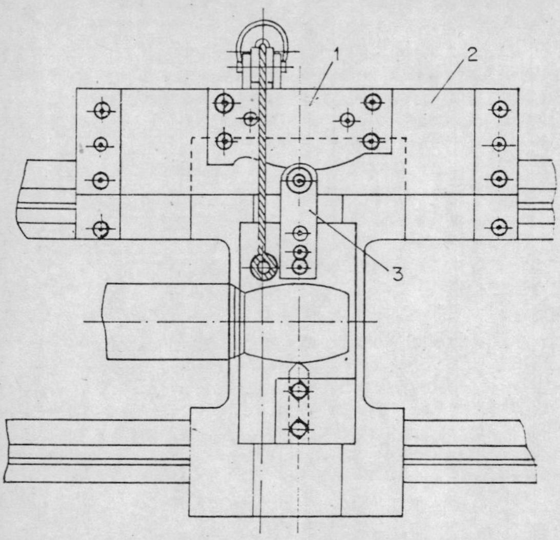

Fig. 9.41. Template profiling – the carriage power-fed along longitudinal axis.

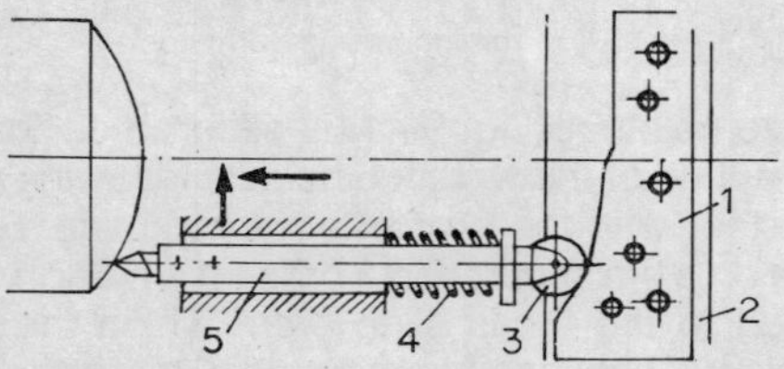

Fig. 9.42. Template profiling with power-fed cross-slide.

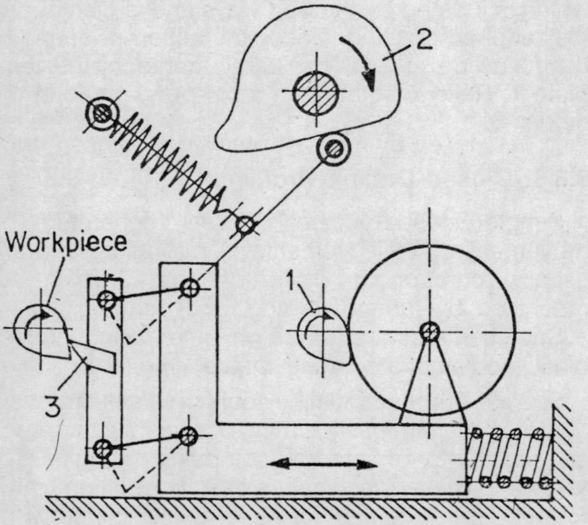

Fig. 9.43. Closed-outline profiling : A cam turning set-up

usually performed by rotation of the workpiece about a vertical axis on vertical milling machines. Profiling in such a manner on an ordinary vertical milling machine is illustrated schematically in Fig. 9.44.

9.8.5. Open-outline Profiling

Open-outline profiles with straight elements may be milled to template on profile milling machines having straight-line table motions. They may also be milled with form-milling cutters or broached with form broaches. In the latter two cases, the profile of the cutting tool will correspond to the profile required on the workpiece. Tracer-controlled milling machines are available in which the required path is imparted to the cutter by means of mechanical, hydromechanical, electromechanical and other tracing devices.

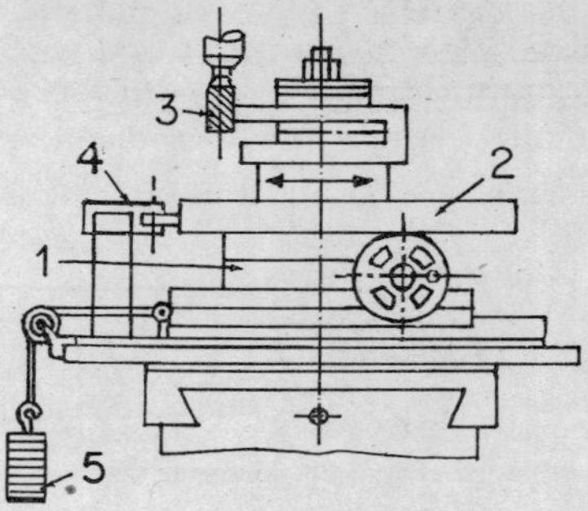

Fig. 9.44. Profile milling of closed-outline surface in a vertical milling machine

9.8.6. Contouring (Three-dimensional Machining)

Three-dimensional shapes are machined, as a rule, not to templates, but to models on tracer-controlled milling machines designed for this purpose. These machines are most frequently equipped with hydromechanical or electromechanical tracing units.

In operation to three-dimensiomal models, tool feed is required, not in two directions (as for profiling to templates) but in three directions (contouring). The following systems are possible for this purpose:

(i) model and workpiece travel in a straight line ;
(ii) model and workpiece rotate;
(iii) model rotates and workpiece travels in a straight line;
(iv) model travels in a straight line and workpiece rotates.

A schematic diagram of contouring to a model or pattern (first method) is shown in Fig. 9.45. Here the model 1, arranged above the workpiece 2, is scanned by a stylus 3. In contouring, the table travels

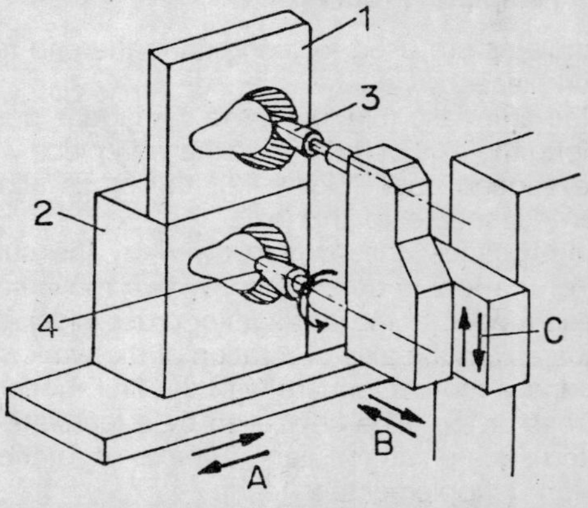

Fig. 9.45: Schematic diagram of 3-D contouring.

in the directions of arrows A with simultaneous movement of the spindle head with the stylus 3 and milling cutter 4 along arrows B. After each pass in the direction of arrows A, the spindle head is led in the direction of one of the arrows C for accomplishing the next pass. After a number of such passes (rows), a contour will have been milled in the workpiece which will duplicate the form of the model or pattern.

9.9. PROCESS-PRODUCT SYNTHESIS OF MACHINING OF GEARS

The two general modes of producing gears are (i) cutting and (ii) generation. Each includes different methods as well, as shown in Fig. 9.46.

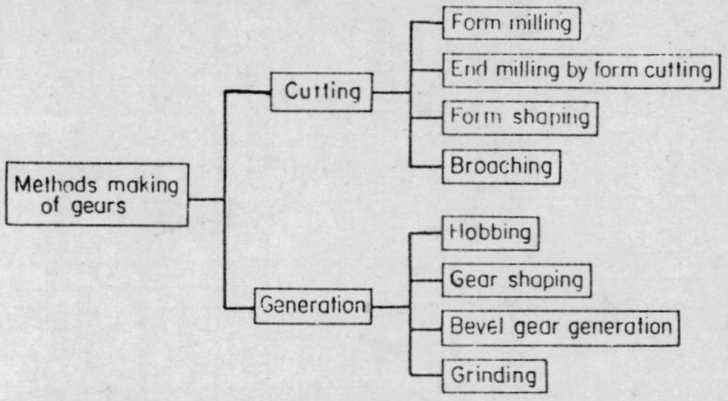

Fig. 9.46. General modes of producing gears.

9.9.1. Gear Milling with Form Cutters

There are three forming methods for cutting gears :

(1) milling the teeth consecutively with a formed cutter of the disc or end-mill types, Fig. 9.47 ;

(2) parallel (simultaneous) shaping of all the teeth on the blank ;

(3) parallel (simultaneous) broaching of all the teeth.

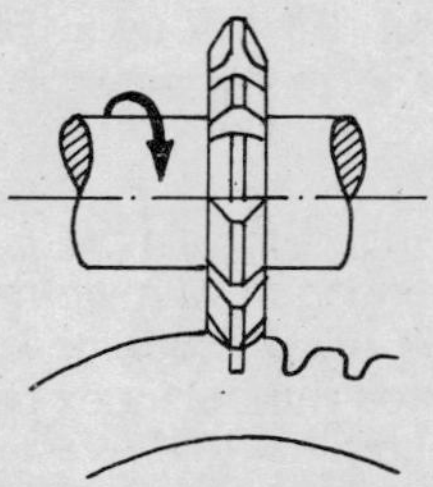

Fig. 9.47: Milling a gear with a form cutter.

Gears are milled with form cutters which have the profile of the tooth space of the corresponding module (Fig. 9.48).

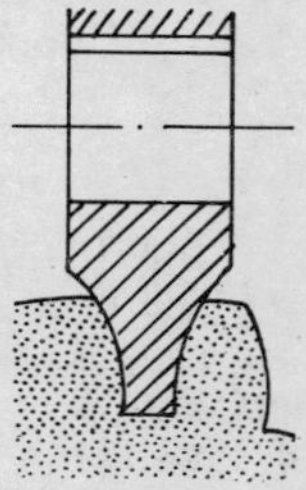

Fig. 9.48: Gear milling cutter in the shape of tooth space.

Theoretically, a different cutter would be required for each number of teeth of gears of the same module. In actual practice, however, a range of numbers of teeth are cut by a single cutter, neglecting the slight errors that may consequently arise. This reduces the number of cutters required to cut all gears of a certain module, up to racks, to 8 or 15 cutters (in certain sets), Table 9.1.

Gears are usually milled by form cutters on plain or universal milling machines with the arrangement illustrated in Fig. 9.49

The gear blank 1 is mounted on an arbour 2 held between the centres of the tailstock 3 and dividing head 4. The tailstock and dividing head are clamped on table 5 of the milling machine.

After milling one tooth space with the working feed *s*, the gear is rapidly returned to its initial position, and is indexed by means of the dividing head to the consecutive tooth space, i.e., through an angle of

$$\alpha = \frac{360°}{z}$$

in which z is the number of teeth on the gear to be cut.

After this, the blank is rapidly advanced to the

TABLE 9.1
SETS OF GEAR CUTTERS FOR INVOLUTE TEETH.

CLASSIFICATION OF THE SET OF 8 CUTTERS

Cutter-No.	1	2	3	4	5	6	7	8
For number of teeth	12 –13	14 –16	17 –20	21 –25	26 –34	35 –54	55 –135	135 up to tooth rack

CLASSIFICATION OF THE SET OF 15 CUTTERS

Cutter-No.	1	1½	2	2½	3	3½	4	4½	5	5½	6	6½	7	7½	8
For number of teeth	12	13	14	15 –16	17 –18	19 –20	21 –22	23 –25	26 –29	30 –34	35 –41	42 –54	55 –80	81 –134	135 up to tooth rack

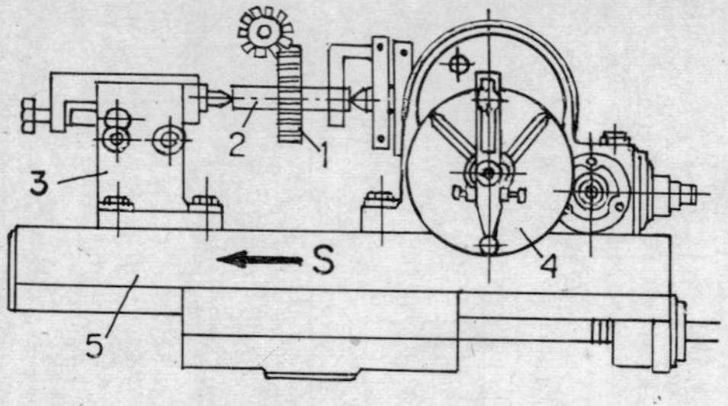

Fig. 9.49: Gear milling setup for the forming method.

cutter, the working feed is engaged, and the next tooth space is cut. All the teeth of the gear are consecutively cut in this manner.

(i) Simple Indexing

Setting up the head for simple indexing: In this case worm z_w is engaged with worm wheel W, and the index plate 1 (Fig. 9.50) or gear z_1 of the differential, on heads without index plates, is fixed in a stationary position.

To rotate the spindle $^1/_z$ part of the circumference, it is necessary to turn the worm through n' revolutions, determined from the formula

$$n' = \frac{N}{z}$$

where N = ratio of the number of teeth on the worm wheel to the number of starts on the worm, called the ratio of the head.

In modern dividing heads, N is usually 40; there are heads, however, in which N = 60 or 80 or even 120.

In the index plate heads index crank 2 is linked to worm z_w through gears with a total gearing ratio of 1 : 1. Therefore, to turn the worm through n' revolutions, it will be necessary to turn the index crank 2 through n′ revolutions where

$$n = n' = \frac{N}{z}$$

In most cases, $n \neq E$, where E is the symbol for an arbitrary whole number. Therefore, it will be necessary to select a whole number B, such that

$$\frac{N}{z} B = E.$$

Thus

$$n = z\frac{N}{Z}$$

If $E > B$,

$$n = C + \frac{A}{B}$$

where C = number of whole turns of index crank 2; B = number of holes in the selected hole circle of the index plate; and A = number of spaces in the selected hole circle of the index plate which the index pin must pass over in addition to C whole turns of the crank.

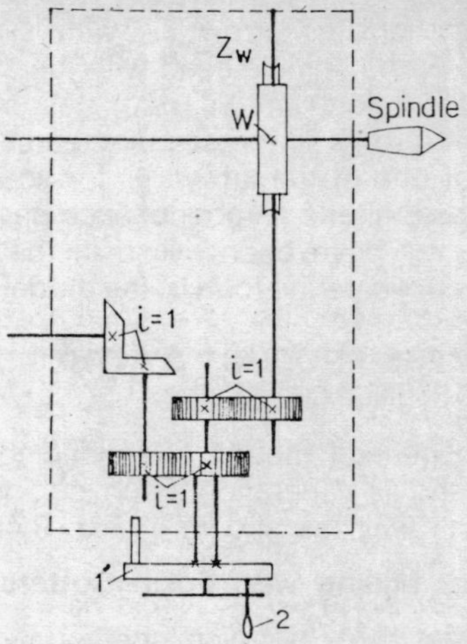

Fig. 9.50. Setting up a head for simple indexing

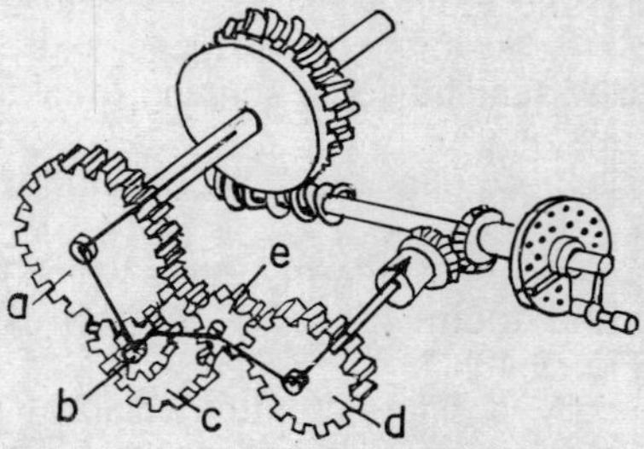

Fig. 9.51: Differential indexing with a~d change gears.

Indexing heads, usually have index plates with blind holes on each side. On one side they have 16, 17, 19, 21, 23, 27, 30 and 31 hole circles, and on the other side they have 33, 37, 39, 41, 43, 47, 49 and 54 hole circles.

(ii) Differential Indexing

Divisions which cannot be performed by direct or compound indexing, will be made possible with the help of change gears by differential indexing, Fig. 9.51.

During differential indexing the index plate is loose. It receives an advancing or retarding movement from the index spindle via the change gears. The number of teeth of the change gears will be ascertained by calculations; usually they are as shown in Table 9.2.

TABLE 9.2
Number of Teeth of the Change Gears.

24	24	28	32	36	40	44
48	56	64	72	86	100	

In differential indexing the worm is engaged to the worm wheel and index plate 1 or gear z_1 (on heads without index plates) must be linked to the spindle of the head through a gear train consisting of the differential quadrant with change gears a, b, c and d in the heads with index plates (Fig. 9.52).

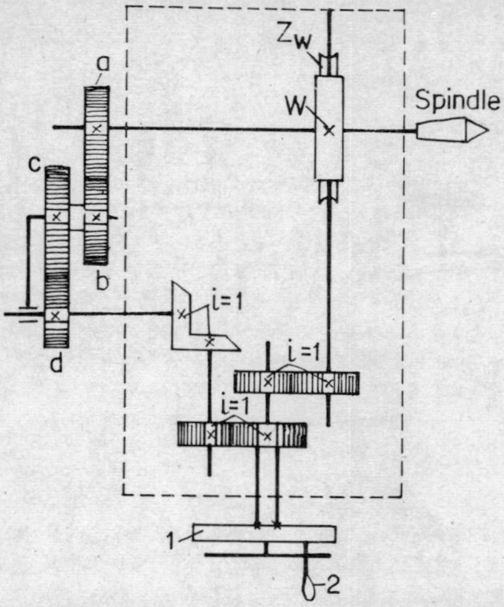

Fig. 9.52. Differtial indexing–Index plate linked to the spindle of the head

(iii) Helical Milling

The helical motion of the work, required to mill the groove, is composed of two simple motions–rotation of the blank about its axis and travel along this axis. The ratio of these two motions should be such that in one revolution of the blank it travels axially a distance equal to the lead of the helical groove to be milled. This ratio is obtained by setting up the gear train which links the lead screw for longitudinal table travel to the spindle of the head. In heads with index plates this gear train includes the following elements (Fig. 9.53): lead screw t_{ls}, screw-cutting quadrant

$y = \frac{a_1}{b_1} \times \frac{c_1}{b_1}$, *constant bevel and spur gearing with*

$i = 1$, index plate 1, index crank 2, worm gearing

$\frac{Z_W}{W} = \frac{1}{N}$, and the spindle.

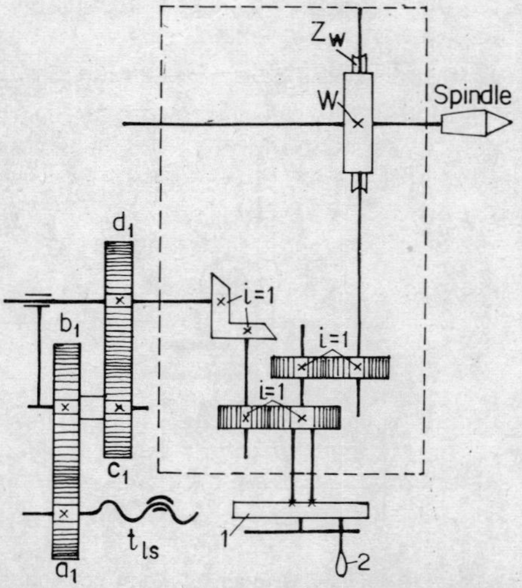

Fig. 9.53. Helical milling of gears–Indexing arrangement

Fig. 9.54. End milling of gears.

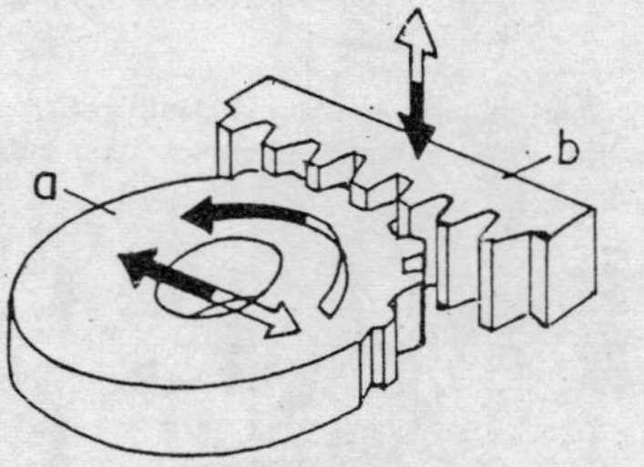

Fig. 9.55 (i) Gear shaping with the rack-shaped cutter. a) Gear blank, b) rack-shaped cutter.

9.9.2. End-milling of Gears

Consecutive milling with end-mill type of gear cutters (Fig. 9.54) is used for cutting large-module gears (from 20 mm and larger) in the heavy engineering industry, where the use of disc type cutters would require very powerful machine tools.

End-mills (formed type) are also applied for cutting large herring-bone gears where it is impossible to use formed gear cutters of the disc type.

9.9.3. Gear Shaping

A gear shaper cutter is shapped like a gear of the same module as the gear to be cut. The shaper cutter and the gear blank rotate together as if they were two gears in mesh. The teeth of the shaper cutter are relieved to obtain the geometrical form required by a metal-cutting tool.

This method is called gear shaping and machines used for this purpose are called gear shapers.

The principal working movements of these machines (Fig 9.55, i-iii) are the following :

(i) rotation of the shaper cutter about its axis ;

(ii) synchronised rotation of the gear blank about its axis ;

(iii) reciprocation of the shaper cutter along its axis (shaping) ;

(iv) infeed of the shaper cutter toward the centre of the gear blank (infeed to the depth of the tooth) ;

(v) withdrawal of the blank from the cutter during the return stroke of the latter (to prevent the flanks of the cutter teeth from rubbing on the machined surfaces of the teeth).

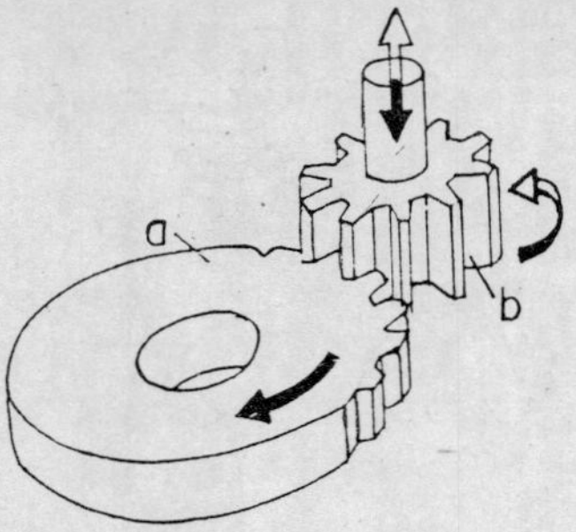

Fig. 9.55 (ii) Gear shaping of external teeth with the pinion type cutter. a) Gear blank, b) pinion type cutter.

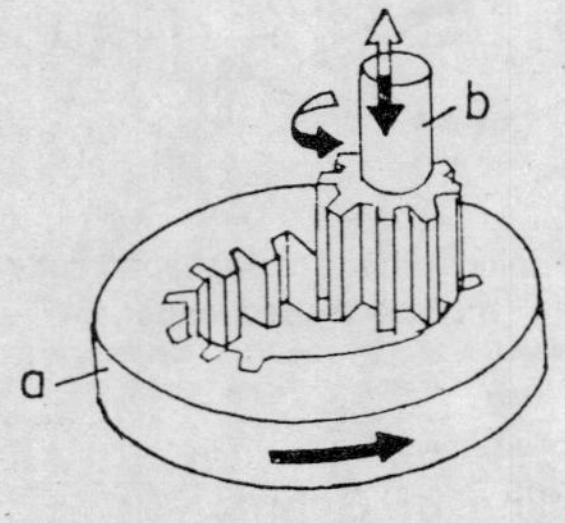

Fig. 9.55 (iii) Gear shaping of internal teeth with the pinion type cutter. a) Gear blank, b) pinion type cutter.

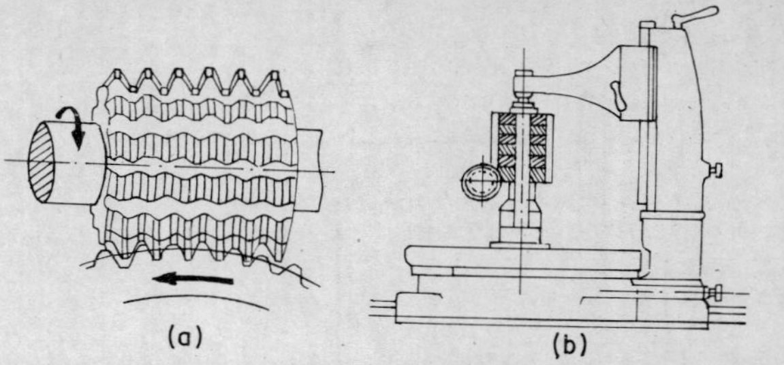

Fig. 9.56: (a) Gear hobbing; (b) hobbing several gears on an arbour

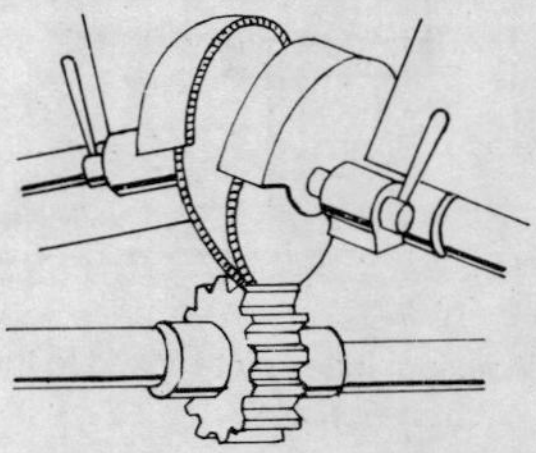

Fig. 9.57: Gear grindling. with dish grinding wheels.

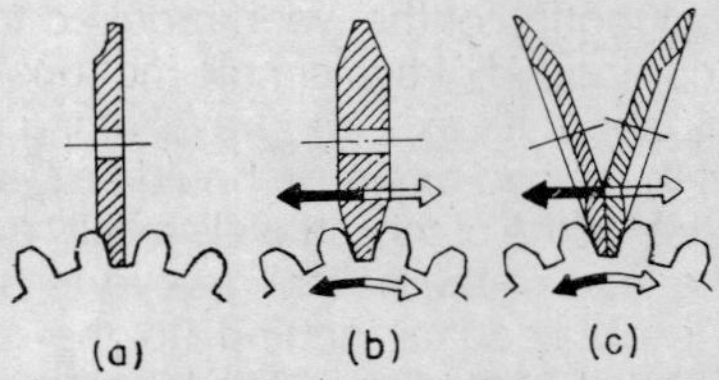

Fig. 9.58: Methods for grinding spur gears.
(a) Profile grinding, (b) grinding with one generating wheel, (c) grinding with dish generating wheels.

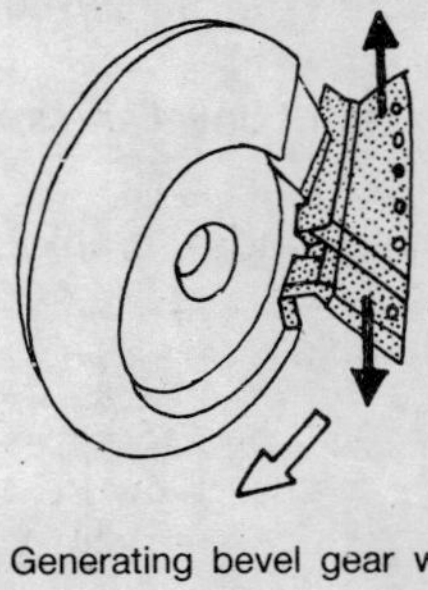

Fig. 9.59: Generating bevel gear with two tools.

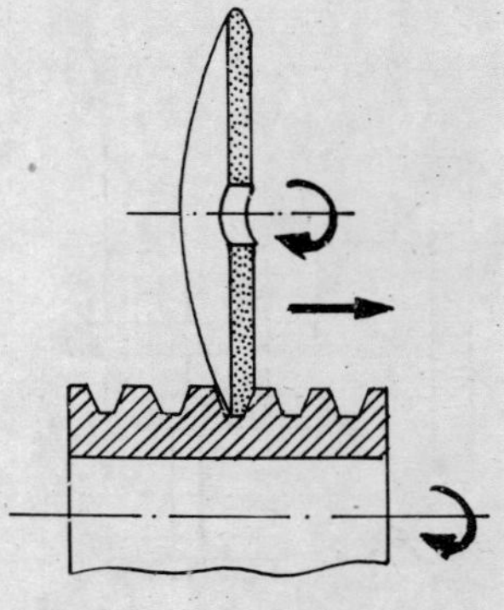

Fig. 9.60: Schematic diagram of worm grinding.

9.9.4. Gear Hobbing

In cutting gears with a hob (Fig. 9.56, a-b) the following working movements are required for the machine: (a) hob rotation at the cutting speed v_h, (b) interacting rotation of the blank (generation), and (c) hob feed along the axis of the blank.

In setting up the hobbing machine, the hob axis is swivelled in reference to the blank axis so that the pitch cylinders of the blank and hob contact at a point where the directions of the gear teeth and hob thread coincide.

The hob, in rotation, cuts metal continuously with its teeth and all the teeth of the gear are consecutively cut, without interruptions in cutting for advancing and withdrawing the hob. Due to this feature, the hobbing of spur and helical gears is one of the most productive gear cutting methods.

A single hob may be used for hobbing any gear of the given module if the following condition has been observed :

$$\frac{n_h}{n_b} = \frac{z}{i}$$

in which z is the number of teeth of the gear being cut ; i is the number of threads (starts) of the hob; n_h is the hob speed in rpm; n_b is the gear blank speed in rpm.

Single - and multiple-thread hobs (two- and three-thread) are used for hobbing spur and helical gears.

A higher rate of production may be obtained using multiple-thread hobs since the blank will rotate at a higher speed in comparison to the hob than for hobbing with a single-thread hob.

9.9.5. Gear Grinding

The accuracy of the tooth shape and the surface quality will be increased by grinding. Hardened gears should be ground to eliminate the distortion, also unhardened gears, if ground, run smoothly (Fig. 9.57 & 9.58).

There are two methods, profile grinding and grinding with generating wheel (generation gear grinding).

9.9.6. Bevel Gear Manufacture

The manufacture of the teeth for bevel gears is difficult since thickness of tooth and flank curvature change towards the cone centre. Bevel gears can be manufactured with the indexing and the generating method.

For the indexing method, four milling cutters are required. Since the width of tooth space is unequal, each tooth flank must be machined individually.

Highly accurate bevel gears can be manufactured by cutting on the bevel gear generator (Fig. 9.59). The machine operates with two tools which cut off chips alternately. When a tooth is completed the gear blank is rotated by one pitch. The generating motion of gear blank and tool is effected by gears. Also spiral bevel gears can be manufactured by the generating method.

9.9.7. Worm Grinding (Fig. 9.60)

External square and trapezoidal threads are usually cut in two operations, *roughing* and *finishing*, in engine lathes. First the groove is cut with a tool slightly narrower than the groove at the root of the thread and then the thread form is obtained by finish tools. Precise threadsare usually finished by tools similar to lathe facing tools which cut the sides of the thread. It is very important to properly set up the tool in reference to the thread. Roughing tools are set with their cutting edge normal to the helical groove. This facilitates chip flow from the tool. A thread is cut in several (5 to 10) passes.

CHAPTER 10

Grinding of Finishing Methods

10.1. GENERAL PRINCIPLES OF GRINDING

Grinding is very similar to other metal cutting operations and can be described as a multitooth operation in which a number of abrasive grains held by a bonding material perform the cutting operation. However, the cutting operation of grits is characterized by several important features :

(a) a varied and indeterminate geometry of the grits,

(b) a random orientation of the grits giving unknown rake angles,

(c) the depth of cut is very small being limited by the capability and geometry of individual abrasive grain,

(d) the very high speed in grinding involving very high rate of strain and a severe temperature gradient.

The process of grinding is dependant on the inherrent properties of the grinding wheel, such as type of abrasive, the size and distribution of grits, amount and type of bond material and volume of pores relative to that of grits and bonds.

10.2. GRINDING WHEEL

The grinding wheel is made up of particles of a hard substance called the *abrasive* embedded in a matrix called the *bond*. These abrasives known as 'grits' provide the sharp edges which do the cutting. Typical abrasives are :

(a) aluminium oxide (alundum) denoted by the letter–*A*,

(b) silicon carbide (carborundum) denoted by the letter–*S*,

(c) diamond–denoted by the letter–*D*.

While both silicon carbide and alumina are hard enough to grind most of the materials that are encountered with, silicon carbide is more suitable for grinding materials of low tensile strength, while alumina is the most efficient when applied to hard, tough materials.

A term 'grit size' is often used to describe the fineness of the mesh used in grading the abrasive grits, the larger the size, the finer the grit. Often grit size S is related to the grit diameter d_g by an approximate formula :

$$S. d_g = 0.6 \qquad \text{Eqn. 10.1}$$

The range of grit sizes is as follows :

Coarse	...	6-24
Medium	...	30-60
Fine	...	70-600

The grits are held by a suitable bonding material, the amount of which decides the grade or hardness of the wheel indicating the relative strength of the holding matrix. Increasing the amount of bonding material thus increases the ratio of the bond material to grit material and makes a wheel appear harder.

The wheels grades are

Very soft	...	*C* to *G*
Soft	...	*H* to *K*
Medium	...	*L* to *O*
Hard	...	*P* to *S*
Very hard	...	*T* to *Z*

The term structure, as shown in Fig. 10.1, refers to the grit spacing and is indicated by a number varying from 0 to 12 with lower numbers signifying closer packed grits.

The following bond materials are generally used:

Vitrified bond	...	*V*
Silicate	...	*S*
Resinoid	...	*B*
Shellac	...	*E*
Oxychloride	...	*O*
Rubber	...	*R*

Using the above notations, a typical grinding wheel might be classified as follows

A-46-*K*-8-*V*

This represents a wheel having aluminium oxide abrasive *A*, grit size of 46 mesh medium strength bond, i.e., hardness or wheel grade *K*; fairly high grit spacing 8, and vitrified bond *V*. Fig. 10.2 shows the efficiency of grinding of different metals when ground with aluminium oxide and silicon carbide.

Opitz and Peklenik have suggested a simple equation for designating grain-bond-void relation,

$$V_k\,(\%) + V_b\,(\%) + V_p\,(\%) = 100\% \qquad \text{Eqn. 10.2}$$

where V_k = % volume of grain, V_b = % volume of bond, V_p = % volume of porosity or voids.

For variation in hardness 'grade', the above equation is modified into

$$V_k + (V_b + \psi_h) + (V_p - \psi_h) = 100\% \qquad \text{Eqn. 10.3}$$

which means more bonding material at the expense of voids for same number of grits. This means increased hardness grade. ψ_h is hardness step.

Similarly, the equation 10.1 is modified by considering the step in structure ψ_s when

$$(V_k - \psi_s) + (V_b + \psi_s) + V_p = 100\% \qquad \text{Eqn.10.4}$$

Fig. 10.1. Structure of grinding wheel showing grits, bonds and voids

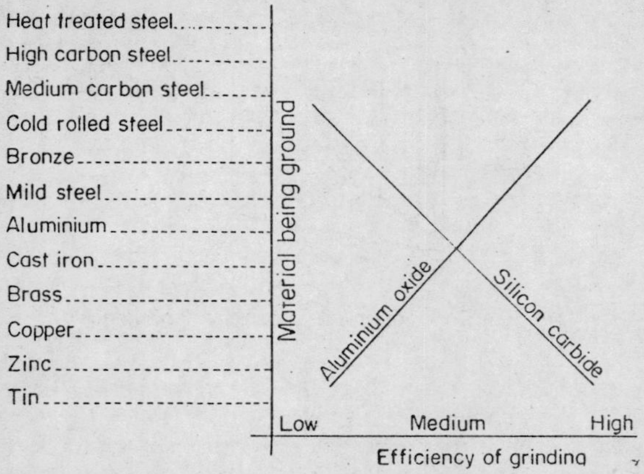

Fig. 10.2. Efficiency of grinding of various metals by AL_2O_3 and SIC Wheels

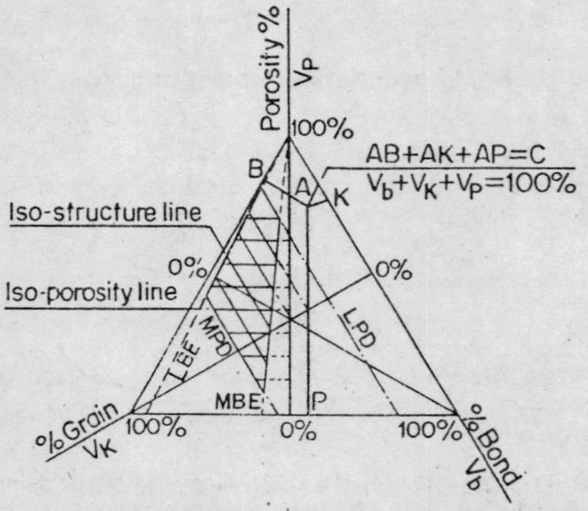

Fig. 10.3 Ternary diagram among porosity (V_P) bond (V_b) and grain (v_k)

This means reduction in grits for same porosity thus changing the structure of the wheel.

A convenient representation of the above equation is the so-called 'ternary diagram' proposed by Peters.

The three quantities, V_k, V_b and V_p can be represented on the axes of an equilateral triangle. Each apex represents 100% of the considered quantity and zero of the two others. Each point similar to *A* within the triangle as shown in Fig. 10.3 represents a wheel composition because the sum of the perpendicular distance of any point within an equilateral triangle equals the unit length of the height, i.e.,

$$AB + AK + AP = \text{constant} = 100\%$$

In Fig. 10.3, the iso-structure lines containing same per cent of grain volume are drawn perpendicular to V_k axis. However, the useful domain is limited by two boundaries: maximum packing density (MPD) line and lowest packing density (LPD) line. These boundaries vary with grain size.

On the other hand, the lateral boundaries of the usable domain are set according to the maximum or minimum volume of bond. Peklenik introduced a very useful notion of 'bond equivalent' which is defined as the average volume of bond per grain and per unit grain surface. For a given grain size, this is represented by lines of same ratio of bond volume to grain volume. On the basis of these values provided by Peklenik as shown in Fig. 10.4, Peters determined the lateral boundaries as lines passing through the point V_p=100%. These lines limit the usable domain as shown in Fig. 10.3.

Based on this 'ternary diagram', manufacturers draw lines of 'iso-hardness' to represent the composition of wheels considered as equally hard although they have a different structure.

Ljumbomudrov uses *iso-porosity* lines as iso-hardness lines. A number of manufacturers, particularly American, use lines with some inclinations. The so-called 'hardness letter' does not represent the real hardness. The co-ordinates of the ternary diagram depends on the manufacturing technique : pressing, baking, temperature, oven atmosphere, etc. Thus, the real operating hardness is often required to be determined experimentally.

10.3. SELECTION OF GRINDING WHEELS

Rules for the selection of the grinding wheels (machine grinding) (Extract from DIN 69 102)

External Grinding Material	Diameter of Grinding Wheel (mm)		
	Up to 350	More than 350...450	More than 450...600
Hardened steel	EK 60 L	EK 50 L	EK 46 L
Unhardened steel	NK 60 M	NK 50 M	NK 46 M
Cast iron	SC, EK 60 I	SC, EK 50 Jot	SC, EK 46 Jot
Internal Grinding	Diameter of Grinding Wheel (mm)		
Material	Up to 16 mm	More than 16...36	More than 36...80
Hardened steel	EK 80 L	EK 60 K	EK 46 Jot
Unhardened steel	NK 80 M	NK 60 L	NK 46 Jot
Cast iron	SC 80 K	SC 60 Jot	Sc 46 I
Surface Grinding	Diameter of Grinding Wheel (mm)		
Material	Straight wheel Up to 200	Cup wheel up to 200	Segments
Hardened steel	EK 46 Jot	EK 36 Jot	EK 30 Jot
Unhardened steel	EK, NK 46 K	EK, NK 46 K	EK, NK 24 K
Cast iron	EK, SC 46 I	EK, SC 46 I	EK, SC 30 Jot

10.4. LAPPING

Lapping is regarded as the oldest method of obtaining a fine finish and is extensively used in engineering for the finishing of gage blocks and flats.

Basically lapping is an abrasive process in which loose abrasives function as cutting points finding momentary support of the laps. Fig. 10.5 depicts the scheme of 'lapping process'. Lapping is characterized by the following features:

(a) use of *loose* abrasives between the lap and the work surface,

(b) usually the lap and the work piece are not positively driven but are guided in contact with each other,

(c) relative motion between the lap and the work surface should be constantly changing as it is desirable that the grains should not follow their own previous path in the machined surface. The most effective path is of cycloidal nature.

10.4.1. Mechanism of Lapping

The material removal rate (MRR), Q, by lapping has been proposed by Matsunaga. Assuming,

(a) scratching with point edges of cone-shaped abrasives,

(b) supporting of the load by abrasives,

(c) independence of hardness H from θ and load, and

(d) satisfying Ehrenberg's experimental equation for scratching, a model for Q during lapping may be constructed.

From the results of Ehrenberg,

$$d^2 = \frac{8\,p}{100\,\pi H}$$

where

d = width of a scratched groove, mm,

p = load on an abrasive grain, kg,

H = indentation hardness, kg/mm^2.

The sum of cross-section is

$$A = \sum \frac{d^2}{4} \cot(\theta/2) = \sum \frac{2p}{100\,\pi H} \cot(\theta/2)$$

where

A = cross-sectional area of scratched grooves,

θ = cross-sectional angle of the points of grains.

Rearranging

$$A = \frac{\cot(\theta/2)}{50\,\pi H} \sum p = \frac{P \cot(\theta/2)}{50\,\pi H}$$

where

P = pressure on the work piece = $\sum p$.

Hence,

$$Q = AGL = \frac{GPL \cot(\theta/2)}{50\,\pi H}$$

where

G = specific density of work piece,

L = lapped distance = vT.

However, experimental results show that Q is proportional to $p^{1.08}$; scratching by rounded edges in wet lapping also need be considered. A modified equation has been proposed and is given by

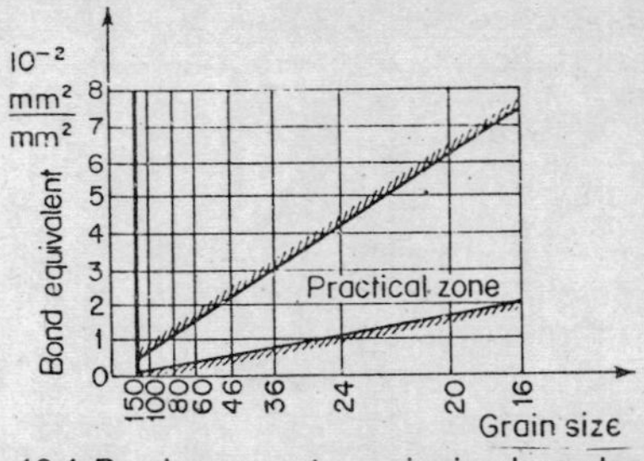

Fig. 10.4. Bond equivalent ~ grain size dependency

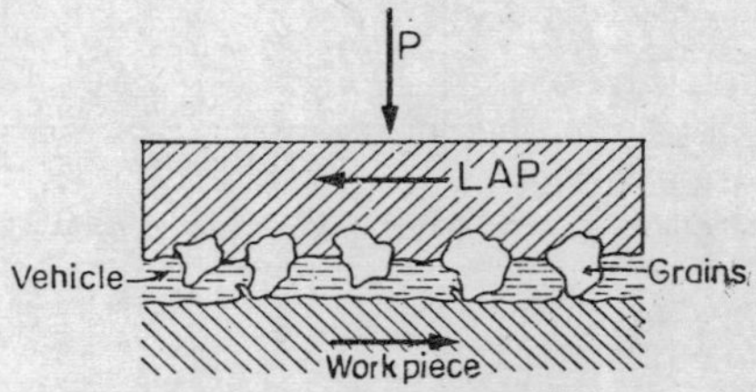

Fig. 10.5. Scheme of lapping process

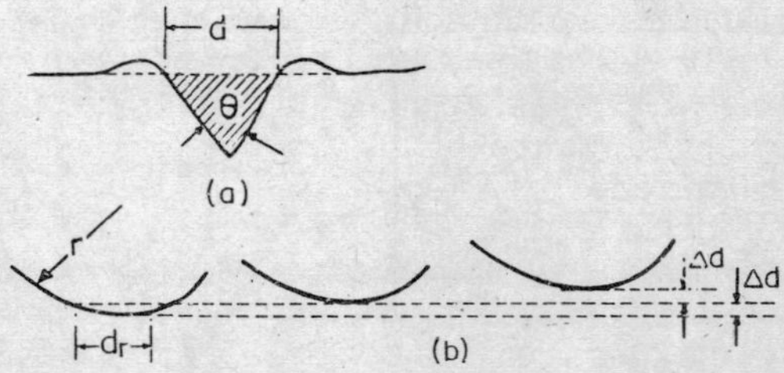

Fig. 10.6. (a) Cross-section of a scratched groove (b) Defining the parameter Δd

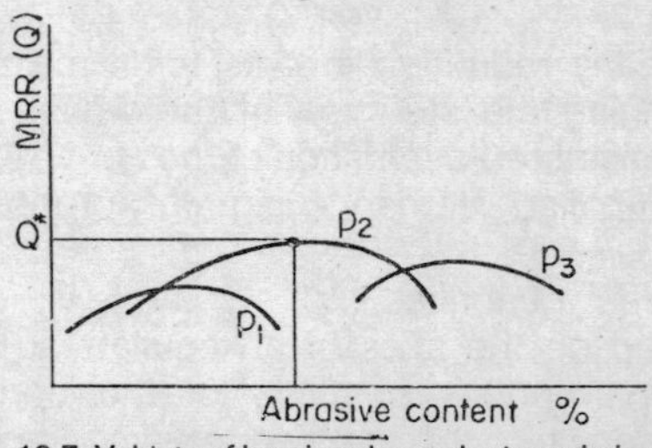

Fig. 10.7. Vehicle of lapping dependent maximised MRR

$$Q = \frac{1.357}{10^3} (\Delta d)^{1/4} \frac{GL}{r^{3/4}} \left(\frac{P}{H}\right)^{5/4}$$

In this equation, the stock removal is proportional to $P^{5/4}$ and the effect of *abrasive to vehicle* ratio and abrasive size is accounted for through the parameter (Δd) (Fig. 10.6).

10.4.2. Laps and Abrasives

The selection of lap material is important for a lapping process. The properties that the lap should have are :

(a) maintenance of ideal shape,

(b) embedding of the abrasive, and

(c) protection of both work piece and lap surfaces.

Thus lap hardness and its surface structure are of special importance. Excessive hardness of a lap causes rapid chipping of grains resulting in faster wear of abrasive mixture and a reduction in Q. Too soft a lap drives the abrasive deeply causing worn corners of grains and results in diminished material removal rate and poorer surface finish.

If the lap material is chosen correctly, it supports the grains temporarily and allow changes of positions of grains enabling each grain to act at several fresh corners. This favours longer abrasive life, increase in metal removal rate and improved surface finish.

Cast iron is mostly used for laps. Ferrite which has considerable hardness and toughness allows maintenance of forms. Graphite in the cast iron serves for embedding of abrasive and for lubricating effect. Besides cast iron, copper alloys, lead, plastics, mirror glass and wood are also used as lap materials. The rule is that the lap material should be of higher hardness if a soft abrasive is used.

Abrasives used are commonly aluminium oxide and silicon carbide. In addition boron carbides are used for lapping sintered carbides, chromium oxide for copper alloys and diamond dust for hard steels and carbides. The grain sizes vary from 1-2 μm for chromium oxide, 0.5-5 μm for diamond dust, 5-60 μm for boron carbide and 5-100 μm for silicon carbide and aluminium oxides.

The *vehicle* of lapping is the lubricating liquids in which the abrasive elements are suspended. Machine oil, rape oil, grease, etc., are used as lapping vehicle. Concentration of mixture of abrasives in lubricants depends also on unit pressure, P. Fig. 10.7 shows that at an optimum pressure an optimum degree of concentration for maximized metal removal rate is obtained.

10.4.3 Technological Parameters Affecting Lapping Process

The important technological parameter that affects material removal rate (MRR) and surface roughness (R) are:

(i) unit pressure, P,

(ii) grain size of abrasives,

(iii) concentration of abrasives in the vehicle of lapping,

(iv) lapping speed.

The effect of unit pressure, P, on MRR and R is shown in Fig. 10.8. An optimum range exists within which a value of P should be chosen. For Al_2O_3, a value of 2-5 kg/cm^2 and for SiC, a value ranging from 0.5 to 2.5 kg/cm^2 are usually chosen for rough lapping. For finish lapping the pressure is kept limited in the range of 0.3-1.2 kg/cm^2.

Grain size affects MRR and surface roughness in the manner shown in Fig. 10.9. As R increases monotonically, the value of R_o at Q_{max} can only be chosen if permissible. Otherwise, a value of $R \leq R^*$ is to be chosen in order to get possible maximum of Q.

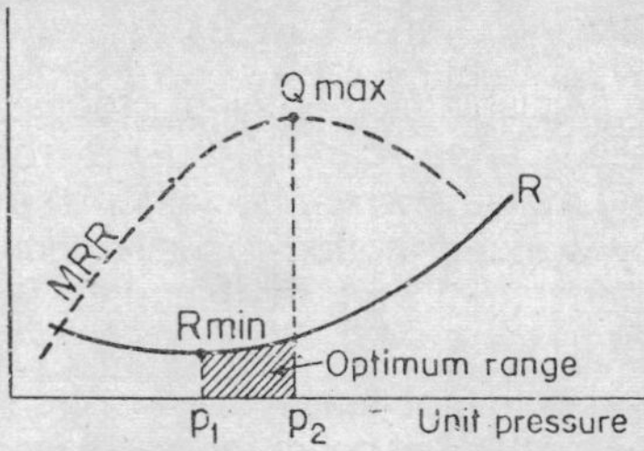

Fig. 10.8. Effect of unit pressure on R and MRR in lapping

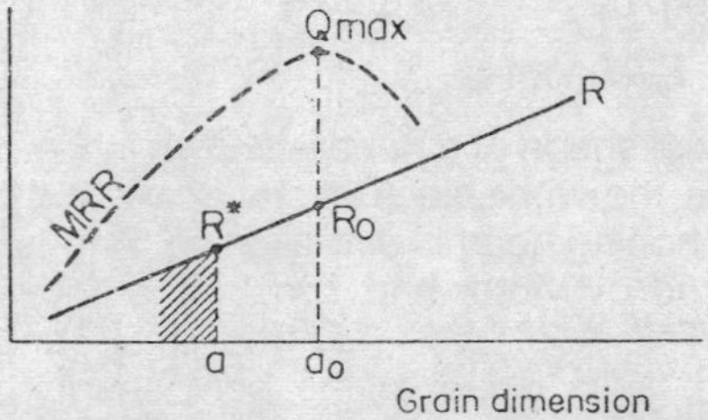

Fig. 10.9. Effect of grain size on R and MRR in lapping

Fig. 10.10 shows the dependence of rate of material removal, Q, and the surface roughness, R as a function of lapping time. Usually, lapping time is decided with respect to permissible surface roughness and linear loss of work piece dimension (L) as shown in Fig. 10.11.

A permissible, L_p, and allowable R decide what τ_p or τ_R should be chosen as lapping time. Fig. 10.11(a) shows $\tau_R < \tau_p$, while in Fig. 10.11(b), $\tau_R > \tau_p$. As τ_R is the desired criterion, the desired characteristic should be as shown in Fig. 10.11(a). Other technological parameters should be so chosen suitably to obtain this condition.

Lapping speed, the speed of the lap relative to workpiece surface, is chosen again considering linear loss of work material (L) and surface roughness (R). Fig. 10.12 shows the dependence of L and R on lapping speed V. The usual lapping speeds are shown in Table 10.1.

TABLE 10.1

LAPPING SPEEDS

Lapping type	Roughness number	Machining allowance	Lapping Speed, V m/minute
Medium accuracy	≤ 10	< 0.5 mm	200-400
Accurate	$10 < R \leq 12$	< 0.25 mm	100-250
Very accurate	$12 < R \leq 14$	< 0.04 mm	10-100

Lap wear increase with lapping speeds. Better cooling at higher lapping speeds should be provided for in order to off-set undesirable rise of temperature.

10.5. HONING

Honing is a finishing (smoothing) process in which a tool called *hone* carries out a rotary and reciprocating motions while the work piece does not perform any working motion. Usually honing stroke is shorter than that of work piece surface length. It is desired that

(a) hone should not leave the work surface,
(b) stroke length must cover the entire work length.

The stroke length is determined from Fig. 10.13 and given by

$$L = l_w - l_h + l_1 + l_2$$

A typical shape of a hone is shown in Fig. 10.14. In this case the work piece is stationary and clamped and the honing head is self-aligning. Another design introduced by Victor and Tönshoff is shown in Fig. 10.15. This tool also centres itself and lesser resulting faults occur in the bore shape.

Usually, for honing steel Al_2O_3, and SiC abrasives are used in honing stick while for honing cast iron and non-ferrous materials green SiC are used. The size of abrasive grains varies from 400 μm on the coarser side to 14 μm on the finer side. Usually phenolic resins are favoured as bond material in

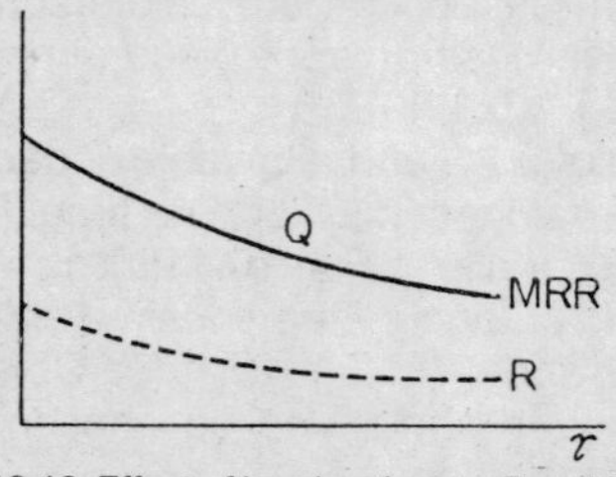

Fig. 10.10. Effect of lapping time on R and MRR

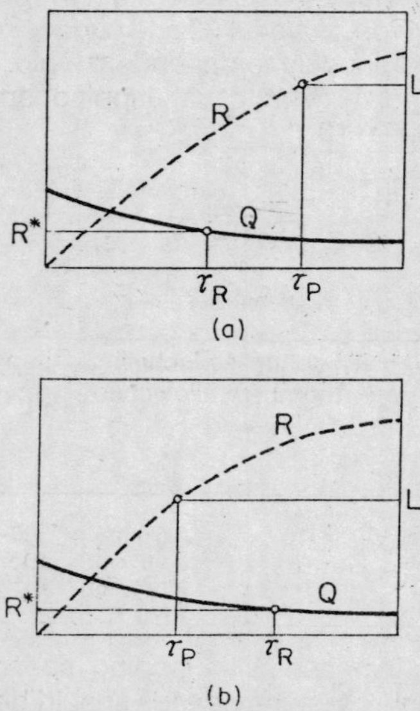

Fig. 10.11. Decision criterion (a) when $\tau_R < \tau_P$ and (b) when $\tau_R < \tau_P$

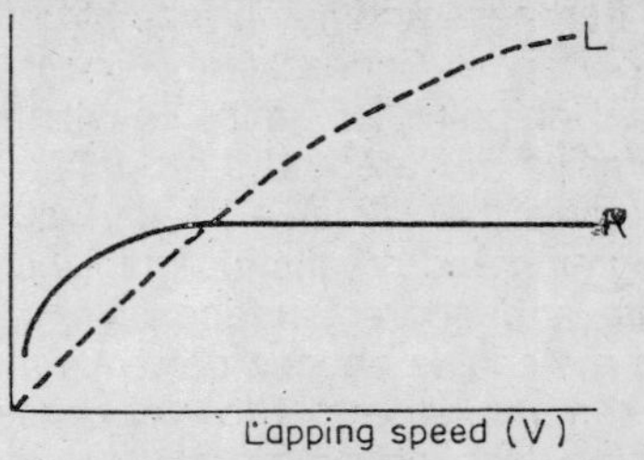

Fig. 10.12. Effect of lapping velocity on L and R

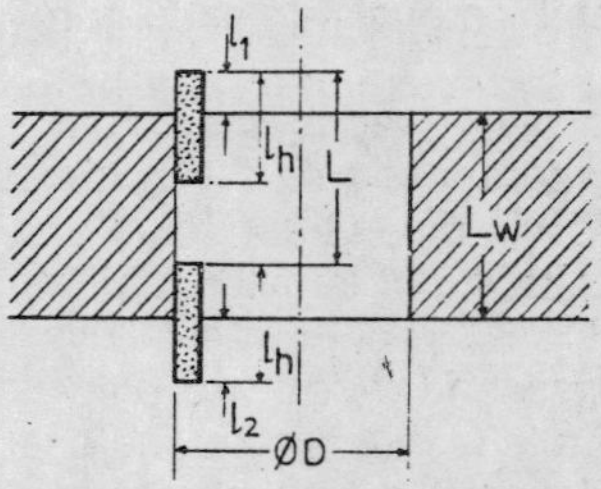

Fig. 10.13. Deciding stroke length in honing

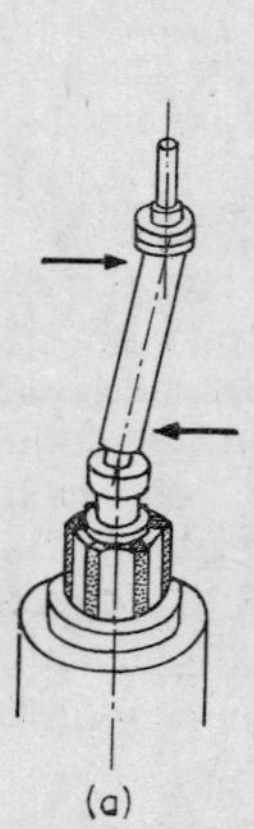

Fig. 10.14. A typical hone

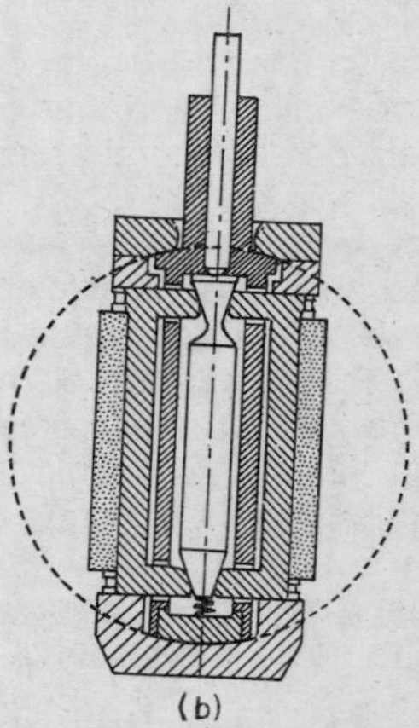

Fig. 10.15. The Victor and Tönshoff hone

hones. Hone hardness ranges from *L* to *P*. For large machining allowances, rigid work pieces and high unit pressure, a higher hardness is chosen. For high accuracy and less rigid work pieces, lower hardness for hones are selected.

Advantages of honing of holes are :

(a) The smooth and quiet course of the process in conjunction with high rigidity of the hone excludes the possibility of vibration and enables extremely high accuracy and surface finish to be achieved,

(b) several hones may be actuated simultaneously on multiple spindle machines,

(c) holes of any diameter or length can be honed,

(d) relatively high productivity and low cost in comparison with other hole finishing methods.

There are certain *disadvantages* in honing which limit its application. These are :

(a) impossibility of improving lack of straightness in holes,

(b) impossibility of honing tough non-ferrous

metals due to glazing or clogging of the pores of the abrasive sticks.

10.5.1. Kinematics of Honing Process

Working motions in honing are composed of (a) a uniform rotary motion with r.p.m. n and peripheral speed $v_p = \pi D n$, (b) reciprocation motion at a frequency f with an average speed $v_s = 2Lf$. The resultant cutting speed is

$$v = \sqrt{v_p^2 + v_s^2}$$

If ϕ is the angle between the tangent to the path of grain motion at the given point and the plane perpendicular to the axis of rotary motion,

$$v_p = v \cos\phi \text{ and } v_s = v \sin\phi$$

In order to avoid retracing of the grain path in subsequent strokes, it is necessary to ensure that the stroke length is not a multiple of the helical pitch of the trajectory. This is determined from the geometry of Fig. 10.16,

$$L = (m + e)\, H = (m + e)\, \pi D \tan\phi$$

where m = an integer and $|e| < 1$.

The track length on the circumferences, S, is given by

$$S = \pi e D \leq \frac{\pi D}{Z} - b$$

when Z = number of hones, and b = hone width. Further,

$$\tan\phi = \frac{L}{(m+e)\pi D} = \frac{v_s}{v_p} = \frac{2Lf}{\pi D n}$$

from which

$$(n/f) = 2(m+e)$$

If (l/e) is made a prime number, grain paths will not be retraced. With L and D *known and* ϕ *assumed* $m + e = \frac{L}{\pi D \tan\phi}$ can be calculated.

10.5.2. Technological Selection of Parameters for Honing

The important parameters that affect honing process include (a) unit pressure, p, (b) peripheral honing speed, v_p, and (c) honing time, τ. The effect of these variables on material removal rate Q and surface roughness height R are shown in Figs. 10.17, 10.18 and 10.19.

From Fig. 10.17, it is seen that the increase of unit pressure results in increased MRR (Q), but after a point p' worsens surface finish. The permissible p lies to the right of p' as further decrease of p does not improve any more the surface finish.

Increased peripheral speed causes increased material removal rate, Q, and decrease in surface roughness, R as shown in Fig. 10.18.

Material removal rate, Q, decreases with honing time, τ. But surface roughness decreases initially and beyond a point τ' increases as depicted in Fig. 10.19. These characteristics are true for all grain numbers in hones.

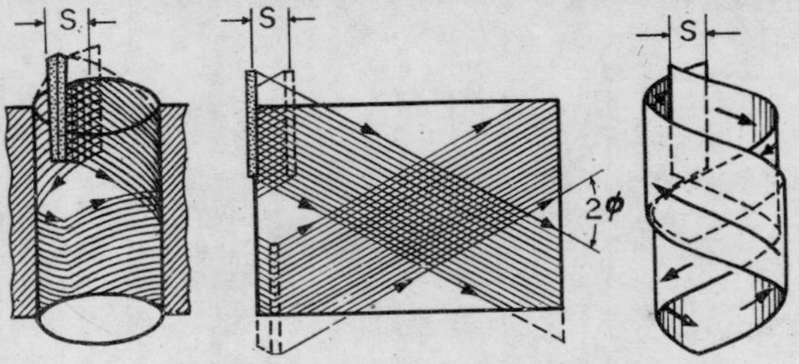

Fig. 10.16. The kinematic path of a hone

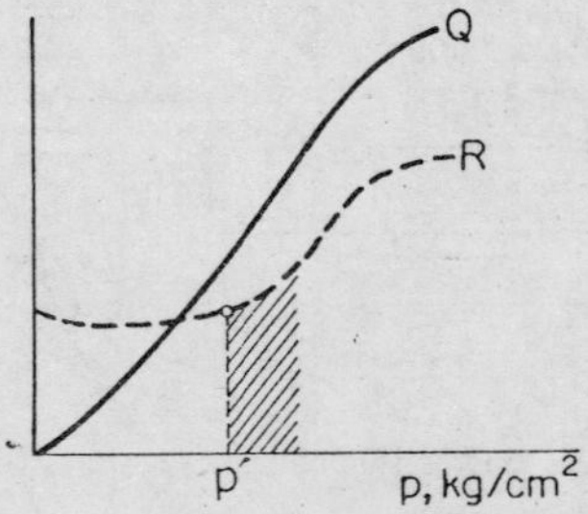

Fig. 10.17 Effect of unit pressure on R and MRR in honing

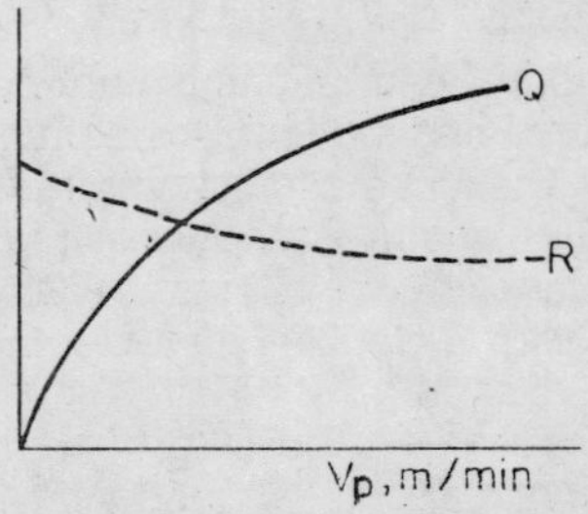

Fig. 10.18. Effect of peripheral honing speed on R & MRR

The angle ϕ depends on work piece materials and the type of cut as shown in Table 10.2.

TABLE 10.2

Work Material	Type of cut	$\frac{V_p}{V_s}$ ratio	$\phi°$
Grey cast iron	Roughing	2-5	26-11
	Finishing	4-8	14-7
Non-hardened steel		2.5-4	22-14
Hardened steel		5-7	11-8
Bronze		10-16	6-35

10.6. SUPERFINISHING

Superfinishing is a special technique for producing high degree of surface finish by using abrasive sticks similar to hones but with a different kinematics for work-tool interaction. The working motions include

(a) oscillatory motion of the tool with short strokes and at a high frequency,

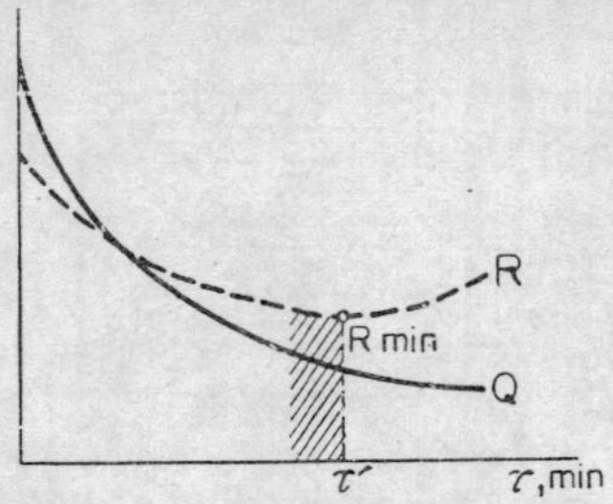

Fig. 10.19. Effect of honing time on R and MRR.

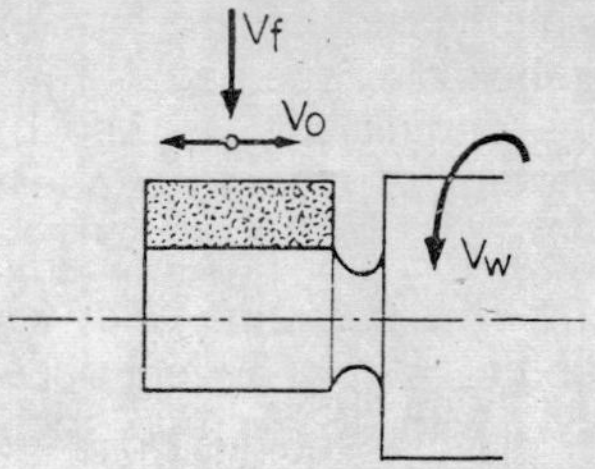

Fig. 10.21. Plunge feed in superfinishing

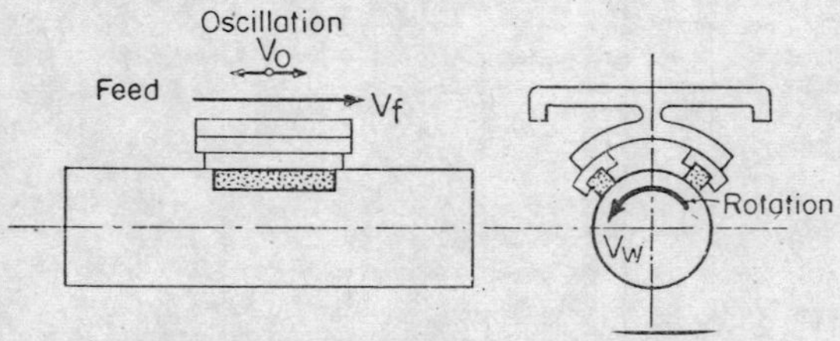

Fig. 10.20. Kinematic scheme of superfinishing a shaft

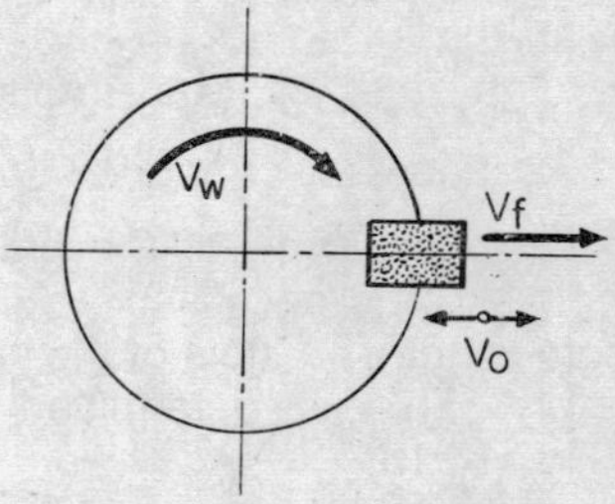

Fig. 10.22. Radial feed in superfinishing

(b) rotation of the workpiece, usually kept low,
(c) feed motion of the tool or of the workpiece.

The direction of oscillations is kept parallel to the axis of rotation. Feed motion can be parallel, radial or plunge type. The kinematic scheme for (a) superfinish process for shafts is shown in Fig 10.20. The other cases are shown in Figs. 10.21 and 10.22.

Superfinishing is applied for external or internal surfaces of cast iron, steel or non-ferrous parts which have been previously ground or precision turned. Though capable of rendering high quality surface finish, macro-geometrical errors cannot be corrected by this technique. The reasons for such product quality obtainable from superfinishing are: (a) small unit pressure, (b) low cutting speeds, (c) oscillation that frees the sticks from chips, (d) complex path of tool allowing abrasive grains to be successfully utilized, (e) low working temperatures due to low heat generation (1-2°C above ambient).

10.6.1. Kinematics of Superfinishing

In Fig. 10.23, the path of superfinishing grains in straight oscillatory motions is projected on a developed work surface. The superfinishing hone starts from an initial position when a large portion of hone length will be in contact with the work piece surface. A part of the hone must protrude beyond this surface. This part l_d should not be smaller than the feed per revolution, i.e., work piece travel in relation to the oscillating part of the superfinishing head during one revolution of the work piece. This condition is recorded as

$$l_d \geq s$$

Combining rotary motion and feed motion the path of grain motion *PR* is obtained. With periodic hone oscillations of amplitude *K* and wave length λ, the

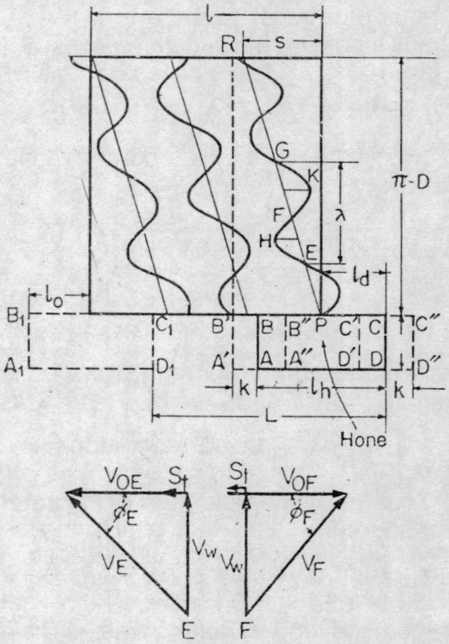

Fig. 10.23. The path of superfinishing grains on the developed surface.

resulting grain path motion *P* will pass through points *E-H-F-K-G-R*. Noting that

$$(m + e)\,\lambda = \pi D,$$

where m = number of full length oscillatory waves,

e = a number given by $0 < e < 1$ $\simeq 0.5$.

Since $f\,(\lambda) = n(m + e)\lambda = \pi D n = v_w$ the frequency of oscillations, f, will be

$$f = n(m + e)$$

These relationships ensure that grains do not move along their previous paths. The cutting speed for straight superfinishing is given by

$$v = \sqrt{v_w^2 + (v_o + s_t)^2}$$

when $v_o = 2\pi f\,K \cos 2\pi ft$

Maximum cutting speed occurs at E where $v_{o\,max}$ agrees with the longitudinal feed.

$$v_{max} = \sqrt{v_w^2 + (2\pi fK + s_t)^2}$$

Minimum cutting speed will occur when $v_o = 0$, thus

$$v_{min} = \sqrt{v_w^2 + (s_t)^2}$$

If oscillatory motion is characterized by the average speed determined by $v_o = 2Kf$, the average cutting speed is

$$v_{av} = \sqrt{v_w^2 + (2Kf + s_t)^2}$$

$$= \sqrt{(\pi Dn)^2 + (2Kf + s_t)^2}$$

Let ϕ be the angle of intersection of grain path with the direction of generation on work surface with hone at the central position. This angle is given by

$$\tan\phi = \frac{v_w}{s_t \pm v_o} \approx \frac{v_w}{v_{o\,max}}$$

CHAPTER 11

The Trident of Speed-Feed-Depth

11.1. CUTTING SPEED

The cutting speed is the rate of cutting on the main motion. It is the cutting length in m per minute (m/minute).

Cutting speed is denoted by the letter v, the diameter of the workpiece in mm by d, the number of revolutions of the workpiece per minute by n.

$$v = \frac{\pi \times d \times n}{1000} \text{ m/min}$$

Example:

Calculate the cutting speed by which a workpiece is to be machined on a lathe.

Given:

Diameter of workplece $d = 50$ mm

$RPM = 160$.

Result:

$$v = \frac{\pi \times d \times n}{1000};$$

$$= \frac{3.14 \times 50 \times 160}{1000} = 25.12 \text{ m/min}$$

The cutting speed cannot be chosen at random. If the cutting speed is too low, the machining time is too long. If it is too high, the cutting edge loses its hardness due to strong heating-up. The cutting edge wears out quickly and must often be reground.

The most favourable cutting speed should be selected (at any time) from Table 11.1.

11.1.1. Calculation of the R.P.M.

Example:

The r.p.m. has to be calculated.

Given : $d = 125$ mm; $v = 20$ m/min

Result:

$$n = \frac{100 \times v}{\pi \times d} = \frac{100 \times 20 \text{ m/min}}{3.14 \times 125 \text{ mm}}$$

$$\approx 51 \text{ r.p.m.}$$

11.2. FEED AND DEPTH OF CUT

Besides the correct selection of r.p.m., the performance of the lathe depends on the feed and the depth of cut.

The Chip Cross-section:

The length (in mm) at which the tool travels at one revolution of the workpiece during longitudinal or transverse turning is called feed. Feed multiplied by depth of cut results in the chip cross-section.

$$(s) \times (a) = (A).$$

Example:

$s = 0.8$ mm/rev, $a = 3$ mm. Calculate A.

Result :

$$A = s \times a = 0.8 \text{ mm} \times 3 \text{ mm} = 2.4 \text{ mm}^2.$$

TABLE 11.1

REFERENCE VALUES OF CUTTING SPEEDS

Material	Tool	Cutting angles			Roughing ∇ Depth of cut a≈4 … 10.s			Finishing ∇∇ Depth of cut a≈2 … 5.s			Coolant and lubricant for	
		α	β	γ	Cutting speed v m/min	Feed s mm/U	Depth of cut a mm	Cutting speed v m/min	Feed smm/U	Depth of cut a mm	Roughing ∇	Finishing ∇∇
Steel	W	8°	62°	29°	14	0.5	4	20	0.2	1		
Strength	HSS				22	1.	10	30	0.5	1	E	E.O.P
50 kg/mm²	H	5°	67°	18°	150	2.5	15	250	0.25	1.5		
	W	8°	68°	14°	10	0.5	4	15	0.2	1		
50-70 kg/mm²	HSS				20	1	10	24	0.5	1	E	E.O.P
	H	5°	75°	10°	120	2.5	15	200	0.25	1.5		
	W	8°	68°	14°	8	0.5	4	12	0.2	1		
70-85 kg/mm²	HSS				15	1	10	20	0.5	1	E	E.O.P
	H	5°	75°	10°	80	2	15	140	0.2	1.5		
	W	8°	76°	6°	6	0.5	3	8	0.2	1		
Tool steel	HSS				12	1	8	16	0.5	1	E	E.O.P
	H	5°	79°	6°	30	0.6	5	50	0.15	1		

W – Tool steel
HSS – High-speed steel
H – cemented carbide
E – Diluted soluble oil
R – Rape-seed oil
P – Kerosene
dr – dry

For thread cutting v will be about $^1/_2$ of the cutting speed for longitudinal turning.

11.3. RPM AND FEED DURING DRILLING

The numbers of revolutions of the drill depend on the cutting speed and on the diameter of the drill.

The cutting speed for drilling is the speed of the periphery of the cutting edge in m/minute.

Frequently, there are diagrams fixed to the drilling machines, from which the numbers of revolutions for each cutting speed and for each diameter of drill can be read.

The feed is expressed in mm per revolution of drill, e.g. 0.2 mm/revolution. Thickness of chip, necessary feed power and surface quality of the hole depend on the feed. The permissible feed depends on the material to be drilled and the diameter of the drill.

Cutting speed (v), feed (s) and coolant for drills of HSS are given in Table 11.2.

Example :

A hole is to be drilled into mild steel flat.

Given : Diameter of hole 14 mm, material M.S. flat St 37.

To find: Numbers of revolutions (n) of the drill.

Result :

Let, a cutting speed of v = 22 m/minute is selected. Diameter of drill d = 14 mm.

$$n = \frac{v \times 1000}{\pi \times d} = \frac{22 \text{ m/min} \times 1000}{3.14 \times 14 \text{ mm}} \approx 501 \text{ r.p.m.}$$

It is assumed that the following numbers of revulutions can be set on the drilling machine: 47.5 – 75 – 118 – 190 – 300 – 475 – 750 – 1180 per minute.

In this case the numbers of revolutions 475 per minute. would have to be selected.

TABLE 11.2

Material		Diameter of drill, mm						Cool-ant
		5	10	15	20	25	30	
Steel	s	0.1	0.18	0.25	0.28	0.31	0.34	
up to 40 kg/mm²	v	15	18	22	26	29	32	E
Steel	s	0.1	0.18	0.25	0.28	0.31	0.35	or
up to 60 kg/mm²	v	13	16	20	23	26	28	S
Steel	s	0.07	0.13	0.16	0.19	0.21	0.23	
up to 80 kg/mm²	v	12	14	16	18	21	23	
Cast iron	s	0.15	0.24	0.3	0.32	0.35	0.38	
up to 18 kg/mm²	v	24	28	32	34	37	39	dr
Cast iron	s	0.15	0.24	0.3	0.33	0.35	0.38	or
up to 22 kg/mm²	v	16	18	21	24	26	27	E
Brass	s	0.1	0.15	0.22	0.27	0.3	0.32	E
up to 40 kg/mm²	v		60	0.70	m/	min		or S
Bronze	s	0.1	0.15	0.22	0.27	0.3	0.32	or dr
up to 30 kg/mm²	v		30	.40	m/min			
Aliminium	s	0.05	0.12	0.2	0.3	0.35	0.4	E
pure	v		80..	.120	m/min			or S
Aliminium	s	0.12	0.2	0.3	0.4	0.46	0.5	or dr
alloy	v		100	..	105 m/min			
Magnesium	s	0.15	0.2	0.3	0.38	0.4	0.45	dr
alloy	v	200 ... 250 m/min						

E = diluted soluble oil, S = cutting and cooling oil, dr = dry

11.4. R.P.M. AND FEED DURING VARIOUS MILLING OPERATIONS

11.4.1. Selection of the R.P.M.

The number of revolutions depend on the permissible cutting speed and the diameter of the cutter. Cutting speed of the cutter means the travel of one cutting tooth in meter/minute. The permissible cutting speed will be taken from the cutting speed table.

Cutting Speed Too High :

Cutter teeth will become blunt prematurely.

Cutting Speed Too Low :

Low cutting capacity.

v = cutting speed in m/min.

d = diameter of cutter in mm.

n = Numbers of revolutions of the cutter per minute.

Revolutions of the milling cutter per min:

$$n = \frac{100 \times v}{\pi \times d}$$

Example :

A plate has to be machined with a plane milling cutter by rough milling. Calculate the number of revolutions of the cutter.

Given :

Material of the plate St 50, diameter of cutter 75 mm.

Result :

Cutting speed 17 m/minute.

$$n = \frac{1000 \times v}{\pi \times d} = \frac{1000 \times 17 \text{ m/minute}}{3.14 \times 75 \text{ mm}} \approx 72 \text{ r.p.m.}$$

As a rule, only certain number of revolutions can be set on a milling machine, e.g. 37 – 49 – 64 – 86 – 113 – 147 – 197 – 260 – 338 – 455 – 600 – 700 per minute

In the present case n = 64 r.p.m. is selected.

The number of revolutions can also be slected from a table, (Table 11.3).

11.4.2. Selection of the Feed

For milling operations, the feed means the rate of feed in mm/minute. It is the distance in mm which the milling machine table and hence the workpiece travels in one min (Fig. 11.1).

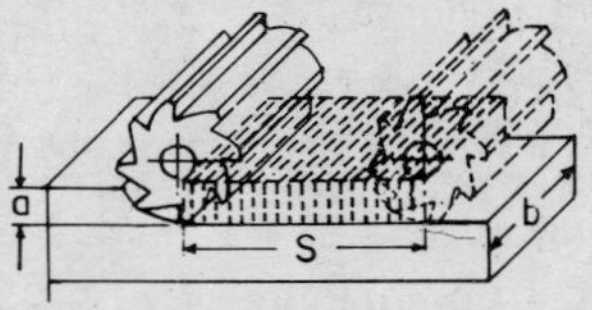

Fig. 11.1: Chips are produced by taking into consideration the following factors:
(a) Dept of cut (in mm), (b) milling width (in mm), s' rate of feed (in mm/min.), V amount of chips.

TABLE 11.3
MILLING CUTTER RPM SELECTION

Materials (Milling width = b) (Depth of cut = a)	Cylindrical Milling Cutter (plain) b = 100 mm				Side Milling Cutter b = 20 mm				Shell End Mill b = 70 mm			
	Roughing a = 5 mm		Finishing a = 0.5 mm		Roughing a = 10 mm		Finishing a = 10 mm		Roughing a = 5 mm		Finishing a = 0.5 mm	
	v	*s′*	*v*	*s′*	*v*	*s′*	*v*	*s′*	*v*	*s′*	*v*	*s′*
Carbon steel up to 65 kg/mm²	17	100	22	60	18	100	22	40	17	100	22	70
Alloy steel, annealed, up to 75 kg/mm²	14	80	18	50	14	80	18	30	14	90	18	55
Alloy steel, tempered, up to 100 kg/mm²	10	50	14	36	12	50	14	25	10	55	14	42
Cast Iron up to 180 Brinell	12	120	18	60	14	120	18	40	12	140	18	70
Brass (Ms 58)	35	70	35	50	36	150	55	75	36	190	55	150
Light metal	200	200	250	100	200	200	250	100	200	250	250	110

	End Milling Cutter b = 25 mm				Inserted Blade Cutter b = 180 mm				Slitting Saws b = 2.5 mm	
	Roughing a = 5 mm		Finishing a = 0.5 mm		Roughing a = 5 mm		Finishing a = 0.5 mm		Roughing a = 10 mm	
	v	*s′*	*v*	*s′*	*v*	*s′*	*v*	*s′*	*v*	*s′*
Carbon steel up to 65 kg/mm²	17	50	22	120	20	20	30	50	45	50
Alloy steel, annealed, up to 75 kg/mm²	15	40	19	100	16	65	23	40	35	40
Alloy steel, tempered, up to 100 kg/mm²	13	20	17	65	14	36	18	30	25	30
Cast Iron up to 180 Brinell	15	60	19	120	16	100	24	90	35	50
Brass (Ms 58)	35	80	55	120	50	200	60	120	350	200
Light metal	160	90	180	120	250	250	300	90	320	180

The rate of feed (s') depends on the cutter, the material of the workpiece, the cutting depth and the required surface quality.

The amount of chips *V* (Fig. 11.1) can also be calculated from the depth of cut (*a*), the milling width (*b*) and the rate of feed (s').

$$V = \frac{a \times b \times s'}{1000} \text{ in cm}^3\text{/min.}$$

By rearranging,

$$\text{Rate of feed in mm/minute, } s' = \frac{V \times 1000}{a \times b}$$

V = biggest possible amount of chips in cm³/min.
V' = permissible amount of chips in cm³/kW min.
P' = machine driving capacity in kW.

The biggest possible amount of chips per minute is equal to the permissible amount of chips per kW in one minute. (cm³/kW minute) multiplied by the machine driving capacity.

Biggest possible amount of chips in cm³/minute is therefore,

$$V = V' \times P$$

Example:

For plain milling of steel of 35...60 kg/mm² strength, the permissible amount of chips amounts to 12 cm³/kW minute.

What amount of chips will be cut off in one minute on a milling machine with 2.5 kW driving capacity?

Result:

$V = V' \times P = 12\ cm^3/kW\ min \times 2.5\ kW$
$= 30\ cm^3/min$

Example:

A plate of St 50.11 has to be machined by plain milling. The cutting depth is 4 mm, milling width 80 mm, and driving capacity of machine 3 kW. Calculate the biggest possible rate of feed.

Result:

1. Biggest possible amount of chips
   $V = V' \times P;\ V' = 12\ cm^3/kW\ min$
   $V = 12\ cm^3/kW\ min \times 3\ kW = 36\ cm^3/min$
2. Rate of feed

$$s' = \frac{V \times 1000}{a \times b} = \frac{36\ cm^3/min \times 1000}{4\ mm \times 80\ mm}$$
$$= 112\ mm/min$$

Usually only certain rates of feed can be set on the milling machine, e.g. 12 – 20 – 33 – 57 – 99 – 167 – 276 – 480 mm/minute.

Thus a rate of feed of 99 mm/minute has to be selected.

11.5. SPEED AND CYCLES (CPM) FOR SHAPING MACHINES

11.5.1. Setting of Stroke (Fig. 11.2).

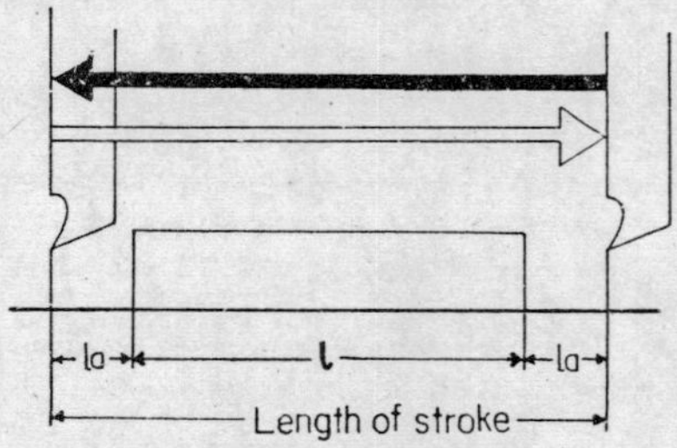

Fig.11.2: Setting of stroke.

11.5.2. Cutting Speeds (m/min).

TOOL MADE OF	STEEL, TENSILE STRENGTH IN KG/MM^3			GRAY CAST IRON	RED BRASS
	40	60	80		
Tool steel	16	12	8	12	20
High-speed steel	22	16	12	14	30

11.5.3. Numbers of Cycles

The number of cycles to be set per minute depends on the permissible cutting speed and the length of the stroke, (Table 11.4).

Example :

A cast iron plate is to be rough machined on the shaper by using a high-speed-steel tool. Length of stroke is 300 mm. Determine the cycles per min.

TABLE 11.4

Number of cycles PER MINUTE	For length of stroke in mm			
	100	200	300	400
Cutting speed	m/min	V_m	m/min	V_m
28	5.3	10.2	14.2	18.2
52	9.8	19	26.2	33.6
80	15.2	29	41	52

Result :

Cutting speed as per table approx. 14 m/min. Cylces as per table 28 per minute.

Calculating the Cycles per Minute:

$$\text{Cycles per min } (n) = \frac{\text{Average speed (in m/min.)}}{\text{Length of cycle (in m)}}$$

$$n = \frac{V_m}{2\ L}$$

Example :

Length of stroke $L = 400$ mm, $v_m = 15$ m/min. Calculate the cycles per min.

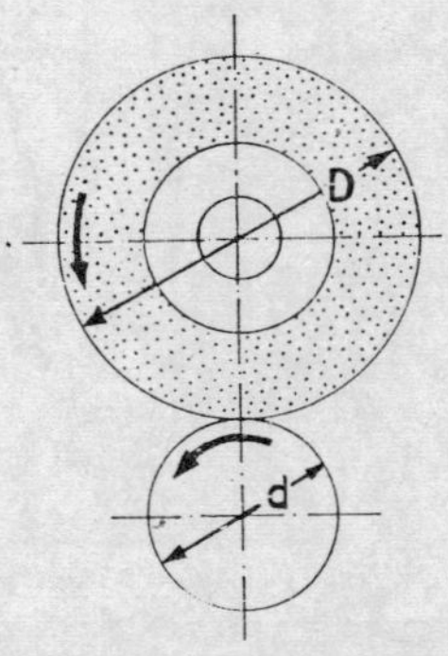

Fig. 11.3: Direction of rotation of grinding wheels (D) and workpiece (d).

Result:

$$n = \frac{V_m}{2\ L} = \frac{15\ m/min}{2 \times 0.4\ m} \approx 20 \text{ cycles per min.}$$

11.5.4. Feed and depth of cut.

Adjusting the Feed and the Cutting Depth:

The amount of feed depends on the type of machining. Chip cross-section = cutting depth × feed; $A = a \times s$. The cross-section of the chip is dependent on the capacity of the machine.

For roughing operations the depth of cut should be 3...5 times bigger than the feed.

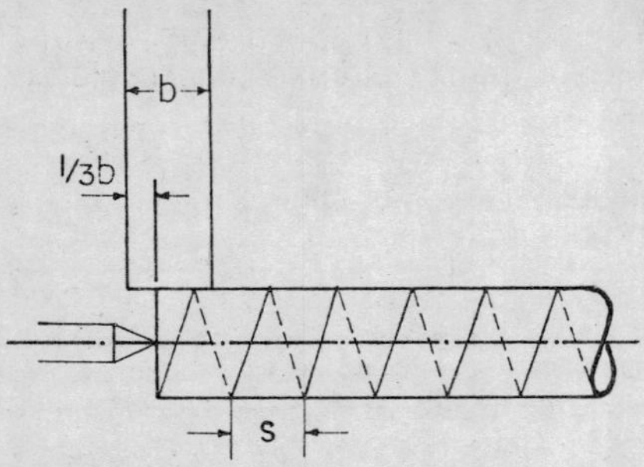

Fig. 11.4: Side feed: (b) Width of grinding wheel, (s) lateral feed in mm per revolution of the workpiece ($^1/_4 ... ^4/_3$ b).

For finishing operations cutting depth and feed should be small.

11.6. CUTTING SPEED AND RPM OF THE GRINDING WHEEL

Cutting Speed and R.P.M. of Grinding Wheel:

The higher the cutting speed the faster the grinding work can be performed; the indicated speeds must, therefore, not fall short of the given instructions. But higher speeds are also to be avoided, since then the wheels clog, get smooth and do not grind any more ; the workpiece overheats and the surface becomes inaccurate; moreover, there is also the danger of accidents. Relative motions are shown in Fig. 11.3.

Circumferential Speed and R.P.M. of the Workpiece:

The ciricumferential speed is designated in m/min. It affects the quality of the cut; if it is low, the cut will be fine, if it is too high, the cut will be coarse.

The wheel width and lateral feed relations are shown in Fig. 11.4.

Calculation of the R.P.M.

v_w = circumferential speed of the workpiece in m/min.

d = diameter of the workpiece in mm,

n_w = R.P.M. of the workpiece per min.

$$\text{R.P.M. of the workpiece } n_w = \frac{v_w \times 1000}{r \times d}$$

Example :

A shaft of St 50 with a diameter of 50 mm shall be ground. n_w is to be calculated.

Result :

v_w = 15 m/min.

$$n_w = \frac{v_w \times 1000}{\pi \times d} = \frac{15 \text{ m/min} \times 1000}{3.14 \times 20 \text{ mm}} \approx 239 \text{ r.p.m.}$$

The Depth of Cut:

Roughing 0.01...0.03 mm ; finishing 0.0025...0.005 mm.

CHAPTER 12

General Rules of Machining Characteristics

12.1. MACHINABILITY AND TOOL LIFE

The 'ease' with which a given material may be worked with a cutting tool is 'machinability'. Machinability depends on

(a) chemical composition of job material,
(b) structure,
(c) mechanical properties,
(d) physical properties,
(e) cutting conditions.

The criterion for judging machinability may be
(a) tool life,
(b) intensity of cutting forces,
(c) quality of surface finish,
(d) form and size of chips,
(e) temperature of cutting.

The main factor to be chosen for judging machinability depends on type of operation and production requirements.

Tool life is the time elapsed between two successive grindings of a cutting tool. There are various ways of expressing tool life, viz.

(i) time of actual operation, viz. the time the tool is in contact with the job,
(ii) total time of operation,
(iii) volume of metal removed,
(iv) number of pieces machined,
(v) equivalent cutting speed.

However, often 'actual cutting time' particularly in turning operations is used to measure tool life.

12.2. FLANK WEAR GROWTH AND 'TOOL LIFE'

The growth of wear land with respect to time is shown in Fig. 12.1. Certain characteristic features of wear growth may be observed from Fig. 12.1. Up to point Q, the region I denotes the zone of initial 'break-in'. The sharp edge rapidly breaks down due to plastic deformation and consequential temperature rise and the point Q is reached. After that the wear process is more or less uniform until a point F has been reached. The Q-F region, shown as region 2 in Fig. 12.1 is the mechanical wear region or wear region under temperature insensitive conditions. Beyond F, the rapid growth of wear process ensues and the cutting tool fails very soon after reaching this point of inflexion F. This point F is often called critical point of flank wear characteristic or simply 'critical flank wear'. Fig. 12.2 shows the effect of cutting speed on critical flank wear. It is observed that the critical point not only shifts to lower time region when cutting speed is increased but the critical value of wear land width is also lowered. Fig 12.3 shows the double-log plot of Fig. 12.2. From Fig. 12.3 it is seen that ABCD forms a critical boundary denoting imminent failure.

Further, $h_f - T$ characteristics bear a relation of the form,

$$h_f = C_1 \, T^{x_1}$$

The exponent x_1 lies between 0.6 and 1.0.

The critical point F can be located by a method indicated by Zorev.

With respect to Fig 12.4 (b), if the radial wear at time T is $(x + A)$, then for a radial wear of H, the time T_o is,

$$T_o = \frac{T\,H}{x + A}$$

$$\text{But, } x = h_f \cdot \frac{\tan \alpha}{\sin \beta} = h_f \, \psi,$$

and A = effect of nose wear = 0.15 to 0.2 mm. Hence,

$$T_o = \frac{TH}{h_f \psi + A}$$

To find maximum T_o with respect to h_f,

$$\frac{dT_o}{dh_f} = \frac{\frac{dT}{dh_f} H}{h_f \psi + A} - \frac{\psi\, T H}{(h_f \psi + A)^2} = 0$$

Hence, the condition is given by

$$\frac{dT}{d\,h_f} = \frac{T}{h_f + \frac{A}{\psi}} \quad \text{or} \quad \frac{dh_f}{dT} = \frac{h_f + \frac{A}{\psi}}{T} \qquad \text{Eqn. 12.1}$$

Thus, the point F is located in Fig. 12.4 whenever the slope at any point satisfies Eqn. 12.1.

The time at which the critical wear value h_f^* is reached is 'tool life' with respect to flank wear growth.

Often a compromise value of h_f^* is chosen as wear limit. Fig. 12.5 (a) shows how 'tool life' values for various cutting speeds are chosen for a limiting value of h_f^*. If the corresponding tool life values for various cutting speeds are plotted, a characteristic curve as shown in Fig. 12.5(b) is obtained. The characteristic of Fig. 12.5(b) can be plotted also in log co-ordinates as indicated in Fig. 12.6.

12.3. TAYLOR'S TOOL LIFE EQUATION

Cutting speed (V_c) is the dominant variable having considerable influence on tool life. Taylor has shown that a relation between cutting speed (V_c) and tool life (T) can be written from Fig. 12.6 in the form :

$$VT^n = \text{constant} = C$$

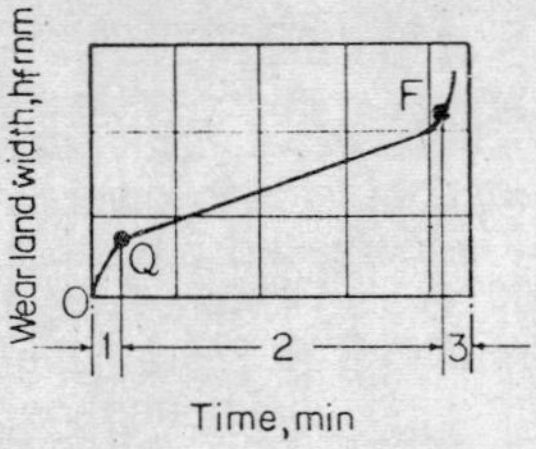

Fig 12.1: Growth of wear land with respect to time.

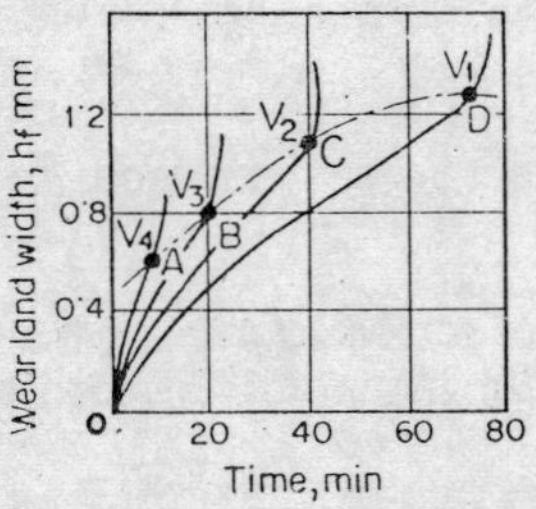

V_4 = 220 m/minute, V_2 = 180 m/minute,
V_3 = 200 m/minute, V_1 = 160 m/minute,

Fig. 12.2: Effect of cutting speed on critical flank wear.

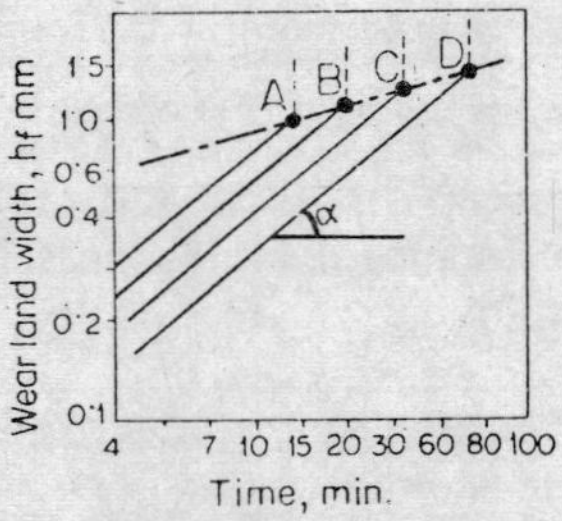

Fig. 12.3: $h_f \sim T$ plot showing critical point locations in double-log co-ordinates.

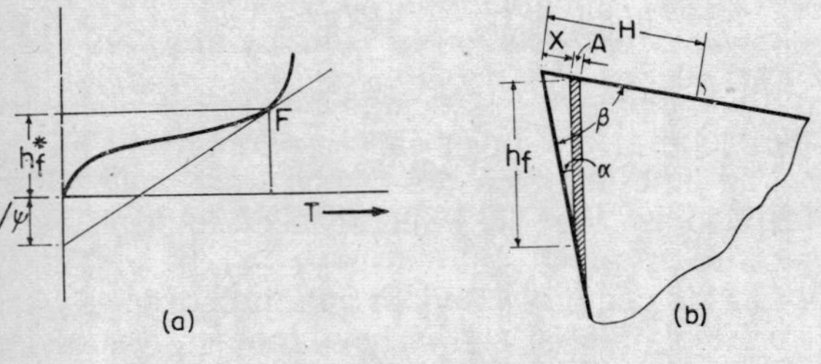

Fig. 12.4: Determination of 'critical flank wear' by method of Zorev.

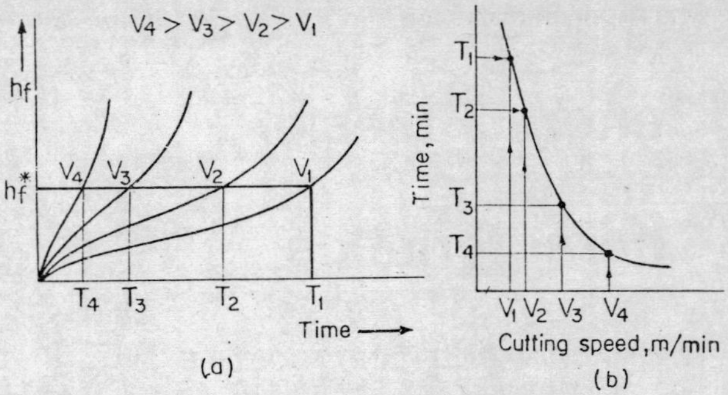

Fig. 12.5: Effect of velocity on tool life.

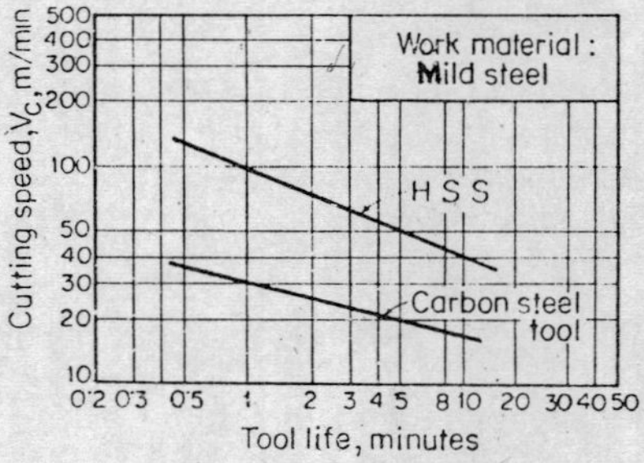

Fig 12.6: A typical tool life characteristic, for given cutting conditions.

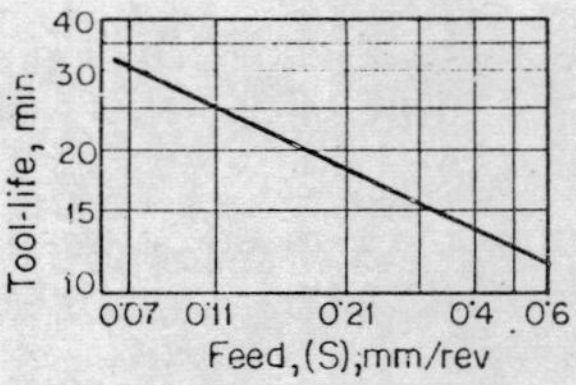

Fig 12.7(a): Effect of feed on tool life
Work material : C-45 steel.
Tool : HSS
Cutting speed : 50 m/minute
Feed : 0-21 mm/revolution.

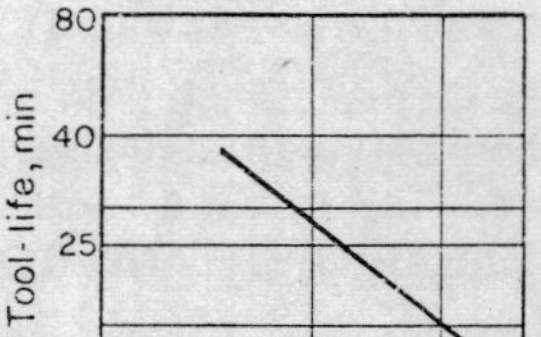

Fig. 12.7(b): Effect of depth of cut on tool life
Work material : C-45 steel
Tool : HSS
Cutting speed : 50 m/minute
Feed : 0-21 mm/revolution.

where T = tool life in minutes,
V = cutting speed, m/minute,
n, C = constants depending on work tool pair, tool shape, cutting environment, etc.

However, from Figs 12.7(a) and 12.7(b), it may be observed that changes in feed and depth of cut also affect tool life.

A generalized Taylor's tool life equation is given by :

$$V = \frac{C_v}{T^{1/m}\, s^{y_v}\, t^{x_v}} \quad \text{Eqn. 12.2}$$

where C_v = modified Taylor constant depending on tool work pair,
$(1/m)$ = Taylor exponent, n,
x_v, y_v = exponents to take care the effect of feed and depth of cut.

The exponents and constants for various tool work pair are shown in Table 12.1

TABLE 12.1

EXPONENTS AND CONSTANTS OF GENERALIZED TAYLOR EQUATION

WORK MATERIAL	TOOL MATERIAL	C_v	X_v	Y_v	N=(1/M)	RESTRICTION
STEEL		273		0.20		$s \leq 0.30$
$\sigma_u = 75$	WTiC	227	0.15	0.35	0.20	$0.3 \leq s \leq 0.75$
kg/mm^2	Co=10%	221		0.45		$s > 0.75$
STEEL		292	0.30	0.15	0.18	$T \leq s$
$\sigma_u = 75$	WTiC	292	0.15	0.30	0.18	$T > s$
kg/mm^2	Co=6%					
CAST IRON :	WC	292	0.15	0.20	0.20	$s \leq 0.4$
BHN=190		243	0.15	0.40	0.20	$s > 0.4$
		324	0.40	0.20	0.28	$T \leq s$
		324	0.20	0.40	0.28	$T > s$

Often, for a given tool life, permissible cutting velocity is sought. For 60 minutes tool life, the Eqn. 12.2 must be modified,

$$V_{60} = \frac{C_v}{[60]^{1/m} s^{y_v} t^{x_v}} = \frac{C_v'}{s^{y_v} t^{x_v}}$$

where $C_v' = \dfrac{C_v}{[60]^{1/m}}$.

12.4. EFFECT OF VARIOUS VARIABLES ON TOOL LIFE

12.4.1. Effect of Mechanical Properties of Work Material

Physical and chemical properties of work materials influence tool life by affecting 'form stability' and rate of wear of tool.

For steel, the permissible cutting speed V^* for a given tool life is given by

$$V^* = \frac{C}{\sigma_u{}^x}$$

where σ_u = ultimate strength of steel in kg/mm^2. The exponent x lies between 1.5 and 1.75.

The effect of the property of material has been investigated by Yanitsky when for a chosen group of materials, an alternative relationship is suggested :

$$V^* = \frac{\text{constant}}{H_B{}^{1.68}\,\psi}$$

where H_B = Brinnel hardness number,
ψ = % reduction.

For cast iron,

$$V^* = \frac{C}{(H_B)^y}$$

where H_B = Brinnel hardness number,
y = exponent, usually about 1.7.

12.4.2. Effect of Tool Material

The effect of tool material on tool life is shown in Fig. 12.8.

For a given cutting speed, more work material is removed by tool materials shown in order–cemented oxide, cemented carbide, HSS. Higher is the cutting speed, more is the gain. An 'ideal' tool material means that at all cutting speeds the maximum and constant volume of work material are removed as indicated in Fig. 12.8.

12.4.3. Effect of Nose Radius

The nose radius tends to improve tool life and is evident from the relation :

$$V T^{0.0927} = 331 \; R^{0.244}$$

where R = nose radius of HSS tool, cutting SAE-2346 steel, inch,
V = cutting speed, f.p.m.,
T = tool life, minute

An alternative equation for a given tool life is in the form :

$$V^* = C r^{0.1} \text{ for steel turning}$$
$$= C r^{0.08} \text{ for cast iron turning.}$$

12.4.4. Optimization of Tool Geometry from Tool Life Consideration

The optimization of tool shape from tool life considerations can be made.

The effect of rake is quite interesting in affecting tool life and is also very complex to be evaluated. The best rake angle lies between – 5° and + 10° for turning austenitic steel by a carbide tool as shown in Fig. 12.9.

However, the optimum rake angle is a function of the properties of work material. Fig. 12.10 shows how the optimum cutting angle, δ, is affected for increase in the ultimate strength of the work material.

In terms of γ, and for materials $\sigma_u < 80$ kg/mm^2

$$\gamma_{opt} = \frac{25 \times 10^4}{\sigma_u{}^{8.0}}$$

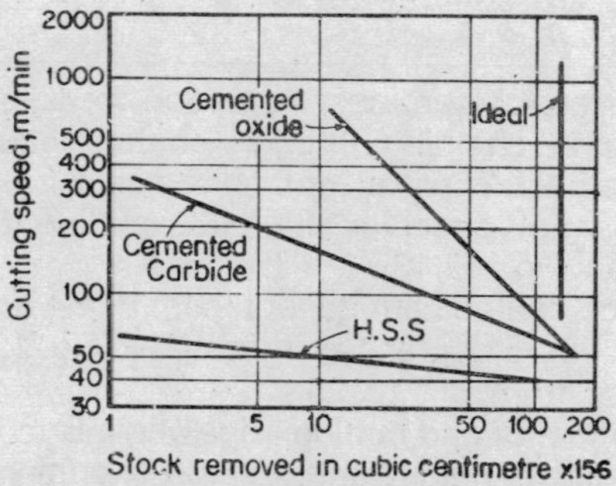

Fig 12.8: Effect of tool material on the $V-T$ characteristic: Work material: Low carbon steel, $\gamma_0 = 10°$.

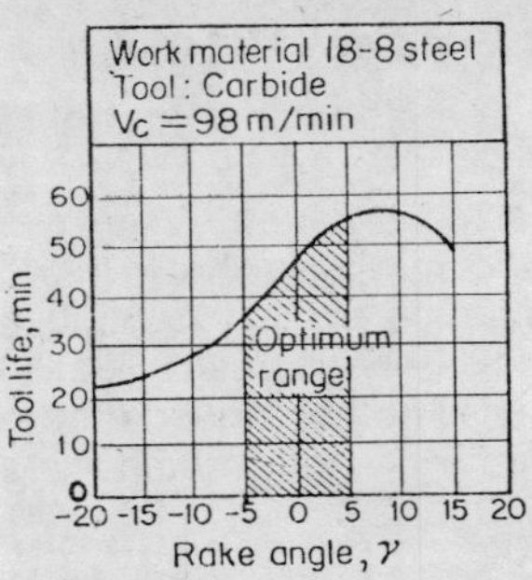

Fig. 12.9: Effect of rake angle on tool life.

For materials with $\sigma_u > 80$ kg/mm^2,

$$\gamma_{opt} = \frac{5 \times 10^{16}}{\sigma_u^{8.0}}$$

The effect of clearance is to improve tool life at first and then tool life decreases because of decreased strength. The optimum clearance angle is between 10 and 15 degrees as shown in Fig. 12.11.

However, change in feed, s, affects the optimum value of clearance angle as seen from Fig. 12.12. More is the feed, less is the value of optimum clearance. This reduction is necessary for providing increased wedge strength and heat dissipative area when load is increased.

Side cutting edge angle (ϕ_s) may improve tool life under non-chatter conditions :

$$VT^{0.11} = 78\ (\phi_s + 15°)^{0.264} \quad \text{(FPS units)}$$

where ϕ_s = complimentary PCEA angle or side cutting edge angle of HSS tool cutting SAE-2345 steel, V = speed, f.p.m. and T = tool life, minute.

Several alternative relationships are found in literature for

$V_c^* = f(\phi) = c/\phi^m$

The exponent m is given in Table 12.2.

TABLE 12.2

Work material	Tool Material	m
Steel	Cemented carbide	0.35
Cast iron	Cemented carbide	0.40
Steel	HSS	0.60
Cast iron	HSS	0.45

However for carbides, reduction in ϕ results in increase in chipping through chatter induced by increased P_y. Thus an optimum exists for carbide which is not observed for HSS cutting tools as shown in Fig. 12.13.

With cemented carbide, optimum PCEA angles (ϕ) is about 60-75 degrees as indicated in Fig. 12.13, i.e. ϕ_s varies from 30° to 25°.

The effect of end cutting edge angle is to improve surface finish, rigidity and equivalent cutting speed. The optimum auxiliary plan angle or end cutting edge angle is 4 to 8° as shwon in Fig. 12.14.

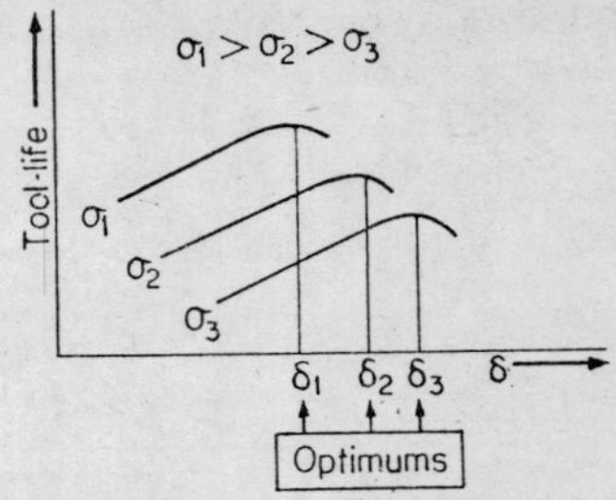

Fig. 12.10: Effect of cutting angle (90° − γ_0) on tool life.

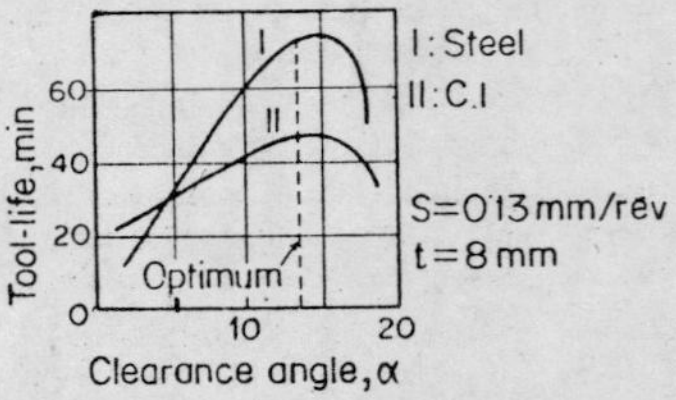

Fig. 12.11: Optimizing clearance angle of a tool.

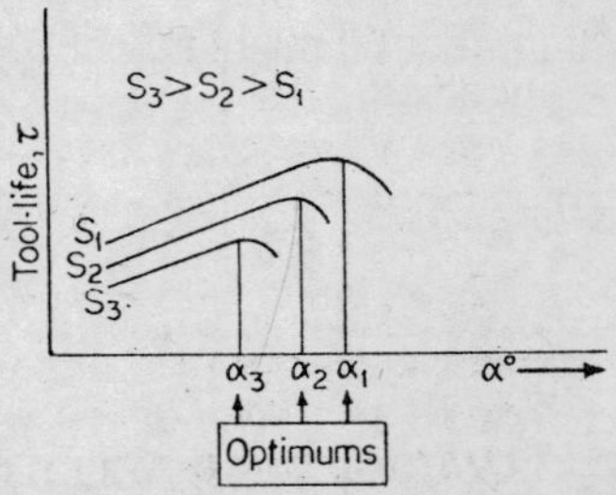

Fig. 12.12: Effect of feed on optimum clearance angle

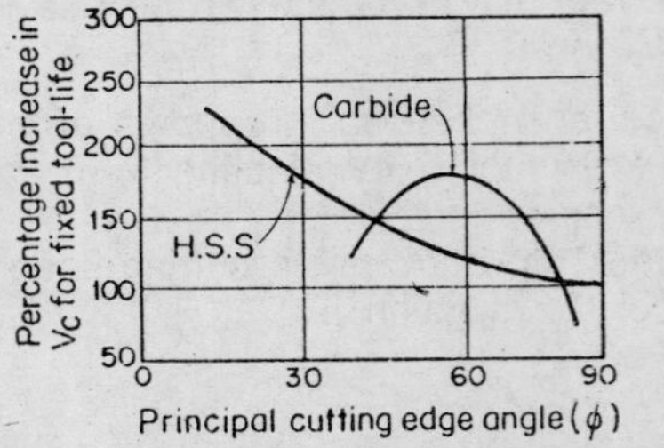

Fig. 12.13: Effect of ϕ on percentage increase in V^*_c.

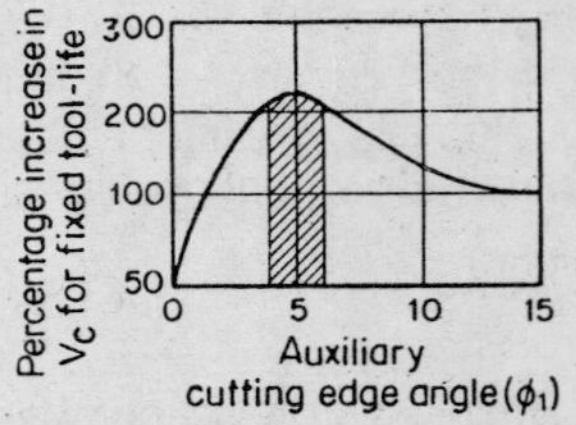

Fig. 12.14: Effect of ϕ_1 on percentage variation in V^*_c.

12.5. LAW OF CUTTING FORCES

12.5.1. First Law of Cutting Force

Kronenberg, Friedrich, Hippler, Okoshi, Optiz and Schallbroch and many others have shown that chip

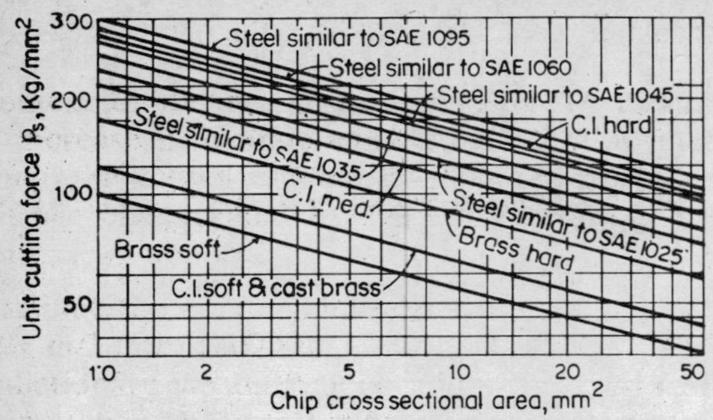

Fig. 12.15: Unit Cutting force ~ area of cut relationship

cross-sectional area, A, during turning affects the unit cutting force, p, related by

$$P = Ap, \text{ kg}$$

where $p = f(A)$, kg/mm².

The first law of cutting force defines the relationship between p and A as given in Table 12.3. :

TABLE 12.3

Author	First Law of Cutting Force	Legend
Friedrich	$p = a + \frac{b}{\sqrt{A}}$	(i) a, b, C_p are constants
Hippler	$p = \frac{C_p}{A^{0.25}}$	(ii) A is the area of cut.
Opitz	$p = \frac{C_p}{A^{0.164}}$	(iii) p is the unit cutting force.
Okoshi	$p = \frac{C_p}{A^{0.197}}$	

The relationship of the unit cutting force with respect to area of cut as determined by Hippler is shown in Fig. 12.15. The experimental results of Opitz and Schallbroch shown in Fig. 12.16 indicate that slenderness ratio has very little effect on the unit cutting force.

In can be seen that the exponent is 0.25 and constant for all materials as observed by Hippler. Schlesinger's Berlin experiments indicate variable values for the exponent for machining various material through turning (Table 12.4).

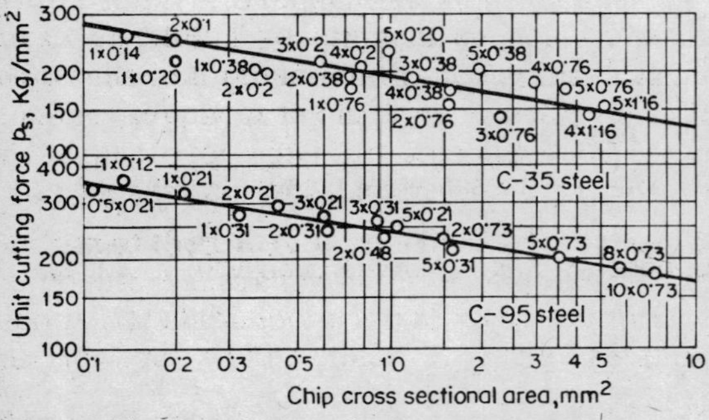

Fig. 12.16: Opitz and Schallbroch experiment

TABLE 12.4
DATA FROM BERLIN EXPERIMENTS
(Schlesinger and Klopstock)

Material	Exponent for First Law of Cutting Force
Cast Iron	0.135
Wrought Iron	0.138
Carbon Steel (σ_v = 75 kg/mm²)	0.197
Cr-Ni Steel	0.198
Copper	0.176
Brass	0.131

The results of Optiz shown in Fig. 12.16 indicate the values of exponent as 0.164 for C-30 steel and 0.182 for C-95 steel.

Kronenberg recommended 0.197 and 0.136 as the best recommended values for exponent to be used in the first law of cutting force for steel and cast iron respectively.

12.5.2. Second Law Of Cutting Force

The cutting forces, under conditions of other variables kept constant, are direct functions of feed (s) and depth of cut (t). Fig. 12.18 shows the variation of P_z as a function of feed during machining of steel (σ_u =45 kg/mm²) with a sintered carbide cutting tool for a constant depth of cut of 2 mm, PCEA (ϕ) of 60° and rake γ_o of 10°. The relationship is given by

$$P_z = C_1 s^{yz}, \text{ kg}$$

From the plotting in double-log co-ordinates, the exponent,

$$yz = \tan a_s = 0.75$$

and $C_1 = 314$

Similarly, the expressions for other components of force are expressible in the forms :

$$P_x = C_2 s^{yx}$$
$$P_y = C_3 s^{yy}$$

Fig. 12.18 shows the variation of tangential cutting force P_z with respect to depth of cut, t, in double co-ordinates, for the same material and cutting conditions as those of Fig. 12.17.

From this figure,

$$P_z = C_4\, t^{xz} = 29\, t^{1.0}$$

The similar expression for these components of force are,

$$P_x = C_5 t^{xx} = C_5 t^{1.1}$$
$$P_y = C_6\, t^{xy} = C_6\, t^{0.9}$$

To combine the relations of feed and depth of cut,

$$P_z = C_4\, t^{xz} = C_z t^{xz}\, s_0^{yz}$$

$$C_z = \frac{C_4}{s_0^{yz}} = \frac{29}{(0.1)^{0.75}} = 157$$

Otherwise, also

$$C_1 s^{xz} = C_z\, t_0^{xz}\, s^{yz}$$

and $C_z = \frac{C_1}{t_0^{xz}} = \frac{314}{2^1} = 157$

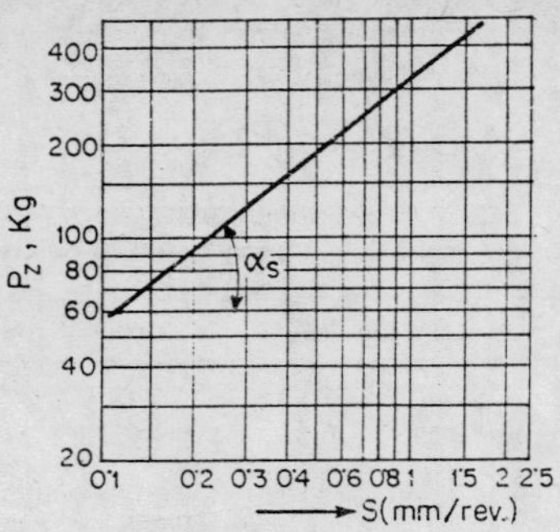

Fig. 12.17: Effect of feed on P_z.

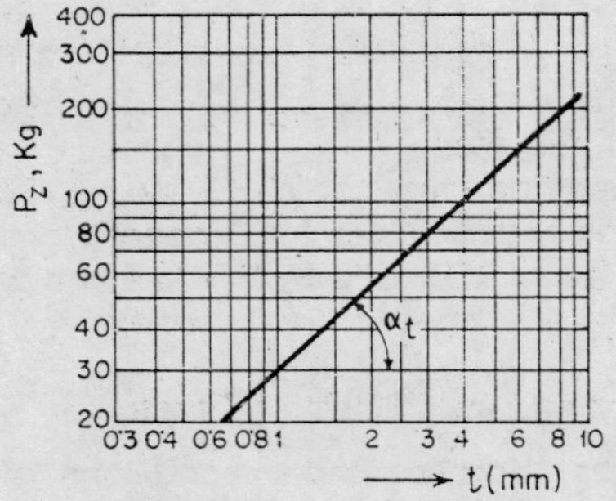

Fig. 12.18: Effect of depth of cut (t) on P_z.

Thus, the equation of P_z with respect to both s and t is given by

$$P_z = 157 \ t \, s^{0.75}$$

Introducing a term, *slenderness ratio, G,* given by the ratio of depth of cut to feed,

$$G = t/s$$

and noting that area of cut, A, is

$$A = s \ t$$

the cutting force equation can be altered.

Known that $P_z = C_z \ t^{xz} \ s^{yz}$

and substituting

$$t = \sqrt{AG}, \ S = \sqrt{(A/G)}$$

$$P_z = C_z \ (AG)^{xz/2} \ (A/G)^{yz/2}$$

$$= C_z.A^{\frac{1}{2}(xz + yz)} \ G^{\frac{1}{2}(xz - yz)}$$

An expression in terms of area of cut (A) and depth of cut feed $\left(\frac{t}{s}\right)$ ratio is derived for P_z, when

$$P_z = C_z A^{\frac{xz + yz}{2}} . \left(\frac{t}{s}\right)^{\frac{xz - yz}{2}}$$

This relationship is called *the second law of cutting force.*

For steel, if $xz = 1.00$ and $yz = 0.75$, the new exponents are calculated

$$\frac{xz + yz}{2} = \frac{1 + 0.75}{2} = 0.85$$

and $\frac{xz + yz}{2} = \frac{1 - 0.75}{2} = 0.825$

Hence, the second law of cutting force for steel is

$$P_z = C_z \ A_0^{0.85} \left(\frac{t}{s}\right)^{0.125} \text{kg}$$

The various constants and exponents are given in Table 12.5 for G varying between 10 and 20 and $\phi = 45°$.

TABLE 12.5

CONSTANTS AND EXPONENTS FOR SECOND LAW OF CUTTING FORCE

Work Material	P_z			P_y			P_x		
	C_z	xz	yz	C_y	xy	yy	C_x	xx	yx
Steel, σ_u = 45 kg/mm^2	157	1	0.78	72	0.90	0.70	44	1.1	0.55
Steel, σ_u = 60 kg/mm^2	171	1	0.78	91	0.90	0.70	55	1.1	0.55
Steel, σ_u = 70 kg/mm^2	184	1	0.78	103	0.90	0.70	63	1.1	0.55
Cast Iron, BHN = 180	98	1	0.73	77	0.90	0.70	38	1.1	0.65
Cast Iron, BHN = 200	105	1	0.73	85	0.90	0.70	42	1.1	0.65

The combined feed and depth-of-cut average characteristics for machining steel is given by (for $\phi = 90°$),

$$P_z = C_z \ t \, s^{0.78}$$
$$P_x = C_x \ t^{1.1} \ s^{0.55}$$
$$P_y = C_y \ t^{0.9} \ s^{0.70}$$

For machining cast iron,

$$P_z = C_z \ t \, s^{0.73}$$
$$P_y = C_y \ t^{0.9} \ s^{0.70}$$
$$P_x = C_x \ t^{1.1} \ s^{0.65}$$

12.5.3. Extended Laws of Cutting Forces

The effect of rake angle, γ, on the cutting force component, P_z, is shown in Fig. 5.51.

Often instead of γ, the relationships are shown in terms of angle of action $\delta = 90° - \gamma$.

Experimental relationships are

$$P_z = C_1 \delta^{x_1}$$
$$P_y = C_2 \delta^{x_2}$$
$$P_x = C_2 \delta^{x_3}$$

The exponents x_1, x_2 and x_3 are usually between

$x_1 = 0.45$ to 0.8,

$x_2 = 3.2$ to 4.5,

$x_3 = 2.8$ to 3.6.

The effect of the principal cutting edge angle, ϕ, can be obtained by substituting

$$a_1 = s. \sin \phi \text{ and } b_1 = \frac{t}{\sin \phi}$$

when

$$P_z = C_p \frac{t^{xz}}{\sin \phi} (s. \sin \phi)^{yz} = C_p \frac{t^{xz} \, s^{yz}}{\sin^{(1-yz)} . \phi}$$

For constant t and s, and for steel with $xz = 1$, $yz = 0.75$,

$$P_z = \frac{K}{\sin^{0.25} . \phi}$$

Fig. 12.20 shows the effect of ϕ on cutting forces during machining of steel and Fig. 5.53 shows the effect of ϕ on cutting forces during the machining of cast iron.

Direct experimental results may also be evaluated to give the relationships in this form

$$P_z = C_{z_1}\,\phi^{z_1}$$

$$P_y = C_{y_1}\,\phi^{y_1}$$

$$P_x = C_{x_1}\,\phi^{x_1}$$

The usual exponents are given in Table 12.6.

TABLE 12.6

EXPONENTS OF ϕ	FOR STEEL		FOR CAST IRON	
	$\phi < 50°$	$\phi > 50°$	$\phi < 45°$	$\phi > 45°$
z_1	− 0.16	+ 0.125	− 0.126	+ 0.126
y_1	− 0.5	− 0.5	− 0.5	− 0.5
x_1	+ 1.08	+ 0.85	+ 1.08	+ 0.35

Inclination angle (λ) also affects cutting forces P_x, P_y and P_z (Fig. 12.22). However, the main cutting force P_z does not change much between $\pm$ 10° of inclination angle.

Nose radius (r) also affects cutting forces. With increase in r, P_z and P_y increases and P_x decreases as shown in Fig. 12.23. The extended laws are

$$P_z = C_{z_2}\,r^{z_2}$$

$$P_y = C_{y_2}\,r^{y_2}$$

$$P_x = C_{x_2}\,r^{x_2}$$

The exponents for force ~ nose radius relationships are shown in Table 12.7.

TABLE 12.7

EXPONENTS OF r	z_2	y_2	x_2
For steel	0.1	0.3	− 0.3
For cast iron	0.07	0.2	− 0.2

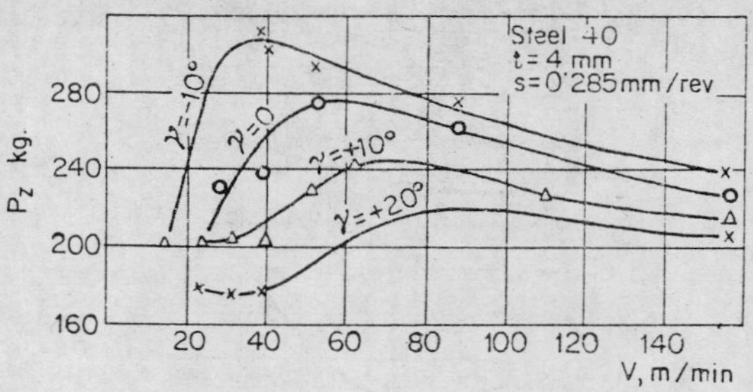

Fig. 12.19: Effect of increasing rake angle on tangential cutting force.

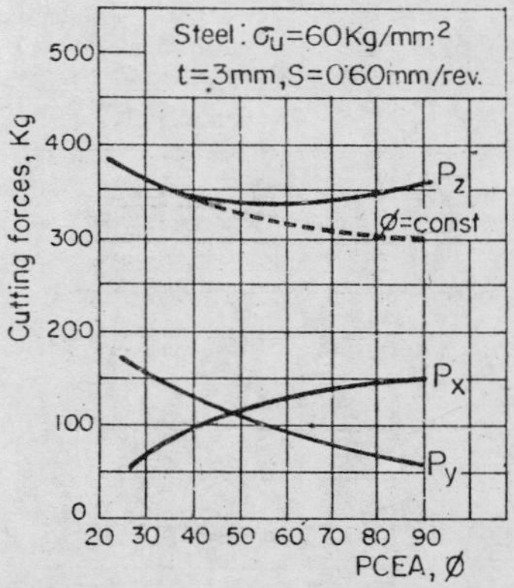

Fig. 12.20: Effect of ϕ on the cutting forces during machining of steel

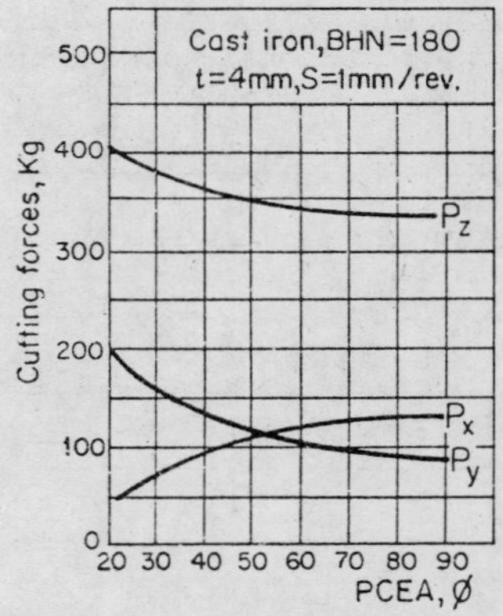

Fig. 12.21: Effect of ϕ on the cutting forces during machining of cast iron.

12.6. CUTTING FLUID ACTION

12.6.1. Purpose and Required Properties of Cutting Fluids

Cutting fluids are used for the following purposes:

(i) cooling purpose–to cool the chip, tool and job;

(ii) friction reduction : to reduce the coefficient of friction at the chip tool interface;

(iii) to protect the finished surface from corrosion and to improve the surface finish, and

(iv) to wash the chips away.

Out of these, the first two functions are very important.

Schallbroch, Schaumann and Wallich gave an empirical formula relating tool life and temperature of cutting tool, which is given by

$$T\theta^n = K$$

where

T = tool life, minute,

θ = temperature at chip tool interface, °C,

n = and exponent, the value of which depends mainly on the tool form and material, and

K = constant.

From the above relation, it has been found that small changes in tool temperature can produce considerable changes in tool life. Hence, the cutting

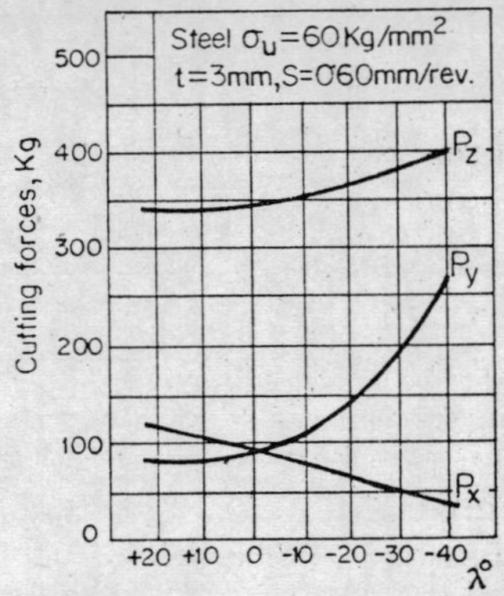

Fig. 12.22: Effect of λ on cutting forces.

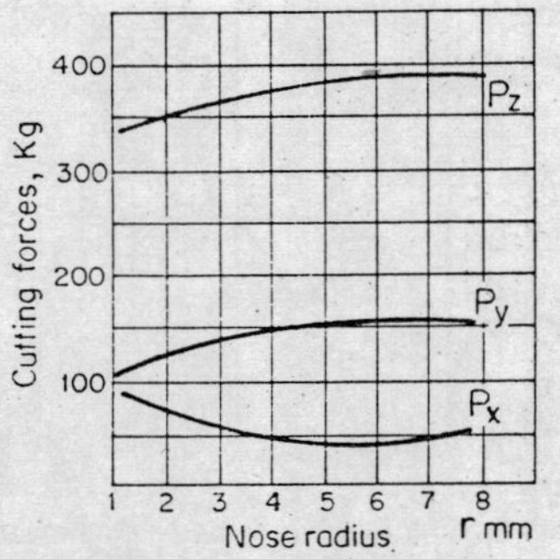

Fig. 12.23: Effect of nose radius on cutting forces.

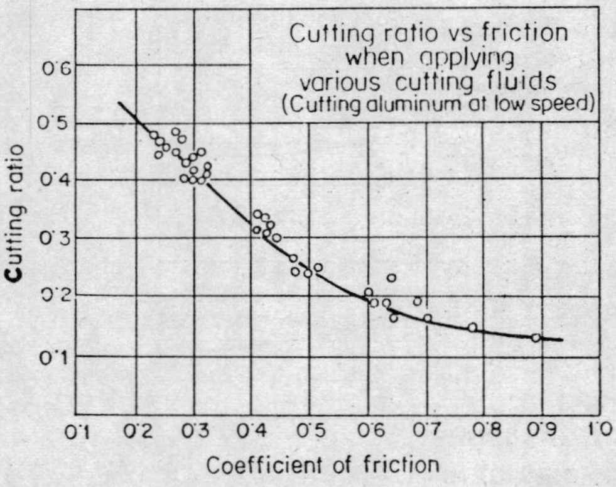

Fig. 12.24: The interrelation between cutting ratio and coefficient of friction.

fluid, which may directly control the amount of heat at the chip tool interface, can play an important role in increasing tool life.

The cooling action of the fluid is by carrying away heat directly thus preventing metallurgical damages. Hence a high specific heat and high heat conductivity together with a high film coefficient for heat transfer are necessary for a good coolant. For cooling ability, water is very effective, but is objectionable for corrosiveness and lack of friction reducing power.

The second basic function of an effective cutting fluid is to reduce kinetic coefficient of friction. Shaw has shown (Fig. 12.24) that cutting ratio ($1/\zeta$) which is a measure of deformation is directly related to chip friction when applying various cutting fluids which indicate that, in most of the cases, cutting fluids reduce friction.

In carrying out only two objectives of friction reduction and cooling, many problems may arise due to the method of application of cutting fluid. Therefore, the fluid must have some auxiliary properties such as it must be stable, non-volatile, odourless, no tendency for gum, foam or smoke formation, anti-corrosive, anti-bactericidal agent, etc.

Merchant has suggested a theory to explain the penetration of cutting fluid. It is assumed that minute capillaries exist at the chip tool interface as shown in Fig 12.25 on a submicroscopic scale. As the chip moves up the tool face, it contacts mainly the tops of the asperities in the point contact zone creating capillaries between the chip and the tool. These capillaries draw in the cutting fluid which chemically reacts to produce a solid low shear strength film. Under the condition of high pressure and temperature at the 'nascent' chip surface the highly reactive chemical action produces relatively weak solid providing a 'sandwich filling' to keep chip and tool apart thereby reducing friction.

As for example a cutting fluid containing chlorine as an additive, coming in contact with the work material, steel, forms iron chloride. As iron chloride is a weak solid with high melting temperature, it provides an excellent sandwich filling to keep chip and tool apart, thereby reducing interface friction.

Therefore, from the above discussion it is found that the cutting fluid should be such selected that it must have the property to penetrate into the capillaries and react in the zone.

The properties of the cutting fluid which are considered to be essential for penetration into the capillaries between the chip and the tool are:

(i) wetting and spreading,
(ii) surface tension, and
(iii) small fat molecules,

to provide a region of contact as shown in Fig. 12.26.

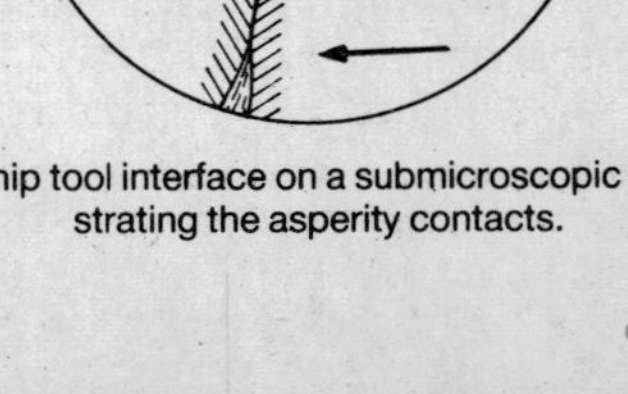

Fig. 12.25: Chip tool interface on a submicroscopic scale demonstrating the asperity contacts.

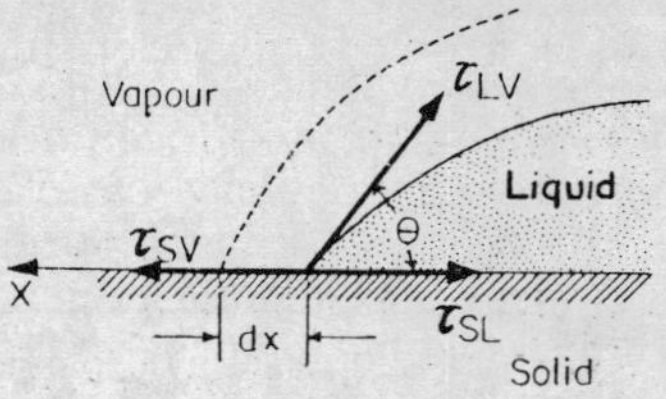

Fig. 12.26: Region of contact between solid, liquid and vapour.

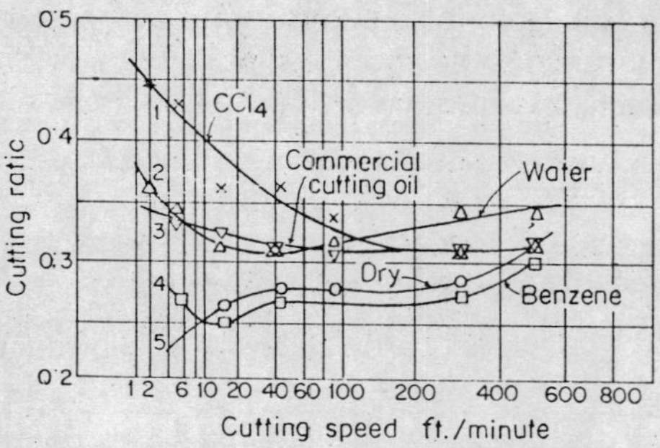

Fig. 12.27: Effect of different cutting fluids at various cutting speeds.

12.6.2. Cutting Fluid Action in the Chip Tool Interface Zone

Reduction of friction between the chip and the tool is very important in cutting operation as reduction in kinetic coefficient of friction not only decreases frictional work, but also decreases the shear work as well due to a resulting increase in the shear angle.

Therefore, cutting fluid should be so chosen that it will react at the interface zone to form a low shear strength solid lubricant.

The fluid when applied finds its way between the asperities at the chip tool interface, near the cutting edge and is subjected to the following condition.

(i) high local temperature on the high points of the surface asperities,
(ii) high local pressures approaching the hardness of metal cut,
(iii) fresh clean surface, and
(iv) highly stressed metal.

Under these circumstances the cutting fluid reacts with the chip to form solid low shear strength lubricant. This thin layer prevents the formation of weld between the chip and the tool, thus reducing kinetic coefficient of friction.

As for example when carbon tetrachloride (CCl_4) is used as a cutting fluid, iron chloride is formed at the chip tool interface which acts as solid low shear strength film.

The nature of cutting fluid action in cutting operation is chemical rather than physical. A number of organic chemicals has been used as cutting fluid to study the effect of chemical structure on the kinetic coefficient of friction. It is found that with increased chain length for all homologous series, the kinetic coefficient of friction decreases.

From Fig. 12.27, the effect of cutting speed on the action of cutting fluid for different agents ranging from dry cutting to cutting with extremely active carbon tetrachloride (CCl_4) may be seen. At lower speeds, the cutting fluid action is 'friction reducing' through a process of chemical action whereby the cutting ratio ($1/\zeta$) is increased reducing the cutting forces. At higher speeds, the action is mostly 'cooling' by reducing the bulk temperature of the chip. Such chilling may improve tool life but will not improve cutting ratio or reduce the cutting force.

With natural contact cutting tools, the contact zone comprises both of interlocked asperities in highly deformed plastic state and welded-on asperities having point-to-point contact. The plastic zone in the case of natural contact cutting tools is less dominant than the other zone where cutting fluid can penetrate and reduce friction by chemical action. But at high speeds, less contact time for stable fluid penetration reduces friction-reducing effect. But in the case of controlled contact cutting tools, restriction imposes a change in the density of the asperities as well as the dominance of plastic zone. This means that the cutting fluids will not be able to penetrate the interlocked asperities except occasionally through random penetrations. This is confirmed by cutting fluid tests performed by Bhattacharyya with controlled contact cutting tests.

CHAPTER 13

Economics of Machining

13.1. CALCULATION OF MACHINING TIMES

The costs may be direct, indirect and overhead. The labour varies from machine to machine and this direct cost is calculated by multiplying the labour rate by the time required for the operation. This time includes time for set-up and operation. Overhead cost is estimated as a product of overhead rate and time. The cost of the major machine and tool may be distributed on hourly basis (Fig. 13.1) or among the pieces produced for the determination of capital cost.

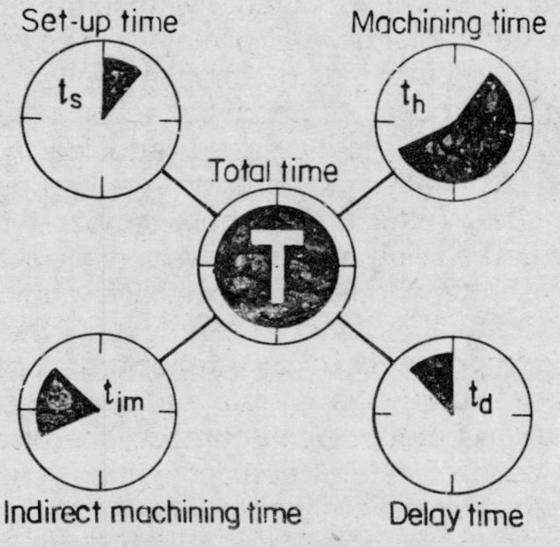

Fig. 13.1: Division of total time.

(i) Set-up time: This is the preparatory time to get tools, to do paper work and to arrange tools on the machine.

Set-up is generally done once for a particular lot.

Set-up and man-time may be estimated from previous performances or the knowledge of motion and time study.

(ii) Man or Handling time: This includes time for measurement, loading and unloading of the piece and manipulation of tools and the machine.

(iii) Machine time: Time during which the machine is busy and the tools are actually cutting.

(iv) Down or Lost time: This is the loss of time due to breakdowns, waiting for the tools, sudden minor accidents, etc.

Indirect machining time is the time spent for operations and operational elements prior to, between, and/or concluding basic elements. Indirect machining time occurs regularly. It includes such operational elements as picking up, positioning, and removing workpiece, measuring, tool sharpening, etc.

Down or lost time being difficult to calculate, the average lost time of the plant is taken into consideration and it is assumed that all the machines are sharing equal down time.

Time for 'approach' and 'overtravel' should be taken into consideration while calculating machine time. 'Approach' is the distance through which the tool is let to move until it starts cutting the full depth while 'overtravel' is the distance of feeding the tool without cutting.

13.2. MACHINING TIME FOR TURNING PROCESS

Machining time can be determined by calculation.

$$\text{Machining time} = \frac{\text{turning length}}{\text{feed/min.}}$$

$$t_h = \frac{L}{s \times n}$$

Designations:

L = turning length [L = length of workpiece (L_1) + starting allowance (l_a) + allowance after turning (l_u)] = $L_1 + l_a + l_u$;

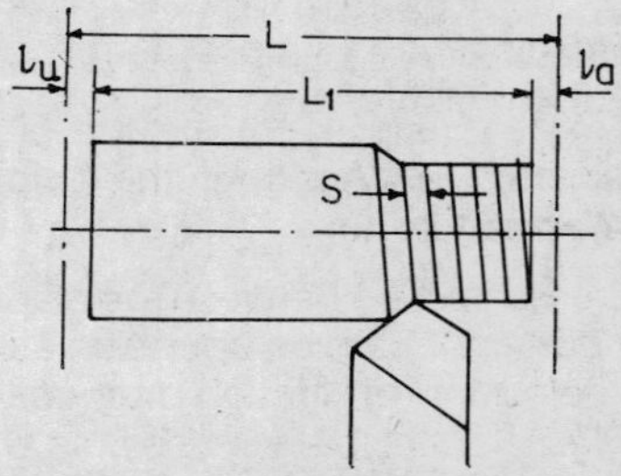

Fig. 13.2: Turning length including allowances

s = feed in mm/rev; n = revs per min.; feed per min.: $s' = s \times n$

Longitudinal Turning:

Example:
d = 80 mm; L_1 = 490 mm; $l_a = l_u$ = 5 mm; v = 20 m/min; s = 0.5 mm/rev

Result:
L = 490 mm + 5 mm + 5 mm;
n = 74 r.p.m.

$$t_h = \frac{L}{s \times n} = \frac{500 \text{ mm}}{0.5 \text{ mm} \times 74/\text{min}} = 13.5 \text{ min.}$$

13.3. MACHINING TIME FOR FACING PROCESS

The facing operation is shown in Fig. 13.3. The turning length L corresponds with the radius plus starting allowance.

$$L = r + l_a.$$

Example:
d = 190 mm; l_a = 5 mm;
v = 20 m/min; s = 0.5 mm/rev

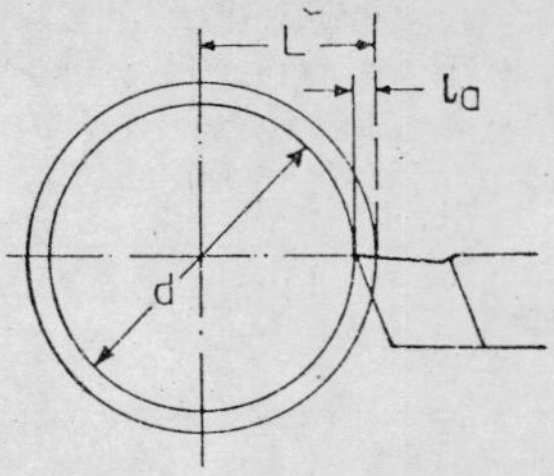

Fig. 13.3: Equivalent turning length in facing

Result:

$$L = \frac{190\text{ mm}}{2} + 5 = 100\text{ mm};$$

$$n = 37\text{ r.p.m.}$$

$$t_h = \frac{L}{s \times n} = \frac{100\text{ mm}}{0.5\text{ mm} \times 37/\text{min.}} = 5.4\text{ min.}$$

13.4. MACHINING TIME FOR DRILLING PROCESS

Machining time (t_h) is the operation time of the machine, i.e. the time during which the cutting edge of the drill cuts off chips (Fig. 13.4).

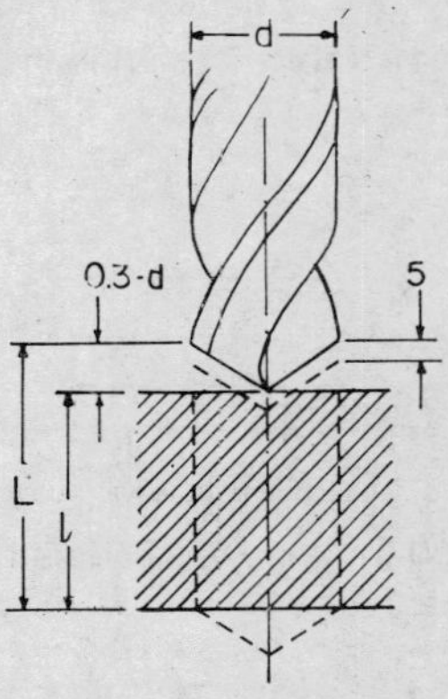

Fig. 13.4: Feed distance of the drill.

Designations :

l = depth of hole

L = feed distance of the drill

= depth of hole + drill point

$L = l + 0.3 \times d$

d = diameter of drill in mm

n = numbers of revolutions of drill per min

s = feed of drill in mm/revolution

Feed/minute = feed/revolution × numbers of revolution/minute

Feed/minute = $s \times n$

$$\text{Machining time} = \frac{\text{Feed distance}}{\text{Feed/minute}}$$

$$t_h = \frac{L}{s \times n}\ \text{min.}$$

Example:

$l = 30$ mm; $d = 18$ mm

$s = 0.2$ mm/rev; $n = 300$ r.p.m.

Result:

$$L = l + 0.3 \times d = 30\text{ mm} + 0.3 \times 18\text{ mm} = 35.4\text{ mm}$$

$$t_h = \frac{L}{s \times n} = \frac{35.4\text{ mm}}{0.2\text{ mm} \times 300/\text{min}} = 0.59\text{ min.}$$

13.5. MACHINING TIME FOR SHAPING PROCESS

L = length of stroke; $L = l + l_a + l_u$;

v_R = return speed in m/min.

v_A = cutting speed in m/min.

s = feed per cycle in mm.

The basic equation for the calculation of the machining time is:

$$\text{Time} = \frac{\text{Distance travelled}}{\text{Speed}}$$

The distance is the length of the stroke. By means of speeds v_A and v_R the time for the working and non-cutting stroke can be calculated. Feed, width etc are shown in Fig. 13.5.

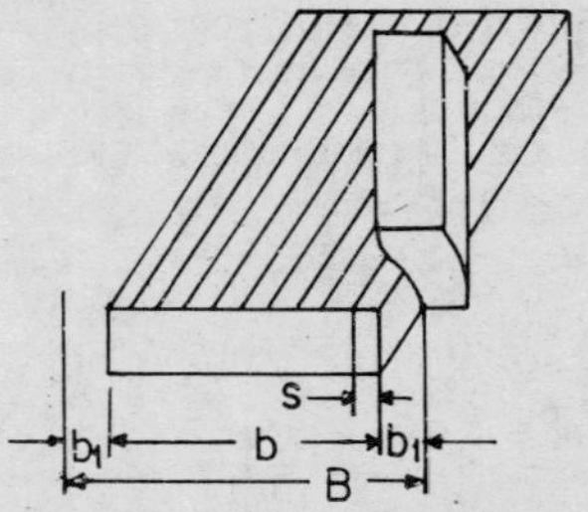

Fig. 13:5: Cycles required for shaping. s) Feed, b_1) feed allowance from the side, b) width of workpiece, B) planing width.

Time for the working stroke

$$t_A = \frac{\text{Length of stroke (m)}}{\text{Cutting speed (m/min)}}$$

$$t_A = \frac{L}{v_A}\ \text{min.}$$

Time for non-cutting stroke

$$t_R = \frac{\text{Length of stroke (m)}}{\text{Return speed (m/min.)}}$$

$$t_R = \frac{L}{v_R}\ \text{min.}$$

Time for cycle t = time for working stroke + time for non cutting stroke

$$t = t_A + t_R$$

For shaping a workpiece, a certain number of cycles is necessary, depending on the amount of feed and the width of the workpiece, plus allowance. The shaping width consists of the width of the workpiece plus an allowance on each side of the workpiece of 5 mm.

Shaping width B = width of workpiece + allowance from both sides

$$B = b + 2 \times 5\text{ mm.}$$

If the shaping width is divided by the feed, the required cycles will be obtained.

Thus number of required cycles

$$Z = \frac{\text{Shaping width}}{\text{Feed}}$$

$$Z = \frac{B}{s}$$

The machining time is calculated by multiplying the number of cycles with the time per cycle.

Machining time t_h = Number of cycles × time per cycle

$$t_h = Z \times t$$

Example :

A plate is to be shaped with roughing cut. Calculate the machining time. Given: Length of plate 260 mm, width 90 mm; $l_a = 30$ mm, $l_u = 10$ mm, $v_A = 10$ m/mm., $v_R = 20$ m/min., feed 1 mm/cycle, allowance from the left and right 5 mm each.

Result :

1. Length of stroke $L = l + l_a + l_u = 260 \text{ mm} + 30 \text{ mm} + 10 \text{ mm} = 300 \text{ mm}$;
2. Time for the working stroke
$$t_A = \frac{L}{v_R} = \frac{0.3 \text{ m}}{10 \text{ m/min.}} = 0.03 \text{ min.};$$
3. Time for the non-cutting stroke
$$t_R = \frac{L}{v_R} = \frac{0.3 \text{ m}}{20 \text{ m/min.}} = 0.015 \text{ min.};$$
4. Time for one cycle $t = t_A + t_R = 0.03 \text{ min.} + 0.015 \text{ min} = 0.045 \text{ min}$;
5. Shaping width $B = b + 2 \times 5 = 90 \text{ mm} + 2 \times 5 \text{ mm} = 100 \text{ mm}$;
6. Number of required cycles
$$Z = \frac{100 \text{ mm}}{1 \text{ mm/cycle}} = 100 \text{ cycles};$$
7. Machining time $t_h = Z \times t = 100 \text{ cycles} \times 0.045 \text{ min/cycle} = 4.5 \text{ min}$.

13.6. CALCULATION OF MILLING TIME

Machining time =

$$\frac{\text{Travelling distance of the milling table (mm)}}{\text{Rate of feed (mm/min)}}$$

$$t_h = \frac{L}{s'} \text{ min.}$$

The travelling distance (L) depends on the length of the workpiece (l), the feed allowance and the over travel (l_a and l_u) (Fig. 13.6).

Example :

A strip of St 42 with a length of 250 mm has to be rough milled by plain milling. Calculate the machining time.

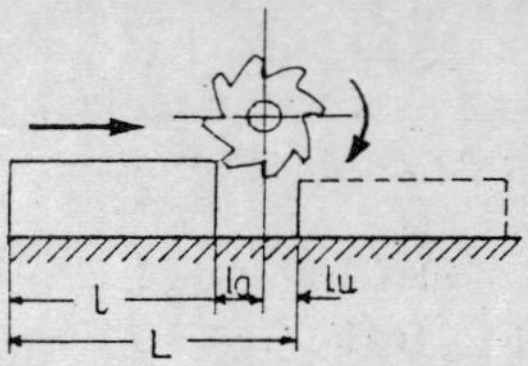

Fig. 13.6: Travelling distance for milling.

Given: $l_a = 30$ mm, $l_u = 5$ mm, rate of feed = 100 mm/min.

Result :

$$L = l + l_a + l_u = 250 \text{ mm} + 30 \text{ mm} + 5 \text{ mm} = 285 \text{ mm},$$

$$t_h = \frac{L}{s'} = \frac{285 \text{ mm}}{100 \text{ mm/min.}} = 2.85 \text{ min}$$

13.7. CALCULATION OF GRINDING TIME

13.7.1. Face Grinding

With reference to Fig. 13.7

l = length of workpiece
L = grinding length ($L = l$ + overtravel)
i = No. of cuts
v = speed of table in m/min.

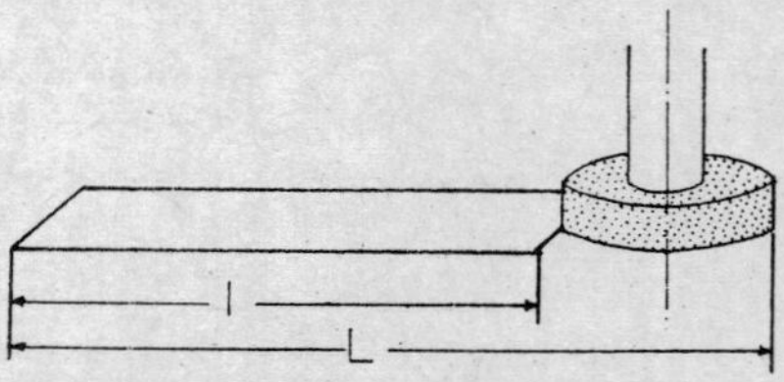

Fig. 13.7: Machining time for face grinding.

It is assumed that one feed adjustment takes place per cycle.

Machining time $t_h = \dfrac{2 \times L \times i}{v \times 1000}$

Example :

A gib of 750 mm length has to be ground. The grinding allowance of 0.6 mm has to be ground down with 4 cuts. Speed of table 1 m/min. The machining time has to be calculated.

Result :

$L = l$ + over-travel;
(over-travel = ϕ of wheel)
$L = 750 \text{ mm} + 150 \text{ mm} = 900 \text{ mm}$;

$$t_h = \frac{2 \times L \times i}{v_m \times 1000} = \frac{2 \times 900 \text{ mm} \times 4}{2 \text{ m/min.} \times 1000}$$

$= 3.6$ min.

13.7.2. Circumferential grinding

With reference to Fig. 13.8

b = width of workpiece
B = grinding width ($B = b$)
s = lateral feed (side feed) in mm/stroke.

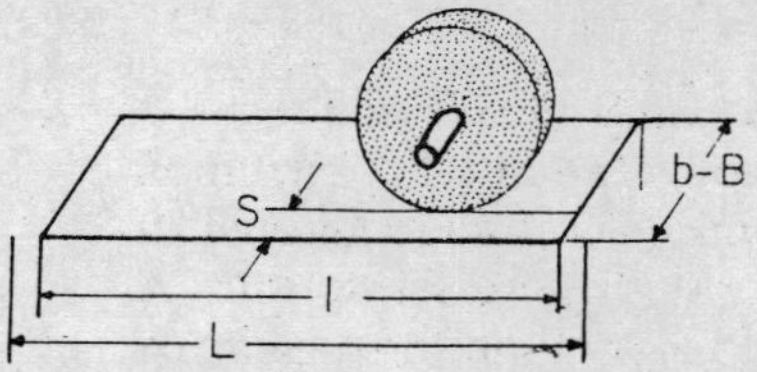

Fig. 13.8: Machining time for circumferential grinding.

Machining time without lateral feed

$$t_h = \frac{2 \times L \times i}{v \times 10000}$$

Machining time with lateral feed

$$t_h = \frac{2 \times L \times B \times i}{v \times 1000 \times s}$$

Example :

A die plate, 190 mm long, 150 mm wide, has to be ground. Grinding allowance = 0.4 mm; number of cuts = 4; width of grinding wheel = 20 mm; lateral feed = 6 mm/stroke; speed of table = 2 m/min. Machining time has to be calculated.

Result :

$B = b = 150$ mm;

$L = l + 2 \times 5$ mm $= 190 + 10$ mm $= 200$ mm;

$$t_h = \frac{2 \times L \times B \times i}{v \times 1000 \times s} = \frac{2 \times 200 \text{ mm} \times 150 \text{ mm} \times 4}{2 \text{ m/min} \times 1000 \times 6 \text{ mm}} = 20 \text{ min.}$$

13.7.3. Cylindrical Grinding Time

With reference to Fig. 13.9

L_1 = length of workpiece in mm,

L = grinding length in mm,

s = feed in mm per revolution of the workpiece,

n_w = No. of revolutions of the workpiece per min.,

i = No. of cuts.

Fig. 13.9: Machining time for cylindrical grinding.

The machining time is calculated when the grinding length is multiplied with the No. of cuts (total distance) and divided by the feed per min.

Feed per min. = feed per revolution of workpiece × r.p.m. of workpiece.

Machining time with feed adjustment at every stroke of the table is

$$t_h = \frac{L \times i}{s \times n_w}$$

Machining time with feed adjustment at every cycle.

$$t_h = \frac{2 \times L \times i}{s \times n_w}$$

Example :

A shaft of St 42, 40 ϕ, 400 mm long, shall be ground. It is supplied with a grinding size of 40.3 ϕ. Machining time has to be calculated.

Particulars: Grinding wheel 40 mm wide, feed adustment per cycle 0.01 mm (return stroke without feed adjustment).

Result :

1. RPM of the workpiece: Circumferential speed from any standard table is 12m/min.

$$n_w = \frac{v_w \times 1000}{\pi \times d} = \frac{12 \text{ m/min.} \times 1000}{314 \times 40 \text{ mm}} \approx 95 \text{ r.p.m.}$$

2. Feed: Selected from table, ½ width of grinding wheel per revolution of workpiece, $s = 40$ mm $\times 0.5 = 20$ mm per revolution of workpiece.
3. Number of cuts: Grinding allowance applied to the radius 0.3:2 = 0.15 mm.

$$i = \frac{\text{Grinding allowance}}{\text{Down-feed adjustment}} = \frac{0.15 \text{ mm}}{0.01 \text{ mm}} = 15.$$

4. Machining time:

$$t_h = \frac{2 \times L \times i}{s \times n_w} = \frac{2 \times 400 \text{ mm} \times 15}{20 \text{ mm} \times 95/\text{min.}} = 6.31 \text{ min.}$$

13.8. GENERAL PRINCIPLES OF ECONOMICS OF MACHINING

Cost conciousness with respect to the metal cutting process is an essential element in efficient manufacturing system. Around 1960, it was reported that in the USA alone, over 15 million tons of metal were converted into chips at a cost of more than 10 billion dollars (100 billion rupees). By 1970, this bill was in excess of 50 billion dollars. Because of the large expenditures involved, it is essential to analyze metal cutting operations in order to operate them at optimum economic conditions.

The money input to produce one work piece is made up of the following items: money spent for machine operation and overhead, and for time spent for loading, for rapid advance, for feed, for rapid return and for unloading.

All these add up to quantity x as shown in Fig. 13.10. Ernst has presented a simple model for estimating machining cost per piece, when

$$\text{cost/piece} = \frac{nx + a}{n} = x + \frac{a}{n}$$

where

n = number of pieces between regrinding,

x = machining and handling cost per piece,

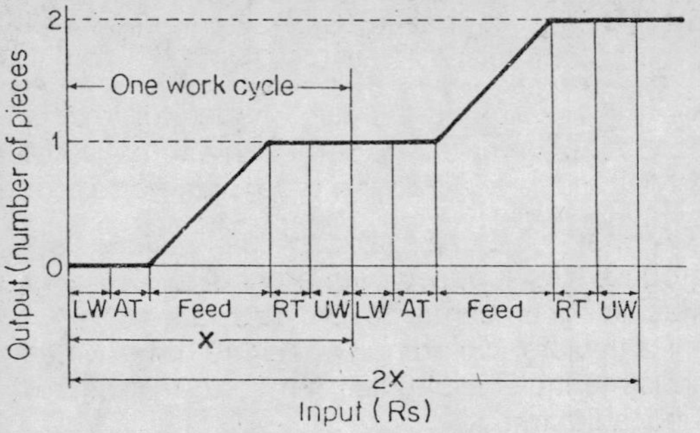

Fig. 13.10 : Money inputs and corresponding work output

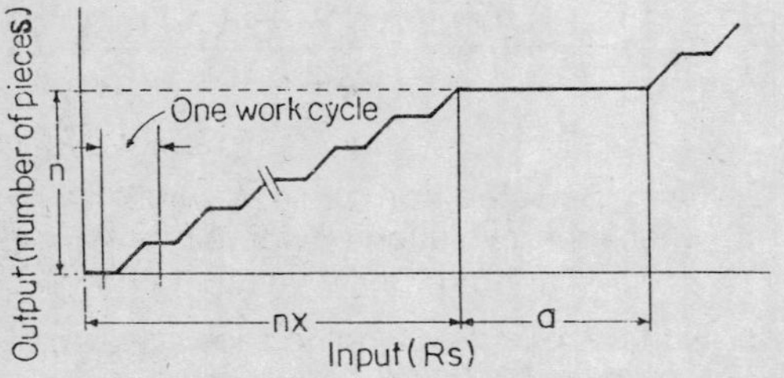

Fig. 13.11 : Total cost picture with n equal pieces

a = total cost of changing and regrinding the tool.

Fig. 13.11 shows the total cost picture with n equaling the number of pieces between tool grindings, x equaling the machining and handling cost per piece and a being the total cost of tool changing and grinding. The sloping parts are the productive inputs and flat parts are non-productive inputs. Each of the small steps in the line represents one work piece cycle.

The basic model provides the average unit cost (K_u) to produce a given work piece configuration as the sum of four costs, namely: (a) the machining cost, (b) the tool cost, (c) the tool-changing cost and (d) the handling cost. Each of these cost elements varies as a funcion of the cutting speed or feed as shown in Fig. 13.12.

Fig. 13.12 (a) depicts the machining cost per piece which goes down with increase in cutting speed while Fig. 13.12(b) shows that the tool cost per piece which increases as the cutting speed increases. The tool-changing cost per piece also increases with increase in cutting speed [Fig. 13.12(c)] while the handling cost remains constant as shown in Fig. 13.12(d).

Assuming that

(i) the work piece material is homogeneous,
(ii) the tool geometry is preselected,
(iii) the feed and the depth of cut are preselected and constant,
(iv) sufficient power is available to undertake the operation,
(v) cost of operating time is independant of actual cutting time.

The unit cost in Rs./piece can be obtained by summing up the individual costs as shown in Fig. 13.13. It is seen that a minimum unit cost is probable under the given characteristic features of the individual cost components.

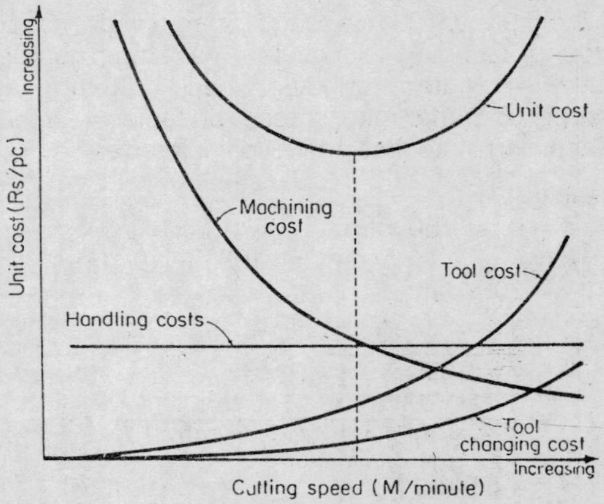

Fig. 13.13 : Unit cost relationship with cutting speed

13.9. ECONOMIC TOOL LIFE : GILBERT'S MODEL

W.W Gilbert has evaluated tool life for (i) minimum cost, and for (ii) maximum production.

Cost per piece = idle cost/piece + cutting cost/piece + tool-changing cost/piece + tool-regrinding cost/piece

Let,

K_1 = direct labour rate + overhead rate, Rs./min,
K_2 = tool cost per grind, Rs./tool,
L = length of machining, mm,
D = diameter of machined part, mm,
V = cutting speed, m/min,

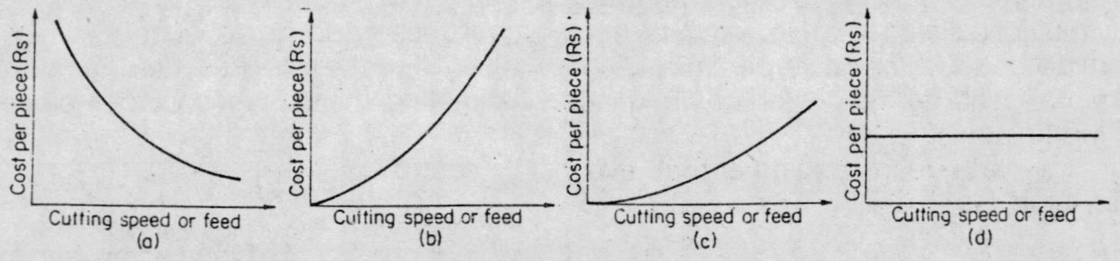

Fig. 13.12: Various cost elements

s = feed, mm/revolution,
VT^n = C, tool life equation,
t_d = TCT=tool-changing time.

Now,

idle cost/piece=K_1 (idle time/piece)

$$\text{cutting cost/piece} = K_1\left[\frac{\pi DL}{1000\,s\,V}\right]$$

Tool-changing cost/piece = K_1 × tool failure per piece × t_d

$$= K_1\left[\frac{L\pi DV^{(1/n)-1}}{1000\,sC^{1/n}}\right] t_d$$

Tool-grinding cost/piece = K_2 × tool failure/piece

$$= K_2\left[\frac{L\pi DV^{(1/n)-1}}{1000\,sC^{1/n}}\right]$$

Thus,

$$\text{cost/piece} = K_1\ (\text{idle time}) + K_1\frac{\pi DL}{1000\,s\,V} + K_1\frac{L\pi DV^{(1/n)-1}}{1000\,sC^{1/n}}\,t_d + K_2\frac{L\pi DV^{(1/n)-1}}{1000\,sC^{1/n}}$$

Differentiating and equating with zero for optimization :

$$\frac{d\left(\frac{\text{cost}}{\text{piece}}\right)}{dV} = 0$$

when V_{opt} the cutting speed for minimum cost per piece is obtained from

$$V_{opt} = C\left(\frac{n}{1-n}\right)^n\left[\frac{K_1}{K_1(t_d)+K_2}\right]^n \qquad \text{Eqn. 13.1}$$

and the tool life for minimum cost is

$$T_{opt} = \left(\frac{1}{n}-1\right)\left[\frac{K_1(t_d)+K_2}{K_1}\right] \qquad \text{Eqn. 13.2}$$

Fig. 13.14 depicts the condition of minimum cost production when cutting speed is increased.

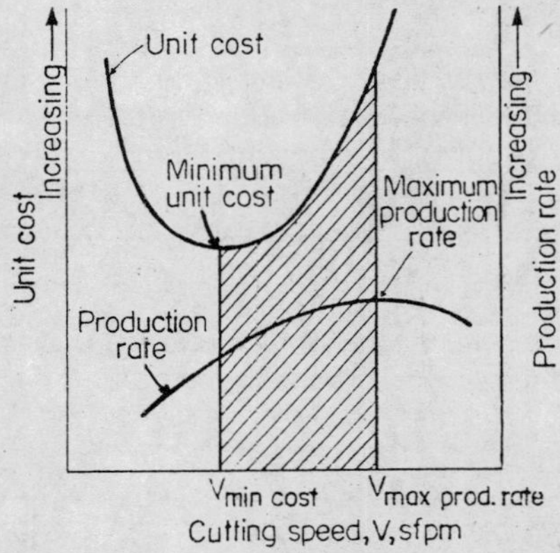

Fig. 13.14 : Condition of minimum cost production

13.10. OPTIMAL CUTTING SPEED FOR MAXIMUM PRODUCTION

If the costs in the basic model for machining economics are not considered, an alternative model can be built for yielding the time required to produce a work piece. The unit time per work piece (t_p) is

$$t_p = t_c + t_d\left(\frac{t_c}{T}\right) + t_l \qquad \text{Eqn. 13.3}$$

where

t_c = machining time,
T = tool life,
t_d = tool-changing time (TCT),
t_l = handling time (idle time).

The minimum unit time will result in maximum production rate as

$$Q = \text{production rate} = \frac{1}{t_p}$$

Thus the maximum production rate will correspond to the minimum production time per piece as obtained by differentiating Eqn. 13.3, with respect to cutting speed after substituting time-cutting speed relationships.

The result can also be obtained by setting $K_2 = 0$ in Eqn. 13.1 when

$$V'_{opt} = \frac{C}{\left[\left(\frac{1}{n}-1\right)(t_d)\right]^n}$$

and tool life for maximum production rate is given by

$$T'_{opt} = \left(\frac{1}{n}-1\right)(t_d)$$

In this case the optimum values are dependent on index, n, and tool-changing time only. Fig. 13.15 schematically shows the V'_{opt} for maximum production. Fig. 13.16 shows the location of V'_{opt} with respect to T'_{opt} for minimum cost and maximum production.

The Hi-E or high efficiency region was proposed

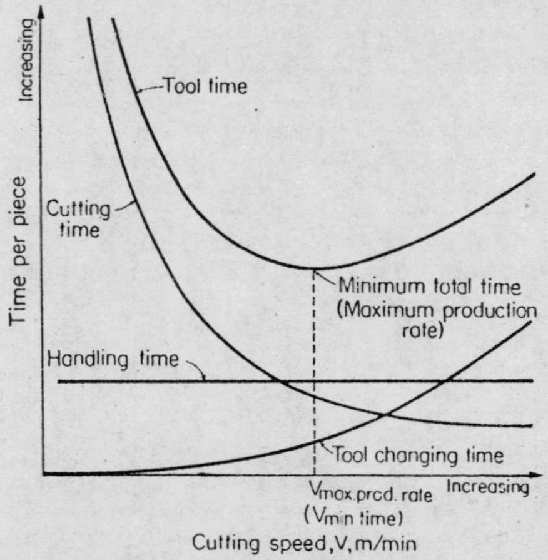

Fig. 13.15 : Optimum cutting speed for maximum production

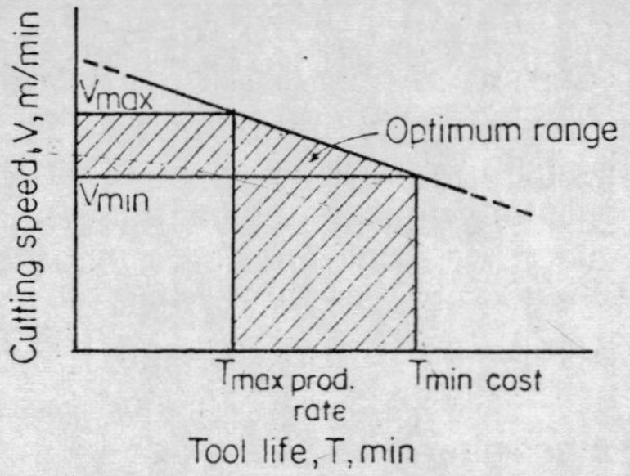

Fig. 13.16 : Optimum cutting speed with respect to optimum tool life

by Gilbert in the cutting speed range between the cutting speed for minimum cost on the lower speed end, and the cutting speed for maximum production rate on the higher speed end. In Fig. 13.17, V'_{opt} for maximum production rate will always be greater than V_{opt} for minimum cost. The reason for this is obvious when the respective equations are compared as the term $[K_1(t_d) + K_2]/K_1$ will always be larger than (t_d) since the ratio (t_d/K_1) is always positive. Operating at cutting speed less than V_{opt} is clearly not desirable since both the unit costs, K_1 and K_2, is increased along with decrease in the production rate. If, however, the cutting speed is between V_{opt} and V'_{opt}, there would be an increase in production rate and the increase in revenue may off-set the increased unit costs. To investigate this effect, an alternative model based on maximum profit criterion has been developed.

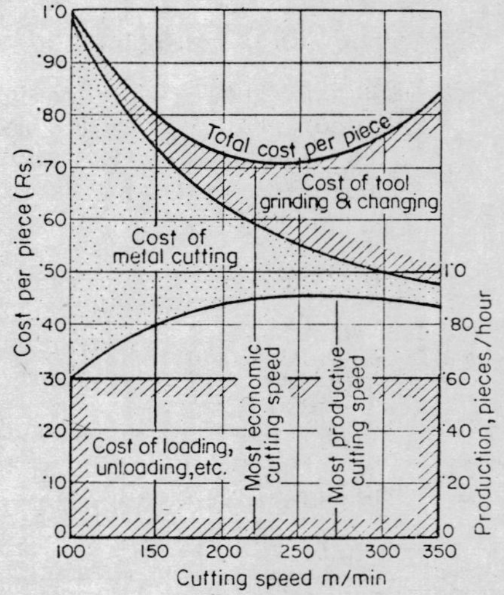

Fig. 13.17 : Comparison of optimum cutting speed for maximum production rate and for minimum cost

CHAPTER 14

Non-Traditional Machining Processes

14.1. NEED FOR NEW MACHINING PROCESSES

Modern machining technology is developing some 'needs' and 'trends' of large-scale and long-range nature which are placing severe demands on the existing system of conventional or traditional machining. This developing challenge of new needs and trends must be met by continuously developing production equipment and processes.

Merchant analysed the needs and trends of future of machining technology by assuming that past and present machining activities can be conveniently projected into the domain of future requirements. According to him, there are several needs of future machining technology out of which the three outstanding ones are:

(i) *Sustained Productivity* in face of rising strength barrier,
(ii) *Higher Accuracy* consistent with the increasing demand for higher tolerances, and
(iii) *Versatility* of automation.

Rising strength barrier poses the most serious problem to the manufacturing engineers and scientists. Metallurgists and materials science people are continuously developing newer and larger varieties of materials having diverse properties and increasingly higher strengths–a trend away from conventional material with conventional properties.

Consider, the strength of alloy steels given in Table 14.1.

TABLE 14.1

STRENGTH OF ALLOY STEELS

Steel AISI Specification	BHN Average
1 006	100
4 340	320
8 650	360
9 440	440
9 850	500

The Table shows a five-fold increase of strength of alloy steels compared to low-carbon steels.

Merchant indicated this rise of strength with respect to year-wise development of materials *vis-a-vis* their use in a given sector of industry, say, aerospace industry. The gradual increase of strength of the materials encountered in aero-space industry alone is indicated in Fig 14.1.

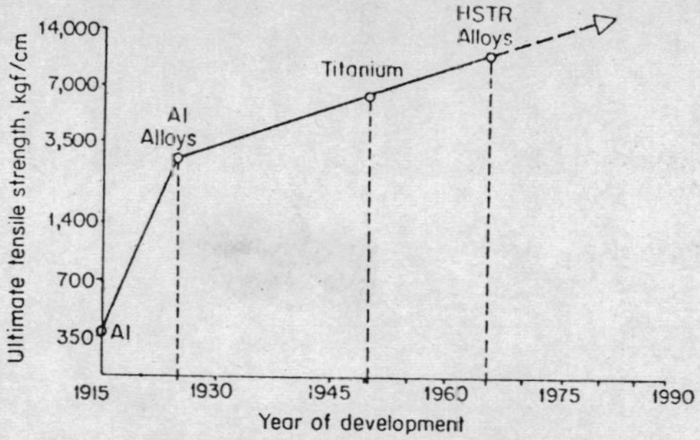

Fig. 14.1: Trand of increase of material strength in aerospace industry

In fact, with the development of new high-strength materials, it has become essential to develop improved cutting tool materials which are necessary for machining the newly developed materials as per requirement of any machining programme in order that the productivity is not hampered by the rising *strength barrier*. This is, however, governed by the fact that the tool mterial must be harder than the workpiece material. Thus, the newly developed cutting tool materials which efficiently fulfil the requirements of any cutting process also cause an improvement in the field of machining technology. Although carbon tool steels were the only tool materials used up to 1870, but tremendous developments have taken place in the field of machining technology because of the use of high-speed steels (HSS), cemented carbides, ceramics and stellites as tool materials nowadays.

In spite of rapid technological advancements in the field of conventional machining processes, the machining of carbides and other hard-to-machine materials has been limited to the diamond wheel grinding for a long time. The process has become costly because of the scarcity and high cost of the abrasives necessary for the diamond wheel. Moreover, the rapid development of the so-called technologically advanced industries like aerospace, nuclear power, etc. has been accompanied by an ever increasing use of the high-strength-temperature-resistant (HSTR) alloys. Besides, the processing of the parts of complicated shapes have been difficult, time consuming and uneconomical by the conventional methods of machining. Thus, neither the conventional techniques of machining are in a position to meet the challenges posed by the new development of materials nor is there any greater scope for their further development.

Rapid growth in the development of difficult-to-machine materials has led to the development of non-traditional machining technologies. The advancing strength level would then otherwise have a catastroplhic effect on the global machining bill (Fig. 14.2) if there was no corresponding improvement in machining technology by employing new techniques.

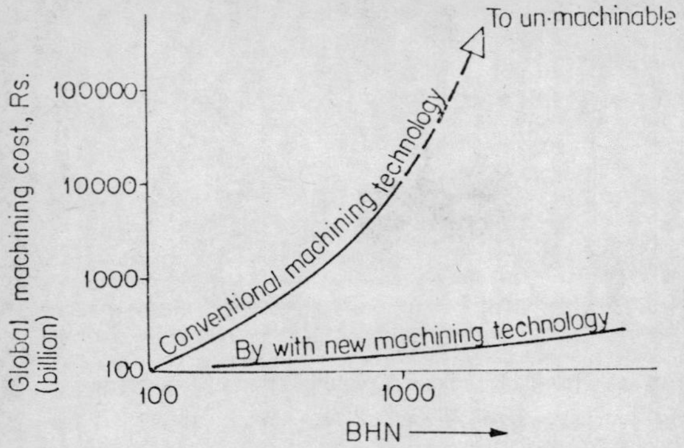

Fig. 14.2: Comparison of machining bills

It is therefore, clear that some new strategies of machining must be developed in order to deal with the problems created by the development and use of the hard-to-machine and high-strength-temperature-resistant (HSTR) alloys. Consequently, some non-traditional techniques of machining have been invented in order to supplement effectively the metal cutting, pressure forming and casting methods, especially when a complicated shape is to be produced on any hard-to-machine and brittle material. These are now a days known as *new machining technologies.*

These new techniques of machining are quite diverse in nature and differ from each other by their characteristic features of design, limitations in operation, complexities in technology and fields of application. Rapid changes are taking place in these areas as regards the bsic theories, requirement of power and energy, design technologies and equipment development.

The classification of the various new machining processes, their historical background and characteristic features have been dealt with in the succeeding sections of this chapter.

14.2. CLASSIFICATION OF NEW TECHNOLOGY

From the preceding discussions, it is clear that new technological processes are those processes which are not affected by hardness, toughness or brittieness of materials and can produce any intricate shape on any workpiece material by suitable control over the various physical parameters of the processes. The underlying principle of the various new machining processes is to apply some sort of energy, namely, mechanical, electrical, chemical, thermal or magnetic, to the workpiece directly and have the desired shape transformation or material removal from the work surface by using the known scientific principles. This is schematically shown in Fig. 14.3. However, these new methods of machining still have got some outstanding problems which, when solved, will bring about extensive application of the processes in the various industrial fields.

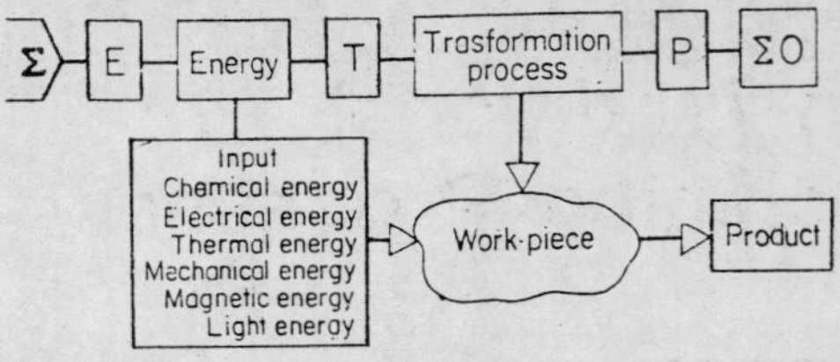

Fig. 14.3: The underlying principle of new machining processes

The new machining processes can be classified into various groups according to the following basic requirements :

(i) Type of energy required, namely, mechanical, electrical, chemical, etc.
(ii) Basic mechanism involved in the processes, like erosion, ionic dissolution, vaporisation, etc.
(iii) Source of immediate energy required for material removal, namely, hydrostatic pressure, high current density, high voltage, ionized material, etc.
(iv) Medium for transfer of those energies, like high velocity particles, electrolyte, electron, hot gases, etc.

In accordance with the above basic requirements, the various processes may be classified as shown in Table 14.2; From the table, it is evident that to cope with the fast changing machining technology and the difficulties caused by the development of hard-to-machine and high-strength-temperature-resistant (HSTR) alloys, the non-traditional techniques of machining like electrical discharge machining (EDM), electro-chemical machining (ECM), electron beam machining (EBM), laser beam machining (LBM), abrasive jet machining (AJM), electro-chemical grinding (ECG), ultrasonic machining (USM), plasma arc machining (PAM), ion beam machining (IBM), chemical machining (CHM) etc. have come about to help the manufacturing engineers.

14.3. HISTORICAL BACKGROUND OF NEW MACHINING PROCESSES

Although, the non-conventional machining processes have created a revolution in the field of machining technology but the development of the ideas of the various processes were initiated as early as in nineteen-twenties in the USSR.

The initiation was first made by Gussev towards the end of 1920 in the USSR. He suggested a method of machining by the combination of chemical and mechanical means and in 1940, he laid down the basis of all the electrochemical machining (ECM) processes known at present. Parallel research had also been made by an American, Burgess in 1941 who had demonstrated the possibilities of the ECM

TABLE 14.2
CLASSIFICATION OF NEW MACHINING PROCESSES

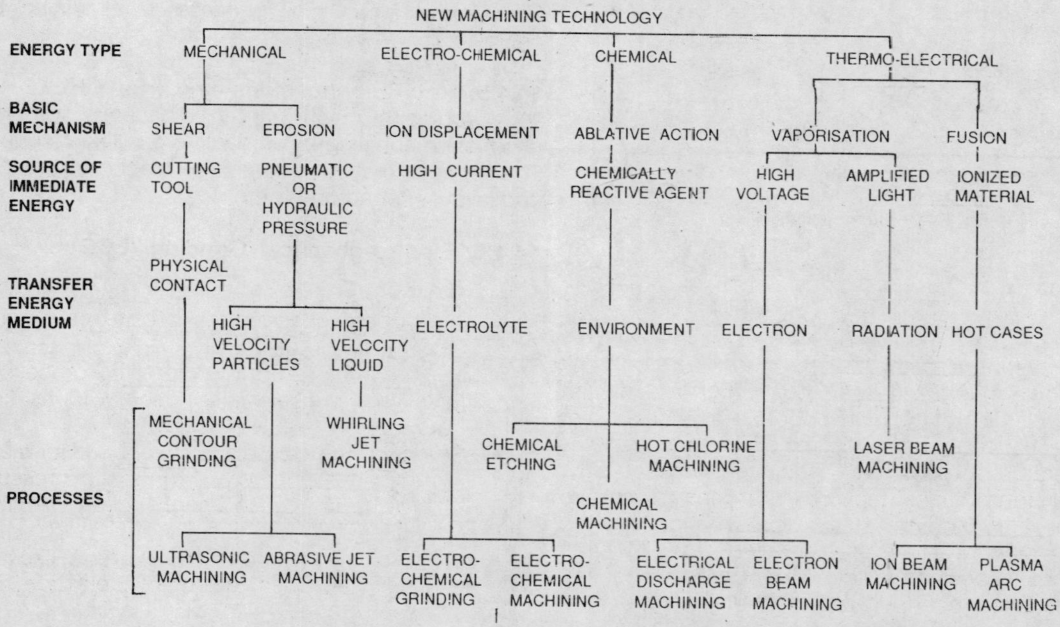

process by drawing a sharp contrast between the mechanical and electrolytic methods of metal removal.

Electrical discharge machining (EDM) was originally developed by B. R. Lazarenko and N. I. Lazarenko in the USSR. They, in the early nineteen-forties (1943), first developed the idea of spark-erosion machining. Advancement also took place parallelly in Europe and the USA. Since then, developments are taking place very rapidly in the theory and technology of the process and hence this technology is gaining much appreciation.

The idea of ultrasonic machining (USM) was invented by Balamuth in the year 1942 at the time of investigating dispersion of solids in liquids with the help of a vibrating magnetostrictive nickel tube. However, the origination of the process was made by Rosenberg.

In the early nineteen-sixties, the idea of ultrasonic machining began to develop widely in the USSR and the basis of this development was laid on extensive investigation that took place in the mechanism of ultrasonic machining and in the design of the magnetostrictive transducers, converters and wave-guides (including the tool) and also on the combination of several ultrasonic processes.

The basis of laser beam machining was established by the processes which were developed by Basov, Prokhorov and Fabrikant in the USSR in 1950. The laser methods which have been developed in the early 1960's, simultaneously in the USA and the USSR, are an application of quantum radio electronics and electron optics. The first application of laser beam machining had been for the combined welding and cutting of components used in the electronic industry and also for drilling small holes.

Electrochemical grinding process has practically been developed in about 1950 and in the early nineteen-sixties the concept of whirling jet machining was innovated.

Many of these new techniques of machining have been developed in the last decade to meet the challenges put forward by the rapid development of the hard-to-machine and high-strength-temperature-resistant (HSTR) alloys. It is anticipated that in the near future, these new technologies will find an ever increasing application in all branches of mechanical engineering industry.

14.4. TOPOLOGY OF NEW MACHINING PROCESSES

14.4.1. Abrasive Jet Machining (AJM)

Abrasive jet machining (Fig. 14.4) is a process of material removal through the action of a focussed stream of fluid with abrasive particles. It is especially used for machining super alloys, ceramics, glass and refractory materials.

14.4.2 Ultrasonic Machining (USM)

Ultrasonic machining (Fig. 14.5) is a process of material removal by the action of abrasives. The material is removed by repetitive impacts at high frequency between the tool and the workpiece with an abrasive slurry inbetween.

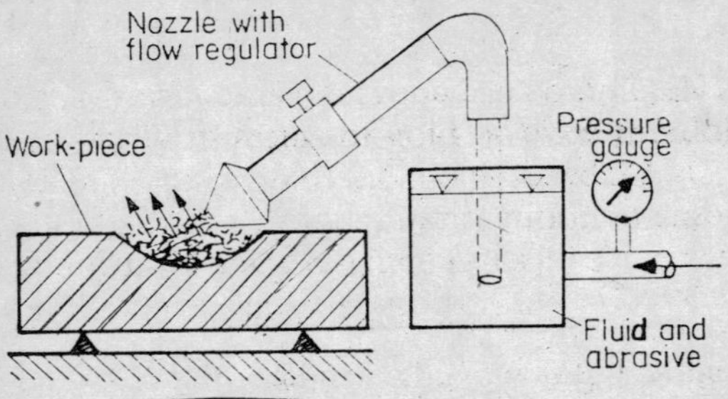

Fig. 14.4: Abrasive jet machining

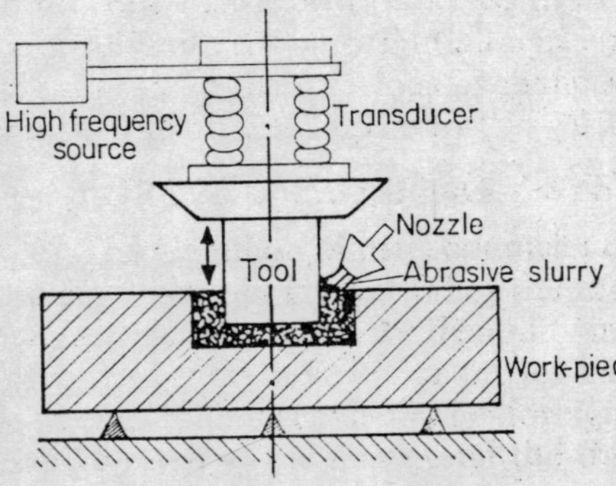

Fig. 14.5: Ultrasonic machining

The process is mainly used for machining semi-conductor elements, brittle ferrite elements used for computors, ceramics, glass and refractory materials.

14.4.3. Chemical Machining (CHM)

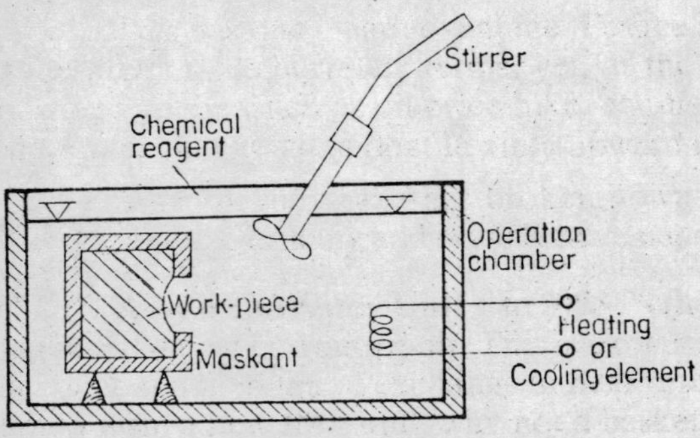

Fig. 14.6: Chemical machining

Chemical machining (Fig. 14.6) is a process of controlled dissolution of material in contact with strong chemical reagents.

The process finds wide applicability in machining metals and alloys like steel and aluminium and also some super alloys and glass.

14.4.4. Chemical Etching (CHE)

Chemical etching or chemical milling is basically a process of metal removal by an etchant solution which may be sprayed to the zone of machining or the workpiece may be dipped in the etchant solution with the unmachined portions protected by a coating of suitable etchant resisting material. The process is used for manufacturing structural members of aerospace equipment, for example, reducing the weight of wing and similar panels by selective metal removal from some areas. The process is a branch of chemical machining.

14.4.5. Hot Chlorine Machining (Hot CHL)

Hot chlorine machining is a gaseous form of chemical machining by a reactive gas like chlorine. The process may be applied for machining metals and alloys, refractories, etc.

14.4.6. Electro-chemical Grinding (ECG)

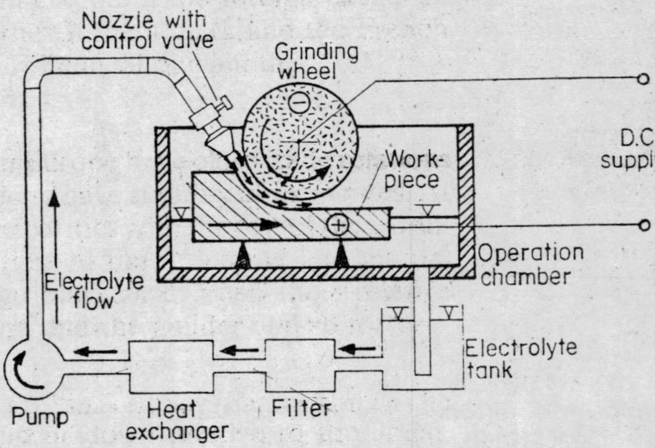

Fig. 14.7: Electro-chemical grinding

Electro-chemical grinding (Fig. 14.7) is an anodic dissolution process of metal removal from a conducting positive workpiece by a rotating conductive abrasive wheel serving as the cathode of an electrolytic cell with a thin film of flowing electrolyte between them.

This is suitable for grinding steels, super alloys, carbides, titanium, etc.

14.4.7. Electro-chemical Machining (ECM)

Electro-chemical machining (Fig. 14.8) is a process of metal removal from a conducting material by

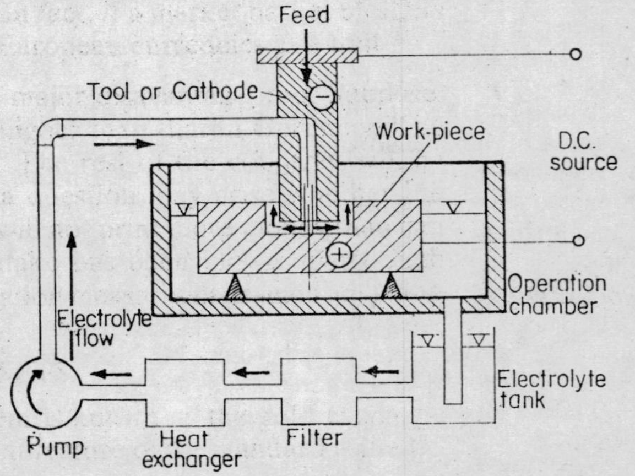

Fig. 14.8: Eletro-chemical machining

the anodic dissolution of the positive workpiece separated from a suitably shaped electrode tool by a thin film of flowing electrolyte.

The process is applicable to any conducting material irrespective of its handness, strength and microstructure. It is especially employed for machining super alloys, steels, etc. Any complicated contour can be produced on the work surface by suitably designing the tool.

14.4.8. Electrical Discharge Machining (EDM)

Electrical discharge machining (Fig. 14.9) is the process of metal removal from a conducting work-

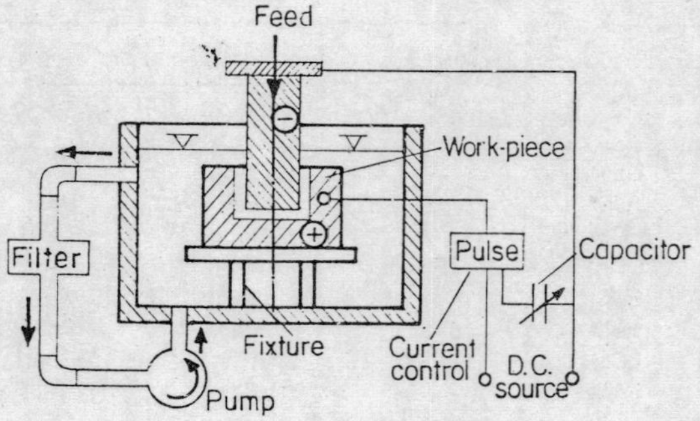

Fig. 14.9: Eletrical discharge machining

piece by a rapid repetitive spark discharge initiated between the negative tool and positive workpiece separated by a flowing dielectric medium.

The process may be applied to machine steels, super alloys, refractories, etc.

14.4.9. Electron Beam Machining (EBM)

Electron beam machining (Fig. 14.10) is a material removal process in which a high velocity focussed

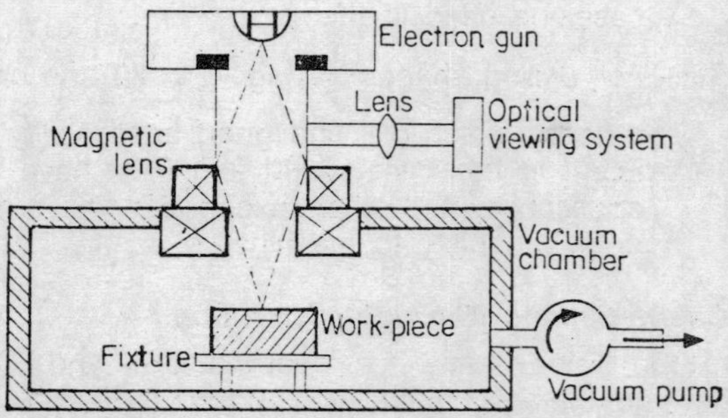

Fig. 14.10: Electron beam machining

stream of electrons hits, melts and vaporizes the work material at the point of bombardment. This process may be suitably applied for machining brittle materials like ceramics, refractories, etc.

14.4.10. Laser Beam Machining (LBM)

In laser beam machining (Fig. 14.11) material is removed by a forcussed mono-frequency collimated

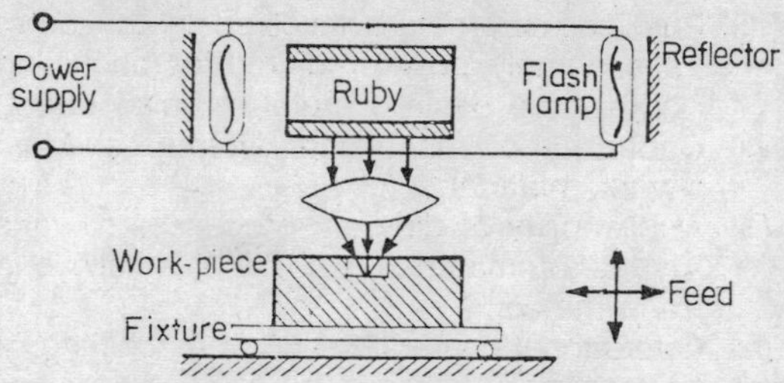

Fig. 14.11: Laser beam machining

light beam which heats, melts and vaporizes the work material at the focussed area.

Laser methods have got wide application for machining ceramics, but may also be used for machining aluminium, steels, super alloys, etc.

14.4.11. Plasma Arc Machining (PAM)

In plasma arc machining (Fig. 14.12), a gas, usually H_2, N_2 etc., is heated by arc to a high temperature to

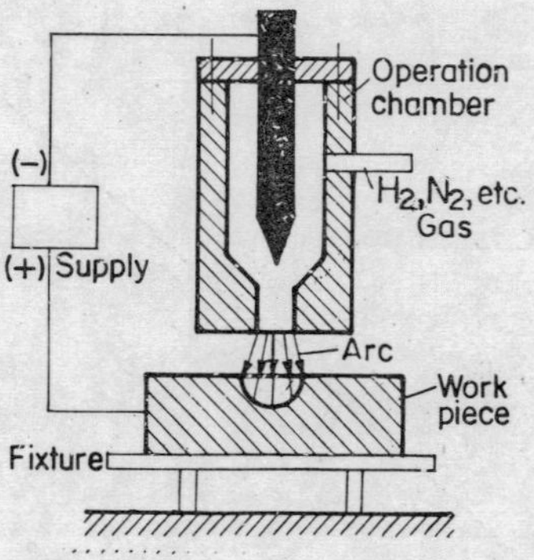

Fig. 14.12: Plasma are machining

become partially ionised. This ionised stream of gas is impinged on the work surface to cause it to melt and erode. This process may be suitably applied to steels, super alloys, etc.

14.4.12. Ion Beam Machining (IBM)

Ion beam machining is a process of sputtering metals by incoming ions.

This process may be employed to machine super alloys, titanium, etc.

14.5. COMPARATIVE ANALYSIS OF NEW MACHINING PROCESSES

In the preceding sections, the role of non-conventional machining processes, their early history as well as their characteristic features have been discussed. In this section, a comparative analysis of the various new technological processes has been made so that a guideline may be drawn for the manufacturing engineers to find the suitability of application of different processes. The analysis has been made from the point of view of:

(i) Physical parameters involved in the processes, namely, power required for machining to obtain the desired metal removal rate;
(ii) Capability of machining different shapes of work material;
(iii) Applicability of different processes to various types of material, *e.g.,* metals, alloys and non-metals;
(iv) Operational characteristics of machining; and
(iv) Economics involved in the various processes.

14.5.1. Physical Parameters

The physical parameters involved in the various processes have got a direct impact on the metal removal rate as well as on the energy consumed in different processes. From a comparative study of the effect of metal removal rate on the power consumed by various new technological processes as shown in Fig. 14.13, it is found that some of the processes (e.g., EBM, ECM) above the mean power consumption line consume a greater amount of power than the processes (e.g., EDM, PAM, MCG) below the mean power consumption line. Thus, the capital cost involved in the processes (EBM, ECM, etc.) lying above the mean line is high whereas for the processes below that line (e.g., EDM, PAM, MCG) is comparatively low. Some of the processes like LBM, USM etc. lie on the mean line power consumption line and the capitial cost is also comparatively low in these cases. However, some of the processes above the mean line, like ECM, have got higher metal removal rate than the processes LBM, USM, EDM etc. lying on or below the mean line.

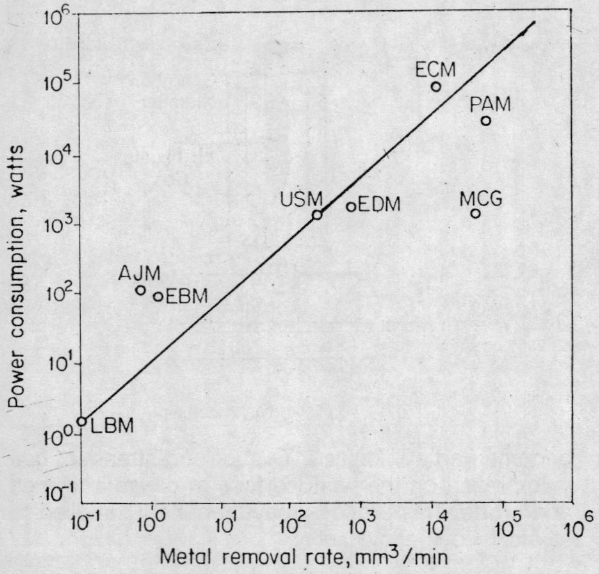

Fig. 14.13: Effect of metal removal rate on power consumption

14.5.2. Capability to Shape

The capability of different processes to perform various machining operations like micro-drilling, drilling, cavity sinking (accurate and standard), pocketing (shallow and deep), contouring a surface, through cutting (shallow and deep), etc can be analysed in the manner as described hereinafter.

Micro-drilling and Drilling Operation

From Table 14.3, the following important inferences may be drawn:

TABLE 14.3

PROCESS	CAPABILITY TO		
		Drill	
	Micro-drill	$l/D<20$	$l/D>20$
USM	–	Good	Poor
AJM	–	Fair	Poor
ECM	–	Good	Good
CHM	Fair	–	–
EDM	–	Good	Fair
EBM	Fair	Fair	Poor
LBM	Good	Fair	Poor
PAM	–	Fair	–

(i) For micro-drilling operation, the only process which has good capability to micro-drill is laser beam machining. However, EBM and CHM processes may also be used.
(ii) For drilling shapes having slenderness ratio, $\frac{l}{D} < 20$, the processes USM, ECM and EDM will be very suitable, whereas other processes like AJM, EBM, LBM and PAM have a fair performance to drill. But, CHM may, in no case, be used in this case.
(iii) For drilling shapes having $\frac{l}{D} > 20$, the only process which will find good applicability is ECM. In this case, CHM and PAM have no applicability and others except EDM have got poor capability to machine.

Cavity-Sinking (through) Operation

From Table 14.4, it is evident that USM and EDM have got the best capability to perform cavity sinking operation.

However, ECM may also be employed with ease for standard operation but has fair performance for accurate operation. Other processes may not be used for cavity sinking operations except AJM and CHM which may sometimes be used for fair performance of standard cavity sinking operation.

Pocketing Operations

Table 14.5 shows that EDM and ECM processes have got **good** capability to make pocketing opera-

TABLE 14.4

Process	Capability To Make	
	Accurate Operation	Standard Operation
USM	Good	Good
AJM	Poor	Fair
ECM	Fair	Good
CHM	Poor	Fair
EDM	Good	Good
EBM	Poor	Poor
LBM	Poor	Poor
PAM	Poor	Poor

TABLE 14.5

Process	Capability to Make	
	Shallow Pocketing	Deep Pocketing
USM	Poor	Poor
AJM	–	–
ECM	Good	Good
CHM	Good	Poor
EDM	Good	Good
EBM	–	–
LBM	–	–
PAM	–	–

tion (shallow or deep) but other processes except CHM (which has got good application for shallow pocketing only) will find poor or no applicability in this type of operation.

Contouring a Surface

From Table 14.6 it is clear that only ECM process has got good capability for the surface contouring operation. Other processes except EDM (which have fair performance) will find no application for the contouring operation.

TABLE 14.6

Process	Performance
USM	Poor
AJM	–
ECM	Good
CHM	–
EDM	Fair
EBM	–
LBM	–
PAM	–

The ultrasonic machining finds in rare occasion application for contouring a surface on some refractory materials.

Through Cutting Operation

Two inferences may be drawn from Table 14.7.

TABLE 14.7

Process	Capability to Have	
	Shallow Cutting	Deep Cutting
USM	Poor	–
AJM	Good	–
ECM	Good	Good
CHM	Good	–
EDM	Poor	–
EBM	Good	Fair
LBM	Good	Fair
PAM	Good	Good

(i) For shallow cutting operation any method except USM and EDM method may be employed to have good cutting performance.
(ii) For deep cutting operation, ECM and PAM have got good cutting performance, whereas other processes except EBM and LBM (which have fair performance) have got no applicability at all.

14.5.3. Applicability to Materials

The applicability of the different processes to different types of material, namely, metals, alloys and non-metals may be analysed in the following way.

Metals and Alloys

From Table 14.8 several important information may be obtained:

TABLE 14.8

Process	Metals & Alloys				
	Aluminium	Steel	Super Alloy	Titanium	Refractory Material
USM	Poor	Fair	Poor	Fair	Good
AJM	Fair	Fair	Good	Fair	Good
ECM	Fair	Good	Good	Fair	Fair
CHM	Good	Good	Fair	Fair	Poor
EDM	Fair	Good	Good	Good	Good
EBM	Fair	Fair	Fair	Fair	Good
LBM	Fair	Fair	Fair	Fair	Poor
PAM	Good	Good	Good	Fair	Poor

(i) USM has got good machining performance for refractory type of material.
(ii) AJM may suitably be employed for machining super alloys and refractory materials.
(iii) ECM has got good applicability for machining steels and super alloys.
(iv) CHM is very useful for aluminium and steel.

(v) EDM may well be adopted for machining steels, super alloys, titanium and refractory materials.

(vi) EBM has found only good machining performance in case of refractory type of materials.

(vii) PAM can be suitably employed for machining aluminium, steels and super alloys.

Non-metals

Table 14.9 provides the following information:

TABLE 14.9

Process	Non-Metals		
	Ceramics	Plastic	Glass
USM	Good	Fair	Good
AJM	Good	Fair	Good
ECM	–	–	–
CHM	Poor	Poor	Fair
EDM	–	–	–
EBM	Good	Fair	Fair
LBM	Good	Fair	Fair
PAM	–	Poor	–

(i) EDM and ECM processes cannot be used for machining non-conducting materials like ceramics, plastics and glass.

(ii) USM and AJM can be suitably employed for machining non-metals like ceramics and glass. They can also be used for machining plastics.

(iii) EBM and LBM have good machining performance for ceramics and fair performance for glass and plastics.

(iv) PAM cannot be used for machining ceramics and glass.

14.5.4. Machining Characteristics

The machining characteristics of different new technological processes can be analysed with respect to

(i) metal removal rate obtained,
(ii) tolerance maintained,
(iii) surface finish obtained,
(iv) depth of surface damage, and
(v) power required for machining.

Table 14.10 shows the machining characteristics of different processes and several important information can be drawn as follows:

(i) The metal removal rate by ECM and PAM is much higher than other non-conventional machining processes when compared with conventional machining. The metal removal rates by ECM and PAM are respectively one-fourth and 1.25 times that of conventional whereas others are only a small fractions of it machining,

(ii) Power requirement of ECM and PAM is also very high when compared with other non-conventional machining processes. This involves higher capital cost for those processes.

TABLE 14.10

Process	MRR, mm^3/min	Tolerance, micron	Surface finish, micron CLA	Depth of surface damage (micron)	Power (watts)
USM	300	7.5	0.2-0.5	25	2,400
AJM	0.8	50	0.5-1.2	2.5	250
ECM	15,000	50	0.1-2.5	5	100,000
CHM	15	50	0.5-2.5	5	—
EDM	800	15	0.2-1.2	125	2,700
EBM	1.6	25	0.5-2.5	250	150 (Average) 2000 (Peak)
LBM	0.1	25	0.5-1.2	125	2 (Average) 200 (Peak)
PAM	75,000	125	Rough	500	50,000
Conventional Machining	50,000	50	0.5-5	25	3,000

(iii) The tolerance obtained by various processes except PAM is within the range of conventional machining which means satisfactory dimensional accuracy can be maintained.

(iv) Satisfactory surface finish are obtained by all the processes except PAM.

(v) Depth of surface damage is very small for AJM, ECM and CHM processes whereas this is very high in case of EDM, EBM, PAM, etc. For this reason, ECM may well be employed for making dies, punches, etc.

14.5.5 Economics of the Processes

The economics of the various processes are analysed by considering the following items:

(i) capital cost,
(ii) tooling cost,
(iii) consumed power cost,
(iv) metal removal rate efficiency, and
(v) wear of tooling.

Table 14.11 shows the economics of the various processes.

TABLE 14-11

Processes	Capital Cost	Tooling Cost	Power Consumption Cost	Material Removal rate Efficiency	Tool Wear
USM	L	L	L	H	M
AJM	VL	L	L	H	L
ECM	VH	M	M	L	VL
CHM	M	L	H*	M	VL
EDM	M	H	L	H	H
EBM	H	L	L	VH	VL
LBM	L	L	VL	VH	VL
PAM	VL	L	VL	VL	VL
MCG	L	L	L	VL	L

* Indicates cost of chemicals

The significance of the various abbreviations are as cited in Table 14.12:

TABLE 14.12

Abbreviation	Significance
V	Very low
L	Low
M	Medium
H	High
VH	Very high

The relative weightages given to different abbreviations are as shown in Table 14.13.

TABLE 14.13

VL	L	M	H	VH
1	2	4	8	16

The following points are clear from Table 14.11

(i) Capital cost for ECM is very high when compared with traditional mechanical contour grinding and other non-conventional machining processes whereas capital costs for AJM and PAM are comparatively low.
(ii) EDM has got higher tooling cost than other machining processes.
(iii) Power consumption cost is very low for PAM and LBM processes whereas it is greater in case of ECM.
(iv) The metal removal efficiency is very high for EBM and LBM than for other processes.
(v) Tool wear is very low for ECM, CHM, EBM, LBM and PAM when compared with other processes whereas it is high in case of EDM process.

In conclusion, it may be stated that the suitability of application of any of the new machining processes is dependent upon various factors as discussed and a manufacturing engineer must judiciously consider all or some of the factors to fulfil his objective before applying new technological principles.